Arthropod Collection and Identification

Field and Laboratory Techniques

Arthropod Collection and Identification Field and Laboratory Techniques

Timothy J. Gibb, Ph.D.
Christian Y. Oseto, Ph.D.

Entomologists
Purdue University

AMSTERDAM • BOSTON • HEIDELBERG • LONDON
NEW YORK • OXFORD • PARIS • SAN DIEGO
SAN FRANCISCO • SINGAPORE • SYDNEY • TOKYO

Academic Press is an imprint of Elsevier

Associate Acquisitions Editor: Kelly Sonnack
Project Manager: Heather Furrow
Marketing Manager: Linda Beattie
Cover Design: Eric DeCicco
Interior Design: Sandra Stephens-Reeves
Composition: Sandra Stephens-Reeves
Cover Printer: Phoenix Color
Interior Printer: RR Donnelley

Academic Press is an imprint of Elsevier
30 Corporate Drive, Suite 400, Burlington, MA 01803, USA
525 B Street, Suite 1900, San Diego, California 92101-4495, USA
84 Theobald's Road, London WC1X 8RR, UK

This book is printed on acid-free paper.

Copyright © 2006, Elsevier Inc. All rights reserved.

No part of this publication may be reproduced or transmitted in any form or by any
means, electronic or mechanical, including photocopy, recording, or any information
storage and retrieval system, without permission in writing from the publisher.

The original edition of this work was published as "Collection and Preservation of Insects"
in 1948 and as U.S. Government material was not subject to copyright protection.

Permissions may be sought directly from Elsevier's Science & Technology Rights
Department in Oxford, UK: phone: (+44) 1865 843830, fax: (+44) 1865 853333,
E-mail: permissions@elsevier.com. You may also complete your request on-line
via the Elsevier homepage (http://elsevier.com), by selecting "Support & Contact"
then "Copyright and Permission" and then "Obtaining Permissions."

Library of Congress Cataloging-in-Publication Data
Gibb, Timothy.
  Arthropod collection and identification : field and laboratory techniques / Timothy Gibb and Christian Oseto.
    p. cm.
  Includes bibliographical references.
  ISBN-13: 978-0-12-369545-1 (pbk. : alk. paper)
  ISBN-10: 0-12-369545-7 (pbk. : alk. paper) 1. Arthropoda--Collection and preservation. 2. Arthropod--
Identification. I. Oseto, C. Y. II. Title.
  QL434.3.G53 2006
  595.7075--dc22
                           2005027226

British Library Cataloguing-in-Publication Data
A catalogue record for this book is available from the British Library.

ISBN 13: 978-0-12-369545-1
ISBN 10: 0-12-369545-7

For all information on all Elsevier Academic Press Publications
visit our Web site at www.books.elsevier.com

Printed in the United States of America
05 06 07 08 09 10 9 8 7 6 5 4 3 2 1

Working together to grow
libraries in developing countries

www.elsevier.com | www.bookaid.org | www.sabre.org

ELSEVIER   BOOK AID International   Sabre Foundation

Contents

PREFACE vii

PART 1: BASIC TOOLS AND GENERAL TECHNIQUES 1

Introduction 5

Chapter 1: Equipment and Collecting Methods 7

1.1 Equipment 7
    1.1.1 Forceps 7
    1.1.2 Sample Vials 8
    1.1.3 Killing Bottles 8
    1.1.4 Small Containers 8
    1.1.5 Small Envelopes 8
    1.1.6 Aspirators 8
    1.1.7 Absorbent Tissue 8
    1.1.8 Notebook 8
    1.1.9 Tools for Cutting or Digging 9
    1.1.10 Brush 9
    1.1.11 Bags 9
    1.1.12 Hand Lens 9
    1.1.13 Summary 9
1.2 Collecting Nets 10
1.3 Killing Containers and Agents 13
    1.3.1 Freezing Insects 13
    1.3.2 Injecting Insects with Alcohol 13
    1.3.3 Killing Jars for Field Collecting 13
    1.3.4 Liquid Killing Agents 14
    1.3.5 Solid Killing Agents 15
1.4 Aspirators and Suction Devices 17
1.5 Other Collection Devices 19
    1.5.1 Beating Sheets 19
    1.5.2 Drag Cloth 19
    1.5.3 Sifters 20
    1.5.4 Separators and Extractors 20
1.6 Traps 21
    1.6.1 Windowpane Trap 22
    1.6.2 Interception Nets and Barriers 22
    1.6.3 Malaise Traps 22
    1.6.4 Pitfall and Dish Traps 23
    1.6.5 Emergence Traps and Rearing Cages 24
    1.6.6 Traps Using Lobster or Eel Trap Principle 24

Arthropod Collection and Indentification Techniques

	1.6.7	Light Traps 25
	1.6.8	Light Sheets 27
	1.6.9	Color Traps 28
	1.6.10	Sticky Traps 28
	1.6.11	Snap Traps 29
	1.6.12	Artificial Refuge Traps 29
	1.6.13	Electrical Grid Traps 29
1.7	Baits, Lures, and Other Attractants 29	
	1.7.1	Sugaring for Moths 30
	1.7.2	Feces 31
	1.7.3	Oatmeal 31
1.8	Pheromones and Other Attractants 31	
	1.8.1	Carbon Dioxide 32
	1.8.2	Sounds 33
1.9	Collecting Aquatic Insects 33	
1.10	Collecting Soil Insects 34	
1.11	Collecting Ectoparasites 34	
1.12	Collecting Regulated Insects 34	
1.13	Collecting Insects for Pest Management Audits 34	
1.14	Collecting Insects for Medico-Criminal Investigations 35	
1.15	Rearing 35	
	1.15.1	Containers for Rearing 36
	1.15.2	Rearing Conditions and Problems 38
		1.15.2.1 Moisture 38
		1.15.2.2 Temperature 39
		1.15.2.3 Dormancy and Diapause 39
		1.15.2.4 Light 40
		1.15.2.5 Food 40
		1.15.2.6 Artificial Diets 41
		1.15.2.7 Special Problems and Precautions in Rearing 41
1.16	Collecting Insects for Molecular Research 42	

Chapter 2: Agents for Killing and Preserving 43

Chapter 3: Storage of Specimens 45

3.1	Temporary Storage 45	
	3.1.1	Refrigeration 45
	3.1.2	Dry Preservation 45
	3.1.3	Papering 46
3.2	Mounting Specimens 47	
	3.2.1	Preparing Dry Specimens for Mounting 48
	3.2.2	Degreasing 50
	3.2.3	Preparing Liquid-Preserved Specimens 51
	3.2.4	Direct Pinning 53
	3.2.5	Double Mounts 55
	3.2.6	Pinning Blocks 59

Table of Contents

- 3.2.7 Mounting and Spreading Boards and Blocks 60
  - 3.2.7.1 Construction of Spreading Boards 60
  - 3.2.7.2 Using the Spreading Boards 62
  - 3.2.7.3 Construction of Spreading Blocks 64
  - 3.2.7.4 Riker Mounts 66
  - 3.2.7.5 Inflation of Larvae 67
  - 3.2.7.6 Artificial Drying 67
  - 3.2.7.7 Embedding 69
- 3.2.8 Mounting Specimens for Microscopic Examination 69
  - 3.2.8.1 Maceration 70
  - 3.2.8.2 Washing 71
  - 3.2.8.3 Dehydration 71
  - 3.2.8.4 Staining 72
  - 3.2.8.5 Bleaching 72
  - 3.2.8.6 Mounting Medium 72
    - 3.2.8.6.1 Mounting 73
    - 3.2.8.6.2 Temporary Mounting 73
    - 3.2.8.6.3 Permanent Mounting Media 75
    - 3.2.8.6.4 Ringing 76
    - 3.2.8.6.5 Curing 77
    - 3.2.8.6.6 Labeling 77
  - 3.2.8.7 Preparation and Storage of Genitalia 77
  - 3.2.8.8 Mounting Wings 81
  - 3.2.8.9 Mounting Larvae of Diptera, Coleoptera, Lepidoptera, and Other Groups 82

3.3 Labeling 83
- 3.3.1 Paper 83
- 3.3.2 Pens 84
- 3.3.3 Ink 84
- 3.3.4 Lettered and Printed Labels 84
- 3.3.5 Size of Labels 85
- 3.3.6 Label Data 85
  - 3.3.6.1 Locality 85
  - 3.3.6.2 Date 86
  - 3.3.6.3 Collector 86
  - 3.3.6.4 Other Data 86
- 3.3.7 Placing the Labels 87
- 3.3.8 Labeling Vials 87
- 3.3.9 Labeling Microscope Slides 88
- 3.3.10 Identification Labels 88

3.4 Care of the Collection 89
- 3.4.1 Housing the Collection 89
- 3.4.2 Protecting Specimens from Pests and Mold 91

3.5. Packaging and Shipping Specimens 92
- 3.5.1 Packing Materials 92
- 3.5.2 Pinned Specimens 93
- 3.5.3 Specimens in Vials 94
- 3.5.4 Loading Cartons 94
- 3.5.5 Shipping Microscope Slides 95
- 3.5.6 Shipping Live Specimens 95

Arthropod Collection and Indentification Techniques

PART 2: CLASSIFICATION OF INSECTS AND MITES 97

Introduction 100

Chapter 4: Classification of Insects and Mites 103

    4.1   Key to Classes of Arthropoda 106
    4.2   Class Arachnida 110
        4.2.1   Key to Orders of Arachnida 110
    4.3   Subclass Acari 114
        4.3.1   Key to Some Primary Groups of the Subclass Acari 115
    4.4   Classes Diplopoda, Chilopoda, Pauropoda, and Symphyla 117
    4.5   Class Crustacea 117
    4.6   Class Hexopoda (Insecta) 117

Chapter 5: Synopsis of Insect Orders 121

    5.1   Subclass Class Entognatha: Primitive Wingless Hexapods 121
    5.2.   Subclass Ectognatha: Primitive Wingless Hexapods 121
    5.3   Subclass Pterygota (Insecta): Winged and Secondarily Wingless Insects 121
        5.3.1   Exopterygota 122
        5.3.2   Endopterygota 124
    5.4   Key to Orders of Hexapoda (Insects) 125

Chapter 6: Descriptions of Hexapod Orders 157

    6.1   Protura 157
    6.2   Diplura 157
    6.3   Collembola 158
    6.4   Microcoryphia 158
    6.5   Thysanura 159
    6.6   Ephemeroptera 159
    6.7   Odonata 160
    6.8   Orthoptera 162
    6.9   Blattodea 163
    6.10   Mantodea 164
    6.11   Phasmatodea 165
    6.12   Gryllblattodea 166
    6.13   Dermaptera 166
    6.14   Isoptera 167
    6.15   Embiidina 169
    6.16   Plecoptera 170
    6.17   Psocoptera 172

6.18 Zoraptera   172
6.19 Mallophaga   173
6.20 Anoplura   174
6.21 Thysanoptera   175
6.22 Hemiptera   176
6.23 Homoptera   177
6.24 Coleoptera   179
6.25 Strepsiptera   181
6.26 Mecoptera   182
6.27 Neuroptera   183
6.28 Tricoptera   184
6.29 Lepidoptera   185
6.30 Diptera   189
6.31 Siphonaptera   192
6.32 Hymenoptera   193

Summary 198

GLOSSARY 199

REFERENCES 255

APPENDIX 287
    I Liquid Preservation Formulas   288
    II Guidelines for Mounting Small and Soft-Bodied Specimens (SEL)   289
    III Directory – State Extension Service Directors and Administrators   291
    IV Submitting Specimens for Identification (SEL) 299

INDEX   305

NOTES:

PREFACE

More than 100 years have passed since C. V. Riley (1892) published ***Directions for Collecting and Preserving Insects***. That work marked the starting point in North America of publications on the collection, preservation, and curation of insects and mites. Many of these works are now unavailable, and some of the concepts contained in them involve obsolete technology.

In 1986, ***Insects and Mites: Techniques for Collection and Preservation*** (Steyskal et. al. 1986) was published by the U.S. Department of Agriculture. The book was written by George Steyskal, a research entomologist specializing in the taxonomy of Diptera; William Murphy, information specialist; and Edna Hoover, a materials processor for the Systematic Entomology Laboratory. Their avid interest in collecting and curating insects is contagious and much of their original work is reflected throughout this book. Michael E. Schauff published an electronic version of the original text together with several useful updates. We recognize his valuable contributions in the printed publication of this text.

In ***Arthropod Collection and Identification Techniques***, we have updated information and techniques as well as added new material to make the information more useable and complete. This book describes effective methods and equipment for collecting, identifying, rearing, examining, and preserving insects and mites and for storing and caring for specimens in collection. It also provides instructions for the construction of many kinds of collecting equipment, traps, rearing cages, and storage units, as well as updated and illustrated keys for identification of the classes of arthropods and the orders of insects. Such information not only aids hobbyists and professionals in preparing insect collections, but it has become essential in documenting and standardizing the collection of entomological evidence in forensic as well as pest management sciences.

We thank the Entomological Society of America for originally encouraging us to prepare this book. A number of colleagues, including Gordon Gordh, David Headrick, James McMurtry, Thomas Baker, and David Hawks provided useful comments on the revised text.

We hope that this work will be a useful reference and resource for anyone fascinated by the world of "bugs."

Timothy J. Gibb and Christian Y. Oseto
Entomologists, Purdue University

NOTES:

Part 1: Introduction

Part 1: Basic Tools and General Techniques

Arthropod Collection and Indentification Techniques

PART 1: BASIC TOOLS AND GENERAL TECHNIQUES

Introduction 5

Chapter 1: Equipment and Collecting Methods 7

- 1.1 Equipment 7
  - 1.1.1 Forceps 7
  - 1.1.2 Sample Vials 8
  - 1.1.3 Killing Bottles 8
  - 1.1.4 Small Containers 8
  - 1.1.5 Small Envelopes 8
  - 1.1.6 Aspirators 8
  - 1.1.7 Absorbent Tissue 8
  - 1.1.8 Notebook 8
  - 1.1.9 Tools for Cutting or Digging 9
  - 1.1.10 Brush 9
  - 1.1.11 Bags 9
  - 1.1.12 Hand Lens 9
  - 1.1.13 Summary 9
- 1.2 Collecting Nets 10
- 1.3 Killing Containers and Agents 13
  - 1.3.1 Freezing Insects 13
  - 1.3.2 Injecting Insects with Alcohol 13
  - 1.3.3 Killing Jars for Field Collecting 13
  - 1.3.4 Liquid Killing Agents 14
  - 1.3.5 Solid Killing Agents 15
- 1.4 Aspirators and Suction Devices 17
- 1.5 Other Collection Devices 19
  - 1.5.1 Beating Sheets 19
  - 1.5.2 Drag Cloth 19
  - 1.5.3 Sifters 20
  - 1.5.4 Separators and Extractors 20
- 1.6 Traps 21
  - 1.6.1 Windowpane Trap 22
  - 1.6.2 Interception Nets and Barriers 22
  - 1.6.3 Malaise Traps 22
  - 1.6.4 Pitfall and Dish Traps 23
  - 1.6.5 Emergence Traps and Rearing Cages 24
  - 1.6.6 Traps Using Lobster or Eel Trap Principle 24
  - 1.6.7 Light Traps 25
  - 1.6.8 Light Sheets 27
  - 1.6.9 Color Traps 28
  - 1.6.10 Sticky Traps 28
  - 1.6.11 Snap Traps 29
  - 1.6.12 Artificial Refuge Traps 29
  - 1.6.13 Electrical Grid Traps 29

Part 1: Introduction

- 1.7 Baits, Lures, and Other Attractants 29
  - 1.7.1 Sugaring for Moths 30
  - 1.7.2 Feces 31
  - 1.7.3 Oatmeal 31
- 1.8 Pheromones and Other Attractants 31
  - 1.8.1 Carbon Dioxide 32
  - 1.8.2 Sounds 33
- 1.9 Collecting Aquatic Insects 33
- 1.10 Collecting Soil Insects 34
- 1.11 Collecting Ectoparasites 34
- 1.12 Collecting Regulated Insects 34
- 1.13 Collecting Insects for Pest Management Audits 34
- 1.14 Collecting Insects for Medico-Criminal Investigations 35
- 1.15 Rearing 35
  - 1.15.1 Containers for Rearing 36
  - 1.15.2 Rearing Conditions and Problems 38
    - 1.15.2.1 Moisture 38
    - 1.15.2.2 Temperature 39
    - 1.15.2.3 Dormancy and Diapause 39
    - 1.15.2.4 Light 40
    - 1.15.2.5 Food 40
    - 1.15.2.6 Artificial Diets 41
    - 1.15.2.7 Special Problems and Precautions in Rearing 41
- 1.16 Collecting Insects for Molecular Research 42

Chapter 2: Agents for Killing and Preserving 43

Chapter 3: Storage of Specimens 45

- 3.1 Temporary Storage 45
  - 3.1.1 Refrigeration 45
  - 3.1.2 Dry Preservation 45
  - 3.1.3 Papering 46
- 3.2 Mounting Specimens 47
  - 3.2.1 Preparing Dry Specimens for Mounting 48
  - 3.2.2 Degreasing 50
  - 3.2.3 Preparing Liquid-Preserved Specimens 51
  - 3.2.4 Direct Pinning 53
  - 3.2.5 Double Mounts 55
  - 3.2.6 Pinning Blocks 59
  - 3.2.7 Mounting and Spreading Boards and Blocks 60
    - 3.2.7.1 Construction of Spreading Boards 60
    - 3.2.7.2 Using the Spreading Boards 62
    - 3.2.7.3 Construction of Spreading Blocks 64
    - 3.2.7.4 Riker Mounts 66
    - 3.2.7.5 Inflation of Larvae 67
    - 3.2.7.6 Artificial Drying 67
    - 3.2.7.7 Embedding 69

Arthropod Collection and Indentification Techniques

      3.2.8   Mounting Specimens for Microscopic Examination   69
            3.2.8.1   Maceration   70
            3.2.8.2   Washing   71
            3.2.8.3   Dehydration   71
            3.2.8.4   Staining   72
            3.2.8.5   Bleaching   72
            3.2.8.6   Mounting Medium   72
                  3.2.8.6.1   Mounting   73
                  3.2.8.6.2   Temporary Mounting   73
                  3.2.8.6.3   Permanent Mounting Media   75
                  3.2.8.6.4   Ringing   76
                  3.2.8.6.5   Curing   77
                  3.2.8.6.6   Labeling   77
            3.2.8.7   Preparation and Storage of Genitalia   77
            3.2.8.8   Mounting Wings   81
            3.2.8.9   Mounting Larvae of Diptera, Coleoptera, Lepidoptera, and Other Groups   82
3.3 Labeling   83
      3.3.1   Paper   83
      3.3.2   Pens   84
      3.3.3   Ink   84
      3.3.4   Lettered and Printed Labels   84
      3.3.5   Size of Labels   85
      3.3.6   Label Data   85
            3.3.6.1   Locality   85
            3.3.6.2   Date   86
            3.3.6.3   Collector   86
            3.3.6.4   Other Data   86
      3.3.7   Placing the Labels   87
      3.3.8   Labeling Vials   87
      3.3.9   Labeling Microscope Slides   88
      3.3.10   Identification Labels   88
3.4 Care of the Collection   89
      3.4.1   Housing the Collection   89
      3.4.2   Protecting Specimens from Pests and Mold   91
3.5. Packaging and Shipping Specimens   92
      3.5.1   Packing Materials   92
      3.5.2   Pinned Specimens   93
      3.5.3   Specimens in Vials   94
      3.5.4   Loading Cartons   94
      3.5.5   Shipping Microscope Slides   95
      3.5.6   Shipping Live Specimens   95

Introduction

Many reasons exist to collect and preserve insects. Hobbyists, nature enthusiasts, amateur collectors, 4-H or high school students, scientists, and criminal investigators each have different purposes to collect insects. Regardless of the purpose, however, insects collected must be preserved and processed according to established protocols. For example, a box of collected insects without accompanying collection information is of little scientific value. On the other hand, properly collected and preserved insects accompanied by collecting data can be invaluable.

Insects have long served as indicators of environmental disturbance or pollution. Types and numbers of insects found in various habitats over time provide a measure of change within a system if documented correctly. For example, aquatic insects are sensitive indicators of water pollution. Collecting and preserving aquatic samples requires different collection techniques than other types of insect collecting. Proper equipment and techniques must be understood and used if results are to be of value. Over time, a historical baseline of insects and arthropod species composition and population numbers can be established. Environmentalists look for significant changes from established records as evidence of changes (both positive and negative) to the environment in which they live.

In urban pest management, insect audits have become an extremely important tool in establishing the relative health of buildings. Pharmaceutical and food manufacturing plants and warehouses are some of the places where insect presence must be carefully monitored. Insect-infested products, where a single insect or even as much as a part of an insect is detected, can mean the loss of millions of dollars in potential revenues. Collecting and identifying insects in these sensitive situations requires great care and precision and a deep and working understanding of the life history and behavior of the insects in question.

New pests continue to threaten invasion. Counties, states, and countries spend many billions of dollars each year to prevent insects or other potentially harmful pests from invading and establishing themselves. Pests can be unintentionally introduced into new areas via commerce or tourism or by natural dispersion. Most often, when pests arrive in a new habitat, they cause extreme damage because natural controls, which may keep the pests in check in the original areas, do not exist in new habitats, thus allowing the pests to flourish unchecked. Intensive eradication efforts are sometimes implemented in an effort to exterminate the pest once detected. Regulatory officials have been more successful by either thoroughly regulating the movement of infested materials, or by monitoring all commodities originating from infested regions. In either case, collecting and preserving insects and related pests at ports of entry can help educate people and provide valuable leads in determining how and where to deny pest entry.

Arthropod Collection and Indentification Techniques

Insects found in food products eaten by unsuspecting people are often the subject of serious litigation, and the outcome of a lawsuit often rests on the collection and preservation of the insect sample as evidence. Likewise, insects that damage structures, agriculture or food or other commodities can result in the loss of millions of dollars. Those who control these pests often come under considerable pressure when insect inspections show infestations. Proper collection and preservation of such insects is essential.

Insects found in hospitals, where sterile environments are critical or where fully dependent patients are housed, can literally mean the difference between life and death in some instances. Insects that bite or sting cause annoyance, human sickness, and/or death. Insect-transmitted diseases continue to be a leading cause of death in the world. Accurate identification, monitoring, and controlling the spread of these insects is assisted by collecting and maintaining arthropods in a proper manner.

Criminal investigations also can hinge on insects as evidence if they are collected and preserved appropriately. Death scene investigators are very aware that natural succession of insects infesting a corpse can provide extremely valuable information in a death investigation. The species and stages of insects found, together with the temperatures in the surrounding areas, can be used to estimate the location as well as the time of death. Variations from the norm can indicate unusual sequences of postmortem events, but they stand up in a court of law only if the evidence is collected and preserved according to established protocol. Opportunities to collect insects at a death scene occur only once, so entomologists must be trained, prepared, and equipped adequately. In these cases, chain of custody and proper labeling of the evidence are especially critical.

Collecting insects from any habitat or for whatever purpose requires an understanding of the insects' specific behavior and ecological needs. A collector must understand why and where an insect lives in order to find and collect it. Various traps, flushing agents, and other technical tools may help in this effort. Rearing adult insects from immature life stages is sometimes appropriate, depending on the purpose of the information. Labeling the collected materials with date, precise location, and collector's name is a minimum. Often, identifying the host or describing the behavior of the insects at the time of collection is also valuable. Sometimes collecting associated materials, such as damaged leaves, wood, cast skins, or fecal material, in addition to the insect is needed. Signs or symptoms of infestations and specific behaviors of the insects just prior to collection are often required. Detailed written accounts or photographs can provide very important information, especially when some time has elapsed. Remember that providing too much information is seldom a problem. Having only partial information to document a situation that occurred in the past is a much more common frustration.

Part 1 of this text is devoted to describing the tools and techniques used to collect arthropods and the proper handling of specimens after they have been obtained. **Suggested Reading** features references that provide further detail for selected topics.

PART 1: BASIC TOOLS AND GENERAL TECHNIQUES

Chapter 1: Equipment and Collecting Methods

Collecting methods may be divided into two broad categories. In the first, a collector actively finds and collects the insects with the aid of nets, aspirators, beating sheets, or whatever apparatus suits particular needs. In the second, a collector participates passively and permits traps to do the work. Both approaches may be used simultaneously, and both are discussed in the following pages. Using as many different collecting methods as possible will permit a collector to obtain the greatest number of specimens in the shortest period of time.

Catching specimens by hand may be the simplest method of collecting; however, this method is not always productive because of the evasive behavior of many insects. Some insects are not active at times and places that the collector finds convenient. Some insects cause injury or discomfort through bites, stings, repulsive chemicals, or urticating setae. Often, particular kinds of equipment and special methods are needed. Equipment and methods described here have general application. Advanced studies of specific insect or mite groups may have developed unique procedures for collecting and surveying. For example, Agosti (2001) outlines procedures for surveying ground-nesting ant biodiversity. Clever collectors will make adaptations to fit their specific purposes and resources.

The equipment used to assemble an insect or mite collection is not necessarily elaborate or expensive. In many instances, a collecting net and several killing bottles will suffice. However, additional items will permit more effective sampling of a particular fauna. Many collectors carry a bag or wear a vest in which they store equipment. The following items usually are included in the general collector's bag.

1.1 Equipment

1.1.1 Forceps

Fine, lightweight forceps are strongly recommended for any collector. Specialized forceps may be selected depending upon individual needs (Fig. 1). Lightweight spring-steel forceps are designed to prevent crushing of fragile and small insects. Extra-fine precision may be obtained with sharp-pointed "watchmaker" forceps; however, care must be taken not to puncture specimens. When possible, grasp specimens with the part of the forceps slightly behind the points. Curved forceps often make this easier. When the forceps are not in use, their tips should be protected. This can be accomplished by thrusting the tips into a small piece of styrofoam or cork, or by using a small section of flexible tubing as a collar.

Figure 1.1
Forceps for insect collecting.
A. fine watchmaker forceps
B. curved metal collecting forceps
C. soft forceps

Arthropod Collection and Identification Techniques

Figure 1.2 Sample Vials.

1.1.2 Sample Vials

Sample vials of various sizes containing alcohol or other preservatives are necessary for collecting many species and life stages of insects and mites. Leak-proof caps are recommended for both field and permanent storage (Fig. 1.2).

1.1.3 Killing Bottles

Killing bottles or killing jars in various sizes are important to preserve specimens quickly (Fig. 1.3).

Figure 1.3 An example of a killing bottle appropriately affixed with a "Poison" label.

1.1.4 Small Containers

Small crush-proof containers are necessary for storing and protecting specimens after their removal from killing bottles (Fig. 1.4). These containers may be made of cardboard, plastic, or metal and should be partly filled with soft tissue paper to keep specimens from becoming damaged. Some collectors do not recommend the use of cotton in storage containers because specimens become entangled in the fibers and may become virtually impossible to extricate without damage. However, some collectors of minute or fragile insects find that specimens stored in a few wisps of cotton are better protected from damage.

1.1.5 Small Envelopes

Small envelopes are useful for temporary storage of delicate specimens (Fig. 1.5). Specially designed glassine envelopes, which prevent undue dislodging of butterfly and moth scales, are available from biological supply houses.

1.1.6 Aspirators

Aspirators are necessary for collecting many kinds of small-bodied or agile insects and mites.

1.1.7 Absorbent Tissue

Absorbent tissue is highly recommended for use in killing bottles and aspirators.

1.1.8 Notebook

A notebook and writing equipment are essential for jotting down notes and label data (Fig. 1.6).

Figure 1.5 Glassine envelope designed for field storage of moths and butterflies.

Figure 1.4 Examples of small crush-proof collecting containers.
A. empty film canister
B. plastic box
C. tin pill box

Figure 1.6 An example of a field journal for recording specimen label data and other notes.

Part 1: Basic Tools and General Techniques

1.1.9 Tools for Cutting or Digging

A knife or plant clippers or both are necessary for opening galls, seed pods, twigs, and other kinds of plant material. In addition, a small gardener's trowel (Fig. 1.7) for some kinds of excavation and a heavy knife or small hatchet may be helpful for searching under bark or in decaying logs.

Figure 1.7
A gardener's trowel for excavating insects.

1.1.10 Brush

A small, fine brush (camel's hair is best) is needed to aid in collecting minute specimens (Fig. 1.8). By moistening the tip, tiny specimens will adhere to it and may be quickly transferred to a killing bottle or vial.

1.1.11 Bags

Bags for retrieving plant material, rearing material, or Berlese samples are a good idea (Fig. 1.9). Remember that samples stored in plastic may decompose within a few short hours. Samples must be transferred to more permanent containers immediately upon returning from the field. For collecting much plant material, a botanist's vasculum (tin box) is advisable.

Figure 1.8
A fine-tip camel's-hair brush.

1.1.12 Hand Lens

A hand lens is helpful and will quickly become an indispensable aid to collectors (Fig. 1.10). A lens worn on a lanyard is convenient and prevents its loss while in the field.

Figure 1.10
A collapsible hand lens.

1.1.13 Summary

This list may be modified according to the special kinds of insects or mites to be collected. For example, a plant press to prepare plant specimens for determination or as voucher specimens, especially when leaf-mining insects are studied, may be needed. When collecting at night, a flashlight or headlamp is essential; the latter is especially useful because it leaves the hands free.

Much of the basic collecting equipment may be obtained from ordinary sources, but equipment especially designed for collecting insects often must be bought from biological supply houses. In some cases, addresses may be found on the Internet or in the yellow pages of telephone directories under "Biological Laboratory Supplies" or "Laboratory Equipment and Supplies." Biological and entomological publications often carry advertisements of equipment suppliers. Because these firms are located in many parts of the country and change names and addresses fairly often, it is not practical to list them here. Biologists at a local university usually can recommend a supplier in their area.

Figure 1.9
Bags for collecting miscellaneous plant or soil materials from the field.

Arthropod Collection and Identification Techniques

Figure 1.11 Insect-collecting nets.
A. aerial or sweeping net
B. aquatic net

Figure 1.12 A common design for the rim of an insect-collecting net made from steel wire.

Figure 1.13 General shape of the bag on an aerial or sweep net.

Figure 1.14 Attachment of the net to the rim of a collecting net.

1.2 Collecting Nets

Collecting nets come in three basic forms: aerial, sweeping, and aquatic (Fig. 1.11). The aerial net is designed especially for collecting butterflies and large-bodied flying insects. Both the bag and handle are relatively lightweight. The sweeping net is similar to the aerial net, but the handle is stronger and the bag is more durable to withstand being dragged through dense vegetation. The aquatic net is used for gathering insects from water and is usually made of metal screening or heavy scrim with a canvas band affixed to a metal rim. A metal handle is advisable because wooden ones may develop slivers after repeated wetting. The net chosen depends on the kind of insects or mites intended for collection.

Several kinds of nets, including collapsible models with interchangeable bags, are available from biological supply houses, but anyone with a little mechanical ability can make a useful net. The advantage of a homemade sweep net is that its size and shape can be adapted to the needs of the user, to the kind of collecting intended, and to the material available. Net-constructing materials include the following.

1. A length of heavy (8-gauge) steel wire for the rim, bent to form a ring 30–38 cm in diameter (Fig. 1.12). Small nets 15 cm or smaller in diameter sometimes are useful, but nets larger than 38 cm are too cumbersome for most collecting.

2. A strong, light fabric, such as synthetic polyester, through which air can flow freely. Brussels netting is best but may be difficult to obtain; otherwise, nylon netting, marquisette, or good-quality cheesecloth can be used. However, cheesecloth snags easily and is not durable. The material should be folded double and should be 1.5–1.75 times the rim diameter in length (Fig. 1.13). The edges should be double-stitched (French seams).

3. A strip of muslin, light canvas, or other tightly woven cloth long enough to encircle the rim. The open top of the net bag is sewn between the folded edges of this band to form a tube through which the wire rim is inserted (Fig. 1.14).

4. A straight, hardwood dowel about 19 mm in diameter and 105–140 cm long (to suit the collector). For attachment of the rim to the handle, a pair of holes of the same diameter as the wire are drilled opposite each other to receive the bent tips of the wire, and a pair of grooves as deep and as wide as the wire are cut from each hole to the end of the dowel to receive the straight part of the wire (Fig. 1.15).

Figure 1.1 How to make a handle for an insect net.

Part 1: Basic Tools and General Techniques

5. A tape or wire to lash the ends of the rims tightly into the grooves in the end of the handle. This may be electrician's plastic tape or fiber strapping tape commonly used for packaging. If wire is chosen, the ends should be bound with tape to secure them and to keep them from snagging. A close-fitting metal sleeve (ferrule) may be slipped over the rim ends and held in place with a small roundheaded screw instead of tape or wire lashing.

Figure 1.16 Proper procedure showing a bag flipped over the rim of a collecting net to prevent escape of the insect.

After the net has been placed on the rim, the ends of the band should be sewn together and the rim ends fastened to the handle. The other end of the handle should be filed to remove sharp edges. The net is then ready for use.

Efficient use of a net is gained only with experience. Collection of specimens in flight calls for the basic stroke: Swing the net rapidly to capture the specimen, then follow through to force the insect into the very bottom of the bag. Twist the wrist as you follow through so that the bottom of the bag hangs over the rim (Fig. 1.16); this will entrap the specimen. If the insect is on the ground or other surface, it may be easier to use a downward stroke, quickly swinging down on top of the specimen. With the rim of the net in contact with the ground to prevent the specimen from escaping, hold the tip of the bag up with one hand. Most insects will fly or walk upward into the tip of the bag, which can then be flipped over the rim to entrap the specimen.

Sweeping the net through vegetation, along the sand and seaweed on beaches, or up and down tree trunks will catch many kinds of insects and mites. An aerial net may be used in this way, but the more durable sweeping net is recommended for such rough usage. After sweeping with the net, a strong swing through the air will concentrate anything into the tip of the bag, and then, by immediately grasping the middle of the net with the free hand, the catch will be confined to a small part of the bag. Only the most rugged sweeping net may be used through thistles or brambles. Even some kinds of grasses, such as sawgrass, can quickly ruin a fragile net. Burrs and sticky seeds are also a serious problem.

Figure 1.17 Technique for paralyzing a butterfly by squeezing its thorax between thumb and index finger.

The catch may be conveyed from the bag to a killing jar in a number of ways. Single specimens are transferred most easily by lightly holding them in a fold of the net with one hand while inserting the open killing jar into the net with the other. While the jar is still in the net, cover the opening until the specimen is overcome; otherwise, it may escape before the jar can be removed from the net and closed. To prevent a butterfly from damaging its wings by fluttering in the net, squeeze the thorax gently through the netting when the butterfly's wings are closed (Fig. 1.17). This will temporarily paralyze the insect while it is

Arthropod Collection and Indentification Techniques

being removed to the killing jar. Experience will teach you how much pressure to exert. Obviously, pinching small specimens of any kind is not recommended. When numerous specimens are in the net after prolonged sweeping, it may be desirable to put the entire tip of the bag into a large killing jar for a few minutes to stun the insects. They may then be sorted and desired specimens placed separately into a killing jar, or the entire mass may be dumped into a killing jar for later sorting. These methods of mass collecting are especially adapted to obtaining small insects not readily recognizable until the catch is sorted under a microscope.

Figure 1.18 Technique of stunning insects while in collecting nets by using the killing jar.

Removal of stinging insects from a net can be a problem. Wasps and bees often walk toward the rim of the bag and may be made to enter a killing jar held at the point where they walk over the rim. However, many insects will fly as soon as they reach the rim, and a desired specimen may be lost. A useful technique involves trapping the insect in a fold of the net, carefully keeping a sufficient amount of netting between fingers and insect to avoid being stung. The fold of the net can then be inserted into the killing jar to stun the insect (Fig. 1.18). After a few moments, the stunned insect may be safely removed from the net and transferred to a killing jar. If the stunned insect clings to the net and does not fall readily into the jar, pry the insect loose with the jar lid or forceps. Do **not** attempt this maneuver with fingers because stunned wasps and bees can sting reflexively.

Several special modifications are necessary to adapt a net for aquatic collecting. Aerial nets made of polyester or nylon may be used to sweep insects from water if an aquatic net is not available. The bag will dry quickly if swept strongly through the air a few times. A wet net should not be employed for general collecting until thoroughly dry, or other specimens, especially butterflies, may be damaged.

For more specialized collecting, nets can be adapted in many ways. Nets can be attached to the ends of beams that are rotated about their midlength by a motor drive. Nets also can be adapted to be towed by or mounted on vehicles (Fig. 1.19).

Suggested Reading:
Collecting Nets
- Riley 1957
- McNutt 1976
- Martin 1977
- Rudd & Jensen 1977
- Dunn & Reeves 1980
- LeSage 1991

Vehicle Mounted Nets
- Traver 1940
- Gressitt et al. 1961
- Sommerman & Simmet 1965
- Sommerman 1967
- Hill 1971
- Almand et al. 1974
- Rudd & Jensen 1977
- Holzapfel et al. 1978
- Kronblad & Lundberg 1978
- Barnard 1979

Figure 1.19 A specialized vehicle-mounted insect net.

Part 1: Basic Tools and General Techniques

1.3 Killing Containers and Agents

Killing an insect quickly is paramount to the preservation of a nice specimen. If an insect is allowed to beat its wings or crawl about for extended periods, it will inevitably harm its wings, break legs and antennae, or lose color.

1.3.1 Freezing Insects

When collecting around the home or school, insects can be killed effectively with minimum damage by placing them immediately into a freezer. This method has two distinct advantages: First, no messy and potentially dangerous chemicals are needed; second, insects may be left in the freezer for long periods of time and need only be thawed before pinning. This convenience alone makes the freezer method attractive to many collectors.

1.3.2 Injecting Insects with Alcohol

Most large insects and especially large moths are best killed with an injection of alcohol using a hypodermic needle and syringe. For most insects, inject into the ventral area of the thorax. For large beetles, inject alcohol into a coxal socket. Less than 1 cm^3 of alcohol is usually sufficient to kill large specimens.

1.3.3 Killing Jars for Field Collecting

Any heavy, widemouthed glass jar or bottle with a tight-fitting stopper or metal screw top may be used as a killing container (Fig. 1.20). Glass is preferred over plastic, and jars with relatively thick glass are preferred over thin, fragile glass, for obvious reasons. Olives frequently are sold in bottles that make convenient killing containers. Tops that may be removed with only a quarter turn often are preferred but may not be obtained readily. This type of lid may be quickly removed and returned to the jar with a minimum of effort. Collectors interested in taking minute or small insects may prefer using small vials that can be carried in a shirt pocket. Parallel-sided vials can be closed with cork stoppers. When collecting small-bodied insects in vials, care must be exercised to ensure that the stopper seats firmly against the wall of the vial. Otherwise, specimens become wedged between the glass and stopper, resulting in damage to some specimens. A crumpled piece of tissue paper placed in the vial helps to maintain the specimens clean and disentangled.

Jars for use with liquid killing agents are prepared in one of two ways. One way is to pour about 2.5 cm of plaster of Paris mixed with water into the bottom of the jar and allow the plaster to dry without replacing the lid. Sufficient killing agent is then added to saturate the plaster; any excess should be removed. The

A

B

Figure 1.20
A homemade killing bottle made from absorbent materials covered by a tight fitting cardboard disk.

lid is then replaced. This kind of jar can be recharged merely by adding more killing agent. A second method is to place a wad of cotton, paper, or other absorbent material in the bottom of a jar, pour enough liquid killing agent into the jar to nearly saturate the absorbent material, and then press a piece of stiff paper or cardboard, cut to fit the inside of the jar tightly, over it (Fig. 1.20). The paper or cardboard acts as a barrier between the insect and the killing agent, preventing the specimen from contacting the agent directly and keeping the agent from evaporating too rapidly.

1.3.4 Liquid Killing Agents

Popular liquid killing agents include ethyl acetate ($CH_3CO_2 \cdot C_2H_5Z$), carbon tetrachloride (CCl_4), ether (diethyl ether, $C_2H_5 \cdot O \cdot C_2H_5$), chloroform ($CHCl_3$), and ammonia water ($NH_4OH$ solution). Ethyl acetate is recommended by many entomologists as the most satisfactory liquid killing agent. The fumes of ethyl acetate are less toxic to humans than the fumes of the other substances. Ethyl acetate usually stuns insects quickly but kills them slowly. Specimens, even though they appear dead, may revive if removed from the killing jars too soon. However, a compensating advantage is that most specimens may be left in an ethyl acetate killing jar for several days and remain pliant. (If the ethyl acetate is allowed to evaporate from the killing jar, specimens will harden.) For these reasons, a killing jar with ethyl acetate is preferred by many entomologists over a cyanide jar, especially when the jar is used infrequently.

Hobbyists and other collectors have found that fingernail polish remover (acetone) also works as an alternative to ethyl acetate and is more commonly available, especially in an emergency situation.

Carbon tetrachloride was once very popular as a liquid killing agent because it is not flammable and was easily obtained as a spot remover for clothes. Carbon tetrachloride is no longer recommended, however, because it is a carcinogen and a cumulative liver toxin. Specimens killed with carbon tetrachloride often become brittle and difficult to pin.

Ether and chloroform are extremely volatile and flammable and should not be used near an open flame or lighted cigarette. Their high volatility makes them serviceable in a killing jar for only a short period of time. Perhaps the greatest hazard with chloroform is that even when stored in a dark-colored jar, it eventually forms an extremely toxic gas called phosgene (carbonyl chloride, $COCl_2$). Chloroform, however, is useful when other substances cannot be obtained. Chloroform stuns and kills quickly, but it has the disadvantage of stiffening specimens.

Part 1: Basic Tools and General Techniques

Ammonia is irritating to humans, does not kill insects very effectively, and spoils the colors of many specimens. However, ammonia is readily available and will serve in an emergency. Ammonium carbonate, a solid but volatile substance, also may be substituted.

Spray-dispensed insecticides may be used, if not to kill specimens, to at least "knock them down" into a container from which they may be picked up. If they are directed into a container topped with a funnel, they may be allowed to revive and treated further as desired (Clark & Blom 1979).

Killing agents may be any of various liquids. Never deliberately inhale the fumes, even momentarily. All killing agents are to some extent hazardous to human health. All killing jars or bottles should be clearly labeled "**POISON**" and kept away from persons unaware of their danger (Fig. 1.21).

Collectors who travel on airlines to collecting sites, especially on international flights, should be aware that the transportation industry now prohibits poisons on commercial passenger airliners in many parts of the world. Security at airports continues to increase, and the likelihood of being questioned about toxicants within carry-on baggage is high. Customs officials may also object to the importation of killing agents such as cyanides and ethyl acetate. When an international collecting trip is anticipated, travel agents and consulate officials should be consulted.

> *Never deliberately inhale the fumes, even momentarily.*
>
> *All killing agents can be hazardous to human health.*
>
> *All killing jars or bottles should be clearly labeled "POISON" and kept away from persons unaware of their danger!*

1.3.5 Solid Killing Agents

The solid killing agents used in killing jars are cyanides: potassium cyanide (KCN), sodium cyanide (NaCN), and calcium cyanide ($Ca(CN)_2$). Potassium cyanide is the compound of choice among many collectors. Sodium cyanide is equally effective but it is hygroscopic (it absorbs water and makes the jar wet). Calcium cyanide is seldom available. Handle all cyanide compounds with extreme care. They are dangerous, rapid-acting poisons with no known antidote. If even a single grain touches the skin, wash the affected area immediately with water. To avoid handling the cyanide and storing or disposing of surplus crystals, you may be able to find a chemist, pharmacist, or professional entomologist who can make the killing jar for you. If this is not feasible, use utmost care in following the instructions given here.

To make a moderate-sized cyanide killing jar, place cyanide crystals about 15 mm deep in the bottom. A smaller amount of cyanide may be used for smaller jars. Cover the crystals with about 10 mm of sawdust and add about 7 mm of plaster of Paris mixed with water to form a thick paste, working quickly before the plaster solidifies. Then add crumpled absorbent paper to prevent water condensation on the inside surface of the glass. Instead of the plaster of Paris, a

Figure 1.21
An example of a properly labeled killing jar.

plug of paper or cardboard may be pressed on top of the sawdust. Be sure that it fits tightly. When the jar is ready for use, place several drops of water on the plaster or paper plug. In an hour or so, enough fumes of hydrocyanic acid will have been produced to make the jar operative. **Do not test this by sniffing the open jar!**

Another substance that has been recommended as a killing agent is dichlorvos (2.2-dichloroethenyl dimethyl phosphate), also called DDVP, Vapona, Nogos, Herkol, and Nuvan. Polyvinyl chloride (PVC) resin impregnated with this chemical and sold commercially as bug strips or No-Pest Strips is long lasting and somewhat less dangerous than other killing agents, but its time-release aspect allows only small quantities of the active agent to be released over time, which may be too small to kill quickly. Therefore, PVC-impregnated dichlorvos is more effective as a killing agent in traps than in killing jars.

Every killing jar should be clearly and prominently labeled "**POISON.**" The date the killing jar was fabricated can also be placed on the poison label. This will give the collector a rough idea of the life of a killing jar. The bottom outside of the jar must be reinforced with tape, preferably cloth, plastic, or clinical adhesive tape. The tape cushions the glass against breakage and keeps its dangerous contents from being scattered if the container breaks.

Laws involving toxic materials are changing, and local officials should be consulted as how best to proceed.

Killing jars or bottles will last longer and give better results if the following simple rules are observed.

1. Place a few narrow strips of absorbent paper in each killing jar to keep it dry and to prevent specimens from mutilating or soiling each other. Replace the strips when they become moist or dirty. This technique is useful for most insects, except Lepidoptera, which are difficult to disentangle without damage.

2. Do not leave killing jars in direct sunlight because they will "sweat" and rapidly lose their killing power.

3. If moisture condenses in a jar, wipe it dry with absorbent tissue.

4. Keep delicate specimens in separate jars so that larger specimens do not damage them.

5. Do not allow a large number of specimens to accumulate in a jar unless it is to be used specifically for temporary storage.

Part 1: Basic Tools and General Techniques

6. Do not leave insects in a cyanide killing jar for more than a few hours. The fumes will change the colors of some insects, especially yellows to red, and specimens will generally become brittle and difficult to handle.

7. If it is necessary to hold insects for more than several hours, place the specimens in another container and store them in a refrigerator.

8. Keep butterflies and moths in jars by themselves so that the setae and scales that they shed will not contaminate other specimens. Contaminated specimens often make viewing of taxonomic features more difficult. Further, removing scales and setae is time-consuming, often tedious, always annoying, and potentially destructive to the specimen.

9. Never test a killing jar by smelling its contents.

10. Old jars that no longer kill quickly should be recharged or disposed of in a legal and responsible manner. A cyanide jar that has become dry may be reactivated by adding a few drops of water.

SUGGESTED READING:
Killing Agents
- Lindroth 1957
- Frost 1964
- White 1964
- Pennington 1967
- Preiss et al. 1970
- Clark & Blom 1979
- Banks et al. 1981

1.4 Aspirators and Suction Devices

The aspirator (Fig. 1.22), known in England as a "pooter," is a convenient and effective device for collecting small or highly mobile insects and mites. (The device was named in honor of Frederick W. Poos, an American entomologist who employed the device to collect Cicadellidae.) The following materials are needed to construct an aspirator.

1. A vial 2.5–5 cm in diameter and about 12 cm long.

2. Two pieces of glass or copper tubing about 7 mm in diameter, one piece about 8 cm long and the other about 13 cm long. Copper tubing can be obtained from hobby shops and has the obvious advantage of being less fragile than glass.

3. A rubber stopper with two holes in which the tubing will fit snugly.

4. A length of flexible rubber or plastic tubing about 1 m long, with diameter just large enough to fit snugly over one end of the shorter piece of stiff tubing.

5. A small piece of cloth mesh, such as cheesecloth, and a rubber band.

Figure 1.22
Components of a vial aspirator.
1. vial
2. glass or copper tubing
3. rubber stopper
4. rubber or plastic tubing
5. mesh cloth and rubber band

Arthropod Collection and Indentification Techniques

Figure 1.23
A blow-type aspirator.

To make an aspirator, bend the glass tubes as in Fig. 1.22. In bending or cutting glass tubes, always protect your fingers by holding the glass between several layers of cloth. Obtain the advice of a chemist or laboratory technician for cutting and bending glass. Moisten one end of the longer tube and insert it through one of the holes in the rubber stopper. Moisten one end of the shorter tube and insert it through the other hole in the stopper. Next, use a rubber band to fasten the cloth mesh over the end that was inserted through the stopper. (This will prevent specimens from being sucked into the collector's mouth when the aspirator is used.) Attach one end of the flexible tubing to the free end of this tube. The length, size and amount of bend in the tubing will vary according to the user's needs. To complete the assembly, insert the rubber stopper into the vial. To use the aspirator, place the free end of the flexible tubing in the mouth, move the end of the longer glass tube close to a small specimen, and suck sharply. The specimen will be pulled into the vial.

Instead of a vial, some workers prefer a tube. In either method, keep small pieces of absorbent tissue in the vial or tube at all times to prevent moisture from accumulating. Note that there is some danger of inhaling harmful substances or organisms when using a suction-type aspirator (Hurd 1954).

Either the vial- or tube-type aspirator (Fig. 1.22) may be converted into a blow-type aspirator by removing the 13-cm glass tube and substituting a T-shaped attachment (Fig 1.23). The flexible tubing is attached to one arm of the T, the opposite arm is left open, and the stem of the T is inserted into the vial and covered with mesh. Upon blowing through the flexible tubing, a current of air passes across the T and creates a partial vacuum in the vial, which produces the suction needed to draw specimens into the vial. This kind of aspirator eliminates the danger of inhaling small particles, fungus spores, or noxious fumes. Aspirators with a squeeze bulb sometimes may be purchased, or, if a valved bulb can be obtained, they may be used with either pressure or suction.

Collection traps also have been devised with the suction feature applied on a much larger scale than with the usual aspirator. These include the handheld converted dustbuster insect vacuums (Fig. 1.24), the D-vac, (which employs a backpack motor fan) (Dietrick et al. 1959, Dietrick 1961); the Insectavac (Ellington et al. 1984a, b); and another device developed for the collection of honey bees (Gary & Marston 1976). Suction produced by a fan has been employed in traps in conjunction with light or other attractants. Some of these traps are described in the following references and in Section 1.b on ***Traps***. Suction is created by a piston in a "slurp-gun" described for aquatic collecting (Gulliksen & Deras 1975). This principle could be adapted for use in air to gather insects and to deposit them in a vial attached to the side of the piston.

Figure 1.24
A motorized suction device (i.e. Dustbuster) modified to collect insects.

Part 1: Basic Tools and General Techniques

SUGGESTED READING:

Aspirators and Suction Devices

- Johnson 1950
- Hurd 1954
- Jonasson 1954
- Johnson & Taylor 1955
- Woke 1955
- Johnson et al. 1957
- Lumsden 1958
- Dietrick et al. 1959
- Dietrick 1961
- Minter 1961
- Taylor 1962a
- Singer 1964
- Wiens & Burgess 1972
- Coluzzi & Petrarca 1973
- Sholdt & Neri 1974
- Bradbury & Morrison 1975
- Evans 1975
- Azrang 1976
- Gary & Marston 1976
- Barnard & Mulla 1977
- Ellington et al. 1984a, b
- Summers et al. 1984
- Holtcamp & Thompson 1985
- Belding et al. 1991
- DeBarro 1991
- Governatori et al. 1993
- Moore et al. 1993
- Wilson et al. 1993
- Arnold 1994

A. *square beating sheet*

1.5 Other Collection Devices

1.5.1 Beating Sheets

A beating sheet should be made of durable cloth, preferably white, attached to a frame about 1 m square, with two pieces of doweling or other light wood crossing each other and fitted into pockets at each corner of the cloth. Variations on the design can include a cloth stretched between two dowels or a triangular cloth that can be operated with one hand (Fig. 1.25). An ordinary light-colored umbrella also may be used as a beating sheet. Place the beating sheet or umbrella under a tree or shrub or limb and sharply beat the branches or foliage with a club or stick. Specimens that fall onto the sheet are easily seen against the light-colored material and may be removed by hand or with forceps, a moistened brush, or an aspirator. Locating specimens on the sheet is sometimes a problem because leaves or other unwanted material also drop onto the sheet. Watching for movement helps to locate specimens. Tilting the sheet displaces debris, leaving the insects and mites clinging to the cloth.

B. *two-handled beating sheet*

Beating sheets are especially useful for collecting beetles and are particularly effective early and late in the day and when the weather has turned cold. A "ground cloth" also is used in sampling crop fields (Rudd & Jensen 1977) in much the same manner as a beating sheet.

1.5.2 Drag Cloth

Figure 1.26
Drag cloth.

A drag cloth or flag (Fig. 1.26) made from durable light-colored cloth attached to a piece of doweling along one edge can be dragged through long grass or small shrubs to collect ticks. A variation on this method is to make a flag on a longer handle and to use it to brush against larger trees and shrubs. Questing ticks dislodge easily from the plants and collect on the cloth.

C. *triangular beating sheet*

D. *how to hold a triangular beating sheet*

Figure 1.25
Beating sheets.

Arthropod Collection and Indentification Techniques

1.5.3 Sifters

Figure 1.27 Sifters.

Sifters serve to collect insects and mites that live in ground litter, leaf mold, rotting wood, mammal and bird nests, fungi, shore detritus, lichens, mosses, and similar material (Fig. 1.27). Leaf litter sifters can be constructed very simply by collecting leaf litter on a container, shaking the contents through a wire screen such as a cooling rack to separate the large leaves and then retaining the reduced sample. Sifters also are especially useful for winter collecting to pick specimens in diapause.

Almost any container with a wire-mesh screen bottom will serve as a sifter. The size of the mesh depends on the size of the specimens sought. For general purposes, screening with 2.5–3 meshes per centimeter is satisfactory. To use the sifter, place the material to be sifted into the container and shake it gently over a white pan or piece of white cloth. As the insects and mites fall onto the cloth, they may be collected with forceps, a brush, or an aspirator.

A similar method is used chiefly to collect mites from foliage. Employing a sifter of 20-mesh screen (about 8 meshes per centimeter) with a funnel underneath that leads to a small vial, beat pieces of vegetation against the screen to dislodge the mites and cause them to fall through the screen and into the vial below.

1.5.4 Separators and Extractors

Somewhat similar to the sifter are various devices designed to separate or extract live specimens from substances in which they may be found, such as leaf mold and other kinds of vegetable matter, shore detritus, and dung. A simple extractor can be constructed from a 2-liter plastic pop bottle (Fig. 1.28). Separators are also used effectively in net sweepings that include so much foreign matter that it is difficult to pick out the insects. These devices usually depend on some physical aid, such as light, heat, or dryness, to impel the insects to leave the substrate.

Figure 1.28 Pop bottle extractor.

One of the simplest such devices is the sweeping separator (Fig. 1.29), which consists of a carton or wooden box with a tight-fitting lid. A glass jar is inserted near the top of the box on one side. If the jar is made with a screw top, a hole of proper diameter cut in the side of the carton will permit the jar to be screwed onto it. The cover ring, without the lid, from a home-canning jar may be nailed to the periphery of a hole in a wooden box and the jar then screwed onto the ring.

Figure 1.29 Sweeping separator box.

The sweepings are dumped into the box and the cover is quickly closed. A flashlight, table lamp, or other light source should be placed close to the jar. The insects in the darkened box soon will be attracted to the lighted glass jar. When all the insects appear to have entered the jar, it can be removed and its contents put into a killing jar. Alternatively, a jar cover containing a piece of blotting paper soaked with xylene may be placed over the jar for a short while to stun the insects, which may then be sorted.

Part 1: Basic Tools and General Techniques

The Berlese funnel (Fig. 1.30) and its modifications are cleaner and more efficient than sifting to separate insects and mites from leaf mold and similar materials. The sample is placed on a screen near the top of a funnel. A light bulb can be placed above the sample to produce heat and light, which drive the insects downward into the funnel. In other designs, heated coils or a jacket around the funnel can be used to dry the sample and make it inhospitable. The insects and mites are directed by the funnel into a container, sometimes containing alcohol, at the bottom of the funnel. Care should be taken not to dry the sample so rapidly that slow-moving specimens die before they can leave the sample. To prevent large amounts of debris from falling into the container, place the sample on the screen before the container is put in place.

Figure 1.30
Berlese funnel and variations.

1.6 Traps

A trap is any device that impedes or stops the progress of an organism. Traps are used extensively in entomology and may include devices used with or without baits, lures, or other attractants. The performance of a trap depends on its construction, location, time of day or year, weather, temperature, and kind of attractant. A little ingenuity coupled with knowledge of the habits of the insects or mites sought will allow for modifications or improvements in nearly any trap or may even suggest new traps. Only a few of the most useful traps are discussed here; however, Peterson (1964) and Southwood (1979) include extensive bibliographies on insect trapping.

SUGGESTED READING:

Separators & Extractors

- Salmon 1946
- Kevan 1955, 1962
- Newell 1955
- Kempson et al. 1962
- Murphy 1962
- Woodring 1968
- Brown 1973
- Merritt & Poorbaugh 1975
- Gruber & Prieto 1976
- Lane & Anderson 1976
- Masner & Gibson 1979
- Arnett 1985
- Zolnerowich et al. 1990
- Williams 1951
- Whittaker 1952
- Dales 1953
- Dodge & Seago 1954
- Williams 1954
- Eastop 1955
- Woke 1955
- Heathcote 1957a
- Banks 1959
- Dodge 1960
- Race 1960
- Fredeen 1961
- Glasgow & Duffy 1961
- Gressitt et al. 1961
- Harwood 1961
- Morris 1961
- Taylor 1962b
- Hollingsworth et al. 1963
- Frost 1964
- Hance & Bracken 1964
- Peterson 1964
- Thorsteinson et al. 1965
- Turnbull & Nicholls 1966
- Bidlingmayer 1967
- Kimerle & Anderson 1967
- Everett & Lancaster 1968
- Hartstack et al. 1968
- Nijholt & Chapman 1968
- Heathcote et al. 1969
- Herting 1969
- Thompson 1969
- Catts 1970
- Dresner 1970
- McDonald 1970
- Clinch 1971
- Gojmerac & Davenport 1971
- Hansens et al. 1971
- Nakagawa et al. 1971
- Emden 1972
- Pickens et al. 1972
- A'Brook 1973
- Ford 1973
- Klein et al. 1973
- Hienton 1974
- Weseloh 1974
- Yates 1974
- Acuff 1976
- Hargrove 1977
- Lammers 1977
- Martin 1977
- Rogoff 1978 *(includes extensive bibliography)*
- Stubbs & Chandler 1978 *(collecting and recording on pp.1-37)*
- Barnard 1979
- Meyerdirk et al. 1979
- Southwood 1979
- Hafraoui et al. 1980
- Howell 1980
- Sparks et al. 1980 *(includes pictures of several types of traps)*
- Banks et al. 1981

Traps

- Barber 1931
- Williams & Milne 1935
- Chamberlin 1940
- Fleming et al. 1940
- Huffaker & Back 1943
- Broadbent et al. 1948
- Johnson 1950

Arthropod Collection and Indentification Techniques

SUGGESTED READING:

Light Traps
- Glick 1939, 1957
- Frost 1957, 1958
- Blakeslee et al. 1959
- Meyers 1959
- Goma 1965
- Coon & Pepper 1968
- Fincher & Stewart 1968
- Stewart & Lam 1968
- Callahan et al. 1972
- Hocking & Hudson 1974
- Roling & Kearby 1975

Windowpane Traps
- Chapman & Klinghorn 1955
- Corbet 1965
- Lehker & Deay 1969
- Wilson 1969

Interception Nets and Barriers
- Parman 1931, 1932
- Gordon & Gerberg 1945
- Leech 1955
- Nielsen 1960
- Merrill & Skelly 1968
- Gillies 1969
- Nielsen 1974
  (*describes catches of insects in funnel traps on trunks of beech trees*)
- Steyskal et al. 1981

The elevation or height above ground at which the trap is placed can be important in affecting the performance of traps, especially light traps. Optimum trap placement is a complex issue with many variables, including the behavior of the targeted insect, specific locality, and size and color of the trap, influencing its performance.

1.6.1 Windowpane Trap

This simple and inexpensive trap (Fig. 1.31) involves a barrier consisting of a windowpane held upright by stakes in the ground or suspended by a line from a tree or from a horizontal line. A trough filled with a liquid killing agent is placed so that insects flying into the pane drop into the trough and drown. They are removed from the liquid, washed with alcohol or other solvent, and then preserved or dried and pinned. This trap is not recommended for adult Lepidoptera or other insects that may be ruined if collected in fluid.

Figure 1.31
Windowpane trap.

1.6.2 Interception Nets and Barriers

A piece of netting about 1.8 m high can be stretched between three trees or poles to form a V-shaped trap, with the wide end of the V open. A triangular roof should be adjusted to slope gently downward to the broad open side of the V. A device of this type will intercept many kinds of flying insects, particularly if the trap is situated with the point of the V toward the side of maximum light and in the direction of air movement. A pair of such nets set in opposite directions or a single net in a zigzag shape will intercept specimens from two directions. Insects flying into such a net tend to gather at the pyramidal apex, so they are easy to collect. The so-called "funnel" or "ramp" traps are interception devices that direct insects to a central point, where a retaining device or killing jar may be placed.

1.6.3 Malaise Traps

A modification of the interception net led to the more complex trap developed by the Swedish entomologist René Malaise that now bears his name. Several modifications of his original design have been published, and at least one is available commercially (Townes 1962, 1972; Steyskal 1981). The trap, as originally designed, consists of a vertical net serving as a baffle, end nets, and a sloping canopy leading up to a collecting

Figure 1.32
Malaise trap.

device (Fig. 1.32). The collecting device may be a jar with either a solid or evaporating killing agent or a liquid in which the insects drown. The original design is unidirectional or bidirectional with the baffle in the middle, but more recent types include a nondirectional type with cross-baffles and with the collecting device in the center. Malaise traps have been phenomenally successful, sometimes collecting large numbers of species that could not be obtained otherwise. Attractants increase the efficiency of the traps for special purposes.

1.6.4 Pitfall and Dish Traps

Another simple but very effective and useful type of interception trap consists of a dish, can, or jar sunk in the ground and is called a pitfall trap (Fig. 1.33). Wandering arthropods fall into the trap and are unable to escape from the partly filled container of 70% ethyl alcohol or ethylene glycol (automobile antifreeze). The latter is preferred because it does not evaporate. A cover must be placed over the open top of the container to exclude rain and small vertebrates while allowing insects and mites to enter. Pitfall traps may be baited with various substances, depending on the kind of insects or mites the collector hopes to capture. Although most specimens that fall into the trap remain there, the trap should be inspected daily because specimens can decompose or become damaged. Specimens should be removed and placed in alcohol or in a killing bottle while they are in their best condition.

Other pitfall trap modifications include placing a bait within the collection container or hanging it directly above. The cereal dish trap, another modified pitfall trap, represents a simple but effective device for obtaining insects attracted to dung. This trap consists of a small dish, preferably with a rim, set in the ground (Fig. 1.33C) and partly filled with 70% ethyl alcohol or ethylene glycol. A piece of stout wire, such as a coat hanger, is bent to form a loop at one end that holds the bait receptacle. A few zigzag bends in the other end of the wire will keep the looped end from swinging after the wire is pushed into the earth. The bait receptacle may be a small plastic or metal cup (such as that used for medicine doses), a coffee creamer, or a cup formed from aluminum foil. When baited with animal or human feces, this trap attracts beetles (typically

SUGGESTED READING:
Pitfall and Dish Traps
- Broadbent et al. 1948
- Broadbent 1949
- Tretzel 1955
- Greenslade 1964
- Luff 1968, 1975
- Alderz 1971
- Briggs 1971
- Greenslade & Greenslade 1971
- Gist & Crossley 1973
- Greenslade 1973
- Schmid et al. 1973
- Loschiavo 1974
- Joossee 1975
- Morrill 1975 *(bibliography)*
- Muma 1975 *(includes bibliography of can and pitfall trapping)*
- Newton & Peck 1975
- Shubeck 1976
- Smith 1976
- Smith et al. 1977 *(describes low-cost carrying cages, pitfall traps, and rearing cages)*
- Thomas & Sleeper 1977
- Houseweart et al. 1979
- Reeves 1980

A. *traditional pitfall trap*

B. *baited pitfall trap*

C. *cereal dish trap*

D. *PVC pitfall trap*

Figure 1.33 Pitfall and dish trap

Arthropod Collection and Indentification Techniques

Figure 1.34
Emergence traps.

Scarabaeidae and Staphylinidae), springtails, ants, earwigs, some parasitic Hymenoptera, and several families of flies, including Phoridae, Sepsidae, and Muscidae. The larger, strong-flying calliphorid and sarcophagid flies seldom fall into the liquid, although they are attracted to the bait. The alcohol fumes may cause smaller flies to drop into it. The trap is made of easily obtained materials, is easily transported, and provides excellent results. It deserves wide use.

Modifications to the basic pitfall trap include a section (up to a meter) of PVC pipe, capped at both ends, and where a small channel is cut lengthwise along the pipe (Fig. 1.33D). Burying the pipe such that the open channel is at the soil surface allows wandering insects to fall into the pipe.

1.6.5 Emergence Traps and Rearing Cages

An emergence trap is any device that prevents adult insects from escaping when they emerge from their immature stages in any substrate, such as soil, plant tissue, or water (Yates 1974). A simple canopy over an area of soil, over a plant infested with larvae, or over a section of stream or other water area containing immature stages of flying insects will secure the emerging adults (Lammers 1977). If the trap is equipped with a retaining device (as in the Malaise trap), the adults can be killed and preserved shortly after emergence. Remember, however, that many insects should not be killed too soon after emergence because the adults are often teneral (soft bodied) and incompletely pigmented. For preservation, specimens must be kept alive until the body and wings harden completely and colors develop fully. Emergence traps and rearing cages (Fig. 1.34) enable the insects to develop naturally but ensure their capture when they mature or when larvae emerge to pupate.

1.6.6 Traps Using the Lobster or Eel Trap Principle

This trap design can be made from any container that has its open end fitted with a truncated cone directed inward (Fig. 1.35). This trap design is commonly used to catch minnows, lobsters, and eels in water. An ordinary killing jar with a funnel fastened into its open end is an example. When the funnel is placed over an insect, the specimen will usually walk or fly toward the light and enter the jar through the funnel. Modified traps of this type include the Steiner and McPhail traps, which are used extensively in fruit fly surveys.

McPhail was the first entomologist to use an invaginated clear glass trap for evaluating the attraction of fruit flies to various baits (Burditt 1982). The McPhail Trap (Fig. 1.36), as it is known today, is used extensively for monitoring fruit fly

SUGGESTED READING:
Lobster/Eel Traps
- Nicholls 1960
- Doane 1961
- Brockway et al. 1962
- Morrill & Whitcomb 1972
- Nielsen 1974
  (describes catches of insects in funnel traps on trunks of beech trees)
- Nakagawa et al. 1975
- Reierson & Wagner 1975
- Broce et al. 1977
- Steyskal 1977
- Becker et al. 1991

Figure 1.35
Eel trap.

populations in the United States and abroad. A homemade version of the McPhail trap was devised by Burditt (1988), and a number of baits such as sucrose, yeast, or ethyl alcohol, can be used to attract and collect many kinds of insects. As an example, the McPhail trap was used in a survey to collect euglossine bees in tropical Amazonian forests (Becker et al. 1991). The inside of the Steiner trap usually has a sticky material containing a pheromone or other lure.

1.6.7 Light Traps

Light traps take advantage of the attraction of many insects to a light source and their attendant disorientation. Using various wavelengths of light as the attractant, a great variety of traps can be devised *using* a basic funnel and collection container design (Fig. 1.37).

Figure 1.36
McPhail trap.

Many traps can be constructed easily from materials generally available around the home. All wiring and electrical connections should be approved for outdoor use. Funnels can be made of metal, plastic, or heavy paper. Traps can be with or without a cover, but if they are operated for several nights, covers should be installed to exclude rain.

The New Jersey trap (Fig. 1.38) includes a motorized fan to force insects attracted to the light into a killing jar. This trap has been especially useful for collecting small, nonscaly insects, such as midges and gnats. This type of light trap, in which the insects fall directly into a killing jar, is not recommended for use with moths because such delicate specimens may be badly rubbed or torn. If only small insects are desired, they may be protected from damage by larger insects by placing a screen with the proper-sized mesh over the entrance. The Minnesota trap is very similar to the New Jersey trap, but it does not include a fan or any motorized method of draft induction.

Figure 1.37
Blacklight trap.

The Wilkinson trap (Fig. 1.39) requires somewhat more effort to construct than the preceding traps, but it has the advantage of confining, but not killing, the trapped insects. Moths, therefore, can be collected in good condition if the trap is inspected frequently and desirable specimens are removed quickly through the hinged top and placed in a killing jar.

Several highly effective but more elaborate devices have been made for collecting moths and other fragile insects in good condition. Basically, they all use the principle of a funnel with a central light source above it and vanes or baffles to intercept the

Figure 1.38
New Jersey trap.

Figure 1.39
Wilkinson trap.

Arthropod Collection and Indentification Techniques

Figure 1.40
Mercury vapor light trap.

approaching insects, which are dropped through the funnel into the container beneath that may hold a killing agent. The nature of the container and the type of killing agent affect the quality of the specimens obtained. Some traps catch the insects alive in a large collection chamber (such as a garbage can) that is filled or nearly filled with loosely arranged egg cartons. Most moths will come to rest in the cavities between the egg cartons and will remain there until removed the next morning.

To prevent rainwater from accumulating in the trap, place a screen-covered funnel inside the collection chamber to drain the water out through a hole in the bottom of the trap. Sometimes a system of separators is added to guide beetles and other heavy, hard-bodied insects into a different part of the container than the moths and other delicate specimens.

The most effective light traps use lamps high in their output of ultraviolet light (Fig. 1.40). An example is the Robinson trap, which employs a 175-watt mercury vapor lamp. Other trap designs use ultraviolet fluorescent tubes (most commonly 15-watt), which also are effective and may be powered by a car battery or other portable source of electricity. High-wattage mercury vapor lamps emit more ultraviolet light than the more common fluorescent "blacklight" tubes, and many more insect specimens come to mercury vapor light setups. However, mercury vapor setups are not as portable and, in the field, require the use of a gasoline-powered generator. Self-ballasted mercury vapor lamps are not as effective on a watt-by-watt basis as those with a separate ballast.

Some collectors report that a few groups of insects respond best to incandescent lamps or even gas camping lanterns that do not emit ultraviolet light. However, for the majority of insects that may be taken with lights, high ultraviolet output is desirable.

SUGGESTED READING:

Light Traps
- Gui et al. 1942
- Mulhern 1942
- Pratt 1944
- Williams 1948
- Davis & Landis 1949
- Frost 1952, 1957, 1958, 1964, 1970
- Meyers 1959
- Fredeen 1961
- Graham et al. 1961
- Hollingsworth et al. 1961
- U.S. Department of Agriculture, Agriculture Research Service 1961
- Barr et al. 1963
- Belton & Kempster 1963
- Breyev 1963
- Kovrov & Monchadskii 1963
- Lowe & Putnam 1964
- White 1964
- Lewis & Taylor 1965
- Belton & Pucat 1967
- DeJong 1967
- Bartnett & Stephenson 1968
- Hardwick 1968
- Stewart & Lam 1968
- Powers 1969
- Wilkinson 1969
- Andreyev et al. 1970
- McDonald 1970
- Miller et al. 1970
- Stanley & Dominick 1970
- Barrett et al. 1971
- Carlson 1971
- Stewart & Payne 1971
- Hollingsworth & Hartstack 1972
- Tedders & Edwards 1972
- Clark & Curtis 1973
- Pérez Pérez & Hensley 1973
- Morgan & Uebel 1974
- Smith 1974
- Nantung Institute of Agriculture 1975
- Apperson & Yows 1976
- Onsager 1976
- Stubbs & Chandler 1978
- Burbutis & Stewart 1979
- Howell 1980
- Hathaway 1981

Part 1: Basic Tools and General Techniques

Except in special cases, light sources should be placed near the ground. This is counterintuitive to some collectors who try to place lights as high as possible. However, unpublished studies suggest that more specimens come to ground-level light sources, and species that tend to be "flighty" more reliably remain near these lights. Of course, specialized light collecting may require the placing of a light trap high in a tree or some other location.

A lightweight, spill-proof 12-volt battery, in which the acid electrolyte is a gel rather than a liquid, is far superior to the standard automotive battery for powering light traps. These new batteries are fairly expensive and require a special charger. Special lightweight, nickel-cadmium battery packs, used to power blacklights for collecting, are marketed by some dealers of entomological equipment.

1.6.8 Light Sheets

Another highly effective method of employing light to attract moths and other nocturnal insects involves a "light sheet" (Fig. 1.41). A particularly effective and efficient light sheet or "light net" design is one made of sturdy netting material with edges of ripstop nylon and having a rectangular cutout near the bottom in which a mercury vapor or other light source is placed. The netting and cutout greatly reduce flapping caused by wind, and the cutout allows one light source to shine in all directions. Ground sheets made of ripstop nylon also are useful, lightweight, easy to clean, and quick-drying.

An alternative light-sheet design employs light sources (ultraviolet fluorescent tubes, gasoline lanterns, or automobile headlights) placed in front of the light sheet. As insects are attracted and alight on the sheet, they are easily captured in cyanide bottles or jars by the collector, who stands in attendance or at least checks the sheet frequently. The sheet may be pinned to a rope tied between two trees or fastened to the side of a building, with the bottom edge spread out on the ground beneath the light. Some collectors use supports to hold the bottom edge of the sheet several centimeters above the ground so that specimens cannot walk into the vegetation under the sheet and be overlooked. Other collectors turn up the edge to form a trough into which insects may fall as they strike the sheet.

Figure 1.41
Light sheet.

The light sheet remains unsurpassed as a method of collecting moths in flawless condition or of obtaining live females for rearing purposes. Its main disadvantage is that species that fly very late or those that are active only in the early morning hours may be missed unless the collector is prepared

Arthropod Collection and Indentification Techniques

Figure 1.42
Pan trap.

Figure 1.43
Manitoba trap.

SUGGESTED READING:
Color traps
- Gui et al. 1942
- Hottes 1951
- Brown 1954
- Granger 1970
- Kring 1970
- Beroza 1972
- Emden 1972
- Kieckhefer et al. 1976
- Hendrix & Showers 1990
- McClain et al. 1990

Sticky traps
- Golding 1941, 1946
- Johnson 1950
- Moreland 1955
- Heathcote 1957b
- Still 1960
- Murphy 1962, *pp. 226–227*
- Taylor 1962b
- Maxwell 1965
- Gillies & Snow 1967
- Lambert & Franklin 1967
- Prokopy 1968, 1973
- Dresner 1970
- Harris et al. 1971
- Mason & Sublette 1971
- Emden 1972
- Buriff 1973

to spend most of the night at the sheet. Many other insects besides moths are attracted to the sheet, and collectors of beetles, flies, and other kinds of insects would do well to collect with this method.

It should be emphasized that the phases of the moon profoundly influence the attraction of insects to artificial light. Attraction is inhibited by a bright moon. The best collecting period each month extends from the fifth night after the full moon until about a week before the next full moon.

1.6.9 Color Traps

Colored objects also serve as attractants for insects. A bright yellow pan (Fig. 1.42) containing water is used to collect winged aphids and parasitic Hymenoptera. The insects are attracted by the color and drown in the water. The trap is more effective when a little detergent has been put in the water to reduce surface tension and thereby cause insects to drown more quickly. Yellow seems to be the best color for traps, but various kinds of insects react differently to different colors. The Manitoba trap (Fig. 1.43) has a black sphere to attract horse flies (family Tabanidae), which are then captured in a canopy-type trap.

1.6.10 Sticky Traps

In its simplest design, this type of trap involves a board, a piece of tape, a pane of glass, a piece of wire net, and a sphere, a cylinder, or some other object (often painted yellow). The trap is coated with a sticky substance and suspended from a tree branch or other convenient object. Insects landing on the sticky surface cannot extricate themselves. The sticky material can be dissolved with a suitable solvent, usually toluene, xylene, ethyl acetate, or various combinations of these solvents. Later, the insects are washed in Cellosolve and then in xylene. Sticky traps should not be used to collect specimens such as Lepidoptera, which are ravaged by the sticky substance and cannot be removed without being damaged.

Various sticky-trap materials are available commercially, some with added attractants, including lights. However, exercise caution in selecting a sticky substance because some are difficult to dissolve. General techniques for cleaning specimens taken with sticky traps have been outlined (Murphy 1985), and a cleaning procedure involving an ultrasonic cleaner has been developed (Williams & O'Keeffe 1990).

Part 1: Basic Tools and General Techniques

1.6.11 Snap Traps

Two kinds of traps designed for quantitative sampling may be termed "snap traps." One (Menzies & Hagley 1977) consists of a pair of wooden or plastic disks, slotted in the center to fit on a tree branch and connected to each other by a pair of rods. A cloth cylinder is affixed at one end to one of the disks and at the other end to a ring sliding on the rods. After the cloth cylinder has been pulled to one end and secured in place, the ring is held by a pair of latches. When insects have settled on the branch, its leaves, or flowers, the latches are released by pulling on a string from a distance, and the trap is snapped shut by a pair of springs on the rods, capturing any insects present. One of the canopy traps (Turnbull & Nicholls 1966) operates in a similar fashion. When a remotely controlled latch is pulled, a spring-loaded canopy is snapped over an area of soil, and insects within the canopy are collected by suction or a vacuum device. This trap was designed for use in grasslands.

1.6.12 Artificial Refuge Traps

Many insects, especially beetles, are found under stones, planks, or rotten logs. Providing refuges (such as pieces of wood, cardboard, or complex traps) is also an effective form of trapping.

1.6.13 Electrical Grid Traps

In recent years, electrocuting insects has been used extensively in pest-control work. The insects are attracted to a device by a chemical (see Section 1.7), light, or other substance placed in a chamber protected by a strongly charged electrical grid (Fig. 1.44). This method is not effective in preserving insect specimens but may be useful for purposes such as surveying the arthropod fauna of an area.

Figure 1.44
Electrical grid trap.

Sticky traps (continued)
- Chiang 1973
- Goodenough & Snow 1973
- Williams 1973
- Evans 1975
- Edmunds et al. 1976 *(methods of collecting and preservation on pp. 8-26)*
- Harris & McCafferty 1977
- Murphy 1985
- Bowles et al. 1990
- Knodel & Agnello 1990
- Vick et al. 1990
- Jenkins 1991
- Miller et al. 1993

Artificial Refuge Traps
- Glen 1976
- Shubeck 1976

Electrical Grid Traps
- Graham et al. 1961
- Hollingsworth et al. 1963
- Mitchell et al. 1972, 1973, 1974
- Goodenough & Snow 1973
- Hienton 1974
- Glen 1976
- Rogers & Smith 1977
- Stanley et al. 1977
- Kogan & Herzog 1980

1.7 Baits, Lures, and Other Attractants

Any substance that attracts insects may function as a bait. Natural products, chemicals derived therefrom or synthesized, and, secretions of insects may all be attractants (Jacobson & Beroza 1964, Jacobson 1972). Exposure of attractive substances is found in a variety of constructed traps.

Arthropod Collection and Indentification Techniques

SUGGESTED READING:
Baits and Lures
- Walsh 1933
- Golding 1941
- Dethier 1955
- Macleod & Donnelly 1956
- Atkins 1957
- Colless 1959
- Rennison & Robertson 1959
- Beroza & Green 1963
- Hocking 1963
- Mason 1963
- Wilton 1963
- Jacobson & Beroza 1964
- Coffey 1966
- Newhouse et al. 1966
- Sanders & Dobson 1966
- Strenzke 1966
- DeJong 1967
- Acree et al. 1968
- Everett & Lancaster 1968
- Fincher & Stewart 1968
- Morris & DeFoliart 1969
- Beroza 1970, 1972
- Nakagawa et al. 1971
- Wellso & Fischer 1971
- Beavers et al. 1972
- DeFoliart 1972
- Knox & Hays 1972
- Roberts 1972
- Beroza et al. 1974
- Debolt et al. 1975
- Howell et al. 1975
- Pinniger 1975
- Shorey & McKelvey 1977
- Bram 1978
- Zimmerman 1978
- Howell 1980
- Laird 1981

1.7.1 Sugaring for Insects

One of the oldest collecting methods involves the use of a specially prepared bait in which some form of sugar is an essential component. The bait may be refined or brown sugar, molasses, or syrup. Such substances often are mixed with stale beer or fermented peaches, bananas, or other fruit. Sugar-baited traps are most often used for moths and butterflies, but they are also effective for some flies (Dethier 1955) and caddisflies (Bowles et al. 1990).

One particularly effective recipe for Lepidoptera uses fresh, ripe peaches; culls or windfalls are suitable. Remove the seeds but not the skins, mash the fruit, then place it in a 4-liter (1-gal) container with a snugly fitting but not airtight cover. Avoid using metal containers that may rust or corrode. Fill each container only one-half to two-thirds full to allow space for expansion. Add about a cup of sugar and place in a moderately warm place for the mixture to ferment. The bubbling fermentation reaction should start in a day or so and may continue for two weeks or more, depending on the temperature. During that time, check the fermentation every day or every other day and add sugar until fermentation appears to have subsided completely. As the added sugar is converted to alcohol, the growth of yeast slows and eventually ceases.

If the mixture is allowed to run low in sugar during the fermentation process, vinegar will be produced instead of alcohol. Remember to smell the bait periodically and to add plenty of sugar to avoid vinegar formation. The amount of sugar consumed will be surprising, usually over 0.4 kg per liter (3.3 lb per gal). After fermentation ceases, the bait should remain stable and should be kept in tightly sealed containers to prevent contamination and evaporation. Canned fruit, such as applesauce, may also be used to make the bait, but such products are sterile and a small amount of yeast must be added to start fermentation. The bait should have a sweet, fruity, winelike fragrance. Although the bait may seem troublesome to prepare, it keeps for years and is available at any time, even when fruit is not in season.

Immediately before use, the bait may be mixed with 30% to 50% molasses, brown sugar, honey, or a mixture of these ingredients. This thickens the bait, retards drying, and makes the supply last longer. Set out the sugar bait during the early evening before dark. Apply the bait with a paintbrush in streaks on tree trunks, fence posts, or other surfaces. Choose a definite route, such as along a trail or along the edge of a field, so that later you can follow it in the dark with a lantern or flashlight. Experienced collectors approach the patches of bait cautiously, with a light in one hand and a killing jar in the other to catch moths before they are frightened off. Some collectors prefer to wear a headlamp, leaving both hands free to collect specimens. Although some moths will escape, a net usually is regarded as an unnecessary encumbrance because moths can be directed rather easily into the jar. Sugaring is an exceptionally useful way to collect noctuid moths, and the bait applied in the evening often will attract various diurnal insects on the following days. This bait has been used in butterfly traps with spectacular results. However, collecting with baits is notoriously unpredictable, being extremely productive on one occasion and disappointing on another, under apparently identical conditions.

1.7.2 Feces

Animal, including human, feces attract many insects. A simple but effective method of collecting such insects is to place fresh feces on a piece of paper on the ground and wait a few minutes. When a sufficient number of insects has arrived, a net with its bag held upward can be brought carefully over the bait about 1 m above it. This will not disturb the insects, nor will they be greatly disturbed when the net is lowered gently about two-thirds of the distance to the bait. At this point, the net should be lowered quickly until its rim strikes the paper. The insects, mostly flies, will rise into the net, which may then be lifted a short distance above the bait and quickly swung sideways, capturing the insects in the bottom of the bag. Feces are most attractive to insects during the first hour after deposition, and many flies can be caught during that relatively short period of time. Because of this, the "baiting with feces" method may be used for quantitative studies.

Other insects may continue coming to the feces for a more extended period and may be captured by placing a canopy trap over the feces or by using the feces with the cereal dish trap (see Section 1.6.4). Emergence traps placed over old feces will capture adult insects emerging from immature forms feeding there. The same methods also may be used with other baits, such as decaying fruit, small dead animal carcasses, and a wide variety of other substances.

1.7.3 Oatmeal

Hubbell (1956) showed that dry oatmeal scattered along a path will attract such insects as crickets, camel crickets, cockroaches, and ants. Some of these insects feed only at night and may be best located by using a flashlight or headlamp. The specimens may be hand-collected, aspirated, or netted and placed into a killing jar.

1.8 Pheromones and Other Attractants

Insects rely heavily upon chemical communication. Over the last several years, the research into insect chemical communication has been phenomenal. New

SUGGESTED READING:
Feces
- Steyskal 1957
- Coffey 1966
- Fincher & Stewart 1968
- Merritt & Poorbaugh 1975

Suggested Reading:

Pheromones
- Beaudry 1954
- Holbrook & Beroza 1960
- Jacobson & Beroza 1964
- Howland et al. 1969
- Beroza 1970
- Campion 1972
- Jacobson 1972
- Goonewardene et al. 1973
- Pérez Pérez & Hensley 1973
- Shorey 1973
- Beroza et al. 1974
- Birch 1974
- Campion et al. 1974
- Peacock & Cuthbert 1975
- Weatherston 1976
- Shorey & McKelvey 1977
- Steck & Bailey 1978
- Neal 1979
- Howell 1980
- Sparks et al. 1980
- Hathaway 1981
- Vite & Baader 1990
- Gray et al. 1991
- Mullen 1992
- Mullen et al. 1992
- Bartelt et al. 1994

Carbon Dioxide
- Reeves 1951, 1953
- Rennison & Robertson 1959
- Takeda et al. 1962
- Whitsel & Schoeppner 1965
- Newhouse et al. 1966
- Snoddy & Hays 1966
- Wilson et al. 1966
- Gillies & Snow 1967
- Morris & DeFoliart 1969
- Hoy 1970
- Stryker & Young 1970
- Davies 1971
- Batiste & Joos 1972
- Blume et al. 1972
- Roberts 1972;
- Wilson et al. 1972
- Debolt et al. 1975
- Gray 1985
- Carroll 1988

terms, new concepts, and the refinement of old ideas have been necessary. For instance, the term *allelochemical* has been proposed for chemicals secreted outside the body of an organism. *Semiochemicals* are compounds that influence insect behavior or that mediate the interactions between organisms. Currently, at least three categories of semiochemicals have been identified. A *synomone* is an interspecific chemical messenger that is beneficial to emitter and receiver. Examples include secondary plant chemicals that attract entomophagous insects to the plant and subsequently to prey or hosts. A flower fragrance that is attractive to a bee through nectar and pollen gained is also beneficial to the flower through pollination. In contrast, an *allomone* is an interspecific chemical that, in a two-species interaction, is beneficial only to the emitter. A *kairomone* is an interspecific chemical that is beneficial only to the receiver. Insect collectors are beginning to appreciate the role that interspecific attractants play in capturing insects, and in the future we will probably rely on them even more often. For instance, "Spanish fly" (cantharidin) is an irritant to vertebrates that is produced by meloid beetles but has recently come into use as an extremely effective attractant for other beetles (such as pedilids) and bugs (such as bryocerines).

Substances naturally produced and emitted by insects that cause a behavioral response in individuals of the same species are known as *pheromones*. Sex pheromones, in which the pheromone is emitted by one sex and attracts individuals of the opposite sex, are often used in traps to aid in controlling pest species. They also are emitted into the air from controlled-release dispersers throughout a particular crop to disrupt mating by misdirecting male insects to these synthetic sources as well as habituating males to the scent of their own females. Most pheromones are blends of several components; the blend is highly specific, attracting only one species. Pheromones are known to attract members of one sex, usually males, from great distances.

1.8.1 Carbon Dioxide

Host animals likewise may be used as bait for various bloodsucking insects, with or without constructed traps. Carbon dioxide (Fig. 1.44) in the form of dry ice, cylinder gas, or marble chips treated with an acid such as vinegar serves as an attractant for certain insects and has been very successful in attracting horse flies to Malaise and Manitoba traps. Dry ice is also an effective attractant for adult mosquitoes, horse flies, and parasites such as ticks.

Figure 1.44
CO_2 trap.

Part 1: Basic Tools and General Techniques

1.8.2 Sounds

Sounds are produced by many insects to attract other members of the same species. These sounds are very specific in pitch, tempo, and duration. Sound sources once tape-recorded are now often synthesized and broadcast using high-tech electronic amplifiers. Photo cells often serve as timing devices to turn sounds on or off, rather than manual switches. Recordings of such sounds, played at the proper volume, have been effective in luring grasshoppers, mole crickets, and other kinds of insects.

1.9 Collecting Aquatic Insects

Insects and mites emerging from water may be collected by some of the same methods as terrestrial insects. Aquatic-insect-collecting dip nets and heavy-duty aquatic nets can be purchased commercially (Fig. 1.45 A&B). Kick screens (Fig. 1.45C) can be constructed easily from a fine-mesh window screen attached to two stout poles. Holding the screen on the bottom of the stream immediately downcurrent from where rocks and other debris are dislodged can yield a large number of aquatic organisms. Flume collecting is an old technique that has recently been revived. The extraction of insects and mites from flumes involves glass-topped sleeve cages, modified Malaise and tent traps, and Berlese photoattractive traps positioned directly above water. Other types of aquatic collecting techniques, such as tow traps and dredges (Fig. 1.46), require specialized equipment that we do not describe here. The collection of aquatic insects is of great importance in public health and general ecological studies.

SUGGESTED READING:
Sound Traps
- Belton 1962
- Cade 1975
- Walker 1982
- Bailey 1991
- Dethier 1992
- Parkman & Frank 1993
- Thompson & Brandenburg 2004

Aquatic Insects
- Hodgson 1940
- Welch 1948 (general)
- Jonasson 1954
- Mundie 1956, 1964, 1966, 1971
- Gerking 1957
- Essig 1958
- Lindeberg 1958
- Grigarick 1959
- Pieczynski 1961
- Sladeckova 1962
- Morgan et al. 1963
- Cushing 1964
- Macan 1964
- Corbet 1965
- Wood & Davies 1966
- Kimerle & Anderson 1967
- Tarshis 1968a, b
- Waters 1969
- Coulson et al. 1970
- Elliott 1970
- Carlson 1971
- Edmondson & Winberg 1971
- Mason & Sublette 1971
- Fahy 1972
- Finch & Skinner 1974
- Langford & Daffern 1975
- Murray & Charles 1975
- Apperson & Yows 1976
- Edmunds et al. 1976
- Landlin 1976
- McCauley 1976
- Masteller 1977
- Edmunds & McCafferty 1978
- LaGasa & Smith 1978
- Pennak 1978 (general)
- Ettinger 1979
- Lawson & Merritt 1979
- LeSage & Harrison 1979
- Wood et al. 1979 *(methods of collecting, preserving, and rearing on pp. 45-53)*
- McCafferty 1981
- Merritt et al. 1984 *(general)*
- Halstead & Haines 1987
- Weber 1987

Figure 1.45
Aquatic nets.
A. aquatic dip net
B. aquatic canvas net
C. kick screen

Figure 1.46
Specialized aquatic-insect-collecting techniques.
A. tow trap, B. dredge, C. modified Birge–Ekman dredge

Arthropod Collection and Indentification Techniques

SUGGESTED READING:
Collecting Soil Insects
- Davidson & Swan 1933
- Barnes 1941
- Salt & Hollick 1944
- Kevan 1955, 1962
- Newell 1955
- Turnock 1957
- MacFadyen 1962
- Murphy 1962
- Teskey 1962
- Brindle 1963 *(includes collecting methods)*
- Evans et al. 1964 *(techniques on pp. 61-88)*
- Kühnelt 1976 *(observation and collecting techniques on pp. 35-65, bibliography on pp. 385-466)*
- Lane & Anderson 1976
- Akar & Osgood 1987

Collecting Ectoparasites
- Banks 1909 *(mostly of historical interest, but describes the old methods and contains much general information about insects)*
- Klots 1932
- Cantrall 1939-1940, 1941
- Comstock 1940
- Chu 1949
- Grandjean 1949
- Lipovsky 1951 *(mites)*
- Cook 1954
- Williamson 1954
- Wagstaffe & Fidler 1955
- Balogh 1958
- Lumsden 1958
- Morgan & Anderson 1958
- Oldroyd 1958
- Fallis & Smith 1964
- Peterson 1964
- Urquhart 1965 *(elementary directions for making insect collections, pp. 1-19)*
- Knudsen 1966, 1972, pp. 128-176
- Norris 1966
- U.S. Department of Agriculture, Plant Pest Control Division 1966-1970
- Watson & Amerson 1967
- Lehker & Deay 1969
- Nicholls 1970

(continued next page)

1.10 Collecting Soil Insects

As with aquatic specimens, insects and mites that live on or under the soil surface require special techniques and equipment for their collection and study. Often separators, extractors and pitfall and Berlese traps can be utilized to collect soil-inhabiting insects. Various floatation techniques have been used to collect small insects and mites from soils and other substrates as well. Many soil-inhabiting species are of great economic importance because they devour the roots of crops. Many of these insects spend their immature stages in soil but emerge and leave the soil as adults. A considerable amount of literature on collecting soil insects has been published, the most useful of which is cited here.

1.11 Collecting Ectoparasites

Ectoparasites are organisms that live on the body of their host. Examples include lice and fleas. Some ectoparasites, particularly those that fly, may be collected in some of the traps discussed previously, using their hosts as bait; others may be collected either by hand or with special devices described in the listed readings.

1.12 Collecting Regulated Insects

Collecting insects or other arthropods at points of entry into the state or country or by way of specific and intensive surveys is an important part of regulatory officials' duties. Extreme care must be exercised when collecting, preserving, and indentifying the invasive organisms. Care must be taken to preserve the sample together with its collection information, in a manner that will allow others to also verify species identity. Both shipping and rearing insects (sections found in this book) are critical for regulatory entomologists to understand and utilize.

1.13 Collecting Insects for Pest Management Audits

Collecting tools and techniques are an important part of a pest management audit, regardless of the type of building or the commodities stored or manufactured in the facility. Intensive inspections require a solid understanding of how and why pests may enter a building. Understanding the importance of food, water, and harborage to pests' establishment is crucial in conducting pest audits. Understanding building construction and maintenance is also invaluable. Insect pest managers can take pest information observed during an audit and subsequently describe a correct course of action to eliminate existing and exclude other potential pests from facilities.

Using insect glue traps as monitors is very important to pest auditors. Because inspections occur only over the course of one or two days, they represent a snapshot in time of what is occurring in the building. Monitors, on the other hand, work 24 hours a day, seven days a week, and provide a record of events over a much longer period of time. Flashlights, screwdrivers, probes, flushing agents, scrapers, mirrors, collecting vials, clipboards, blueprints of the building, knee pads, and bump caps are all part of an auditor's routine inspection tool bag. On occasion, more sophisticated equipment is required to listen for or observe pests in areas where they hide. Auditors must be ever alert to signs of pest infestations, even when the pest cannot be found. Often alerting the owners to "conducive conditions" (those conditions that might enable a pest to enter or to flourish if introduced) can be of as much value as finding the pest.

1.14 Collecting Insects for Forensic or Medico-Criminal Investigations

Use of arthropods in forensic or legal investigations has become standard practice in recent years. Insects and mites are often the subject of lawsuits especially where contamination of products, misapplication of pesticides, transmission of disease, or human myiasis is concerned. Collection and documentation of specimens as evidence requires great attention to procedure and protocol.

Many different collecting techniques can be employed in death scene investigations. Direct observation or sweep net sampling can often provide important information of the adult stages of insects above the body if taken before the body is disturbed. These data are often lost evidence if not taken immediately. Direct collecting of insects from on, in, and around the body is one of the most effective means of obtaining reliable data. Often, collecting various stages of insects and recording the exact location of the collections are vital. Rearing a portion of the immature stages to adulthood can provide important times and identifications of insects. For detailed protocols about collecting and preserving insects and related organisms in forensic cases, please consult the cited references.

1.15 Rearing

Collectors should take every opportunity to rear insects and mites. Reared specimens generally are in the best condition. Further, rearing provides life stages that otherwise might be collected only rarely or with great difficulty. With respect to parasitic insects, such as some Hymenoptera and Diptera, rearing provides unambiguous proof of host associations and other scientifically important information.

Ectoparasites (continued)
- Brown 1973
- Ford 1973
- Cogan & Smith 1974, p. 152
- Thompson & Gregg 1974
- Cheng 1975
- McNutt 1976
- Service 1976 *(includes extensive bibliographies)*
- Stein 1976
- Martin 1977
- Bland & Jacques 1978
- Edmunds & McCafferty 1978
- Lincoln & Sheals 1979
- Southwood 1979 *(includes extensive bibliographies)*
- Banks et al. 1981
- Eads et al. 1982
- Pritchard & Kruse 1982
- Arnett 1985
- Stehr 1987, 1991
- Borror et al. 1989
- Dindal 1990
- Stehr 1991
- Dryden & Broce 1993
- Kasprzak 1993

Forensic Entomology
- Erzinclioglu 1983
- Anderson 1999
- Castner & Byrd 2001
- Greenberg & Kunish 2002
- Haglund & Sorg 2002
- Saferstein 2004
- James & Nordby 2005

Arthropod Collection and Indentification Techniques

Figure 1.47
Rearing chambers.
A. *screen cylinder capped on both ends*
B. *glass jar with ventilated lid*

By preserving one or more specimens from each of the stages as they are reared, the collector can obtain a series of immature stages along with associated adults (Clifford et al. 1977). Such series are desirable, especially for species in which the adult is known but the immature stages are unknown or difficult to identify. Often, the converse is also true. Some species of insects, such as stem-mining flies, are fairly abundant in the larval stage. Sometimes we do not know whether the specimens represent a species that has been described and named from an adult whose life history is unknown. Because adults of these flies are seldom found, the easiest way to obtain the stage necessary for specific determination is to rear the larvae or pupae.

If only a few specimens are reared, then the shed "skins" (exuviae) and pupal cases of puparia should be preserved because they are valuable when properly associated with the reared adult. Do not preserve a pupa or puparium with an adult unless you are positive that the association is correct. Put pupae in separate containers so that adults or parasites that emerge are associated with certainty. When feasible, the parasite's host should be preserved for identification. Keep careful notes throughout the rearing so that all data relative to the biology of the species are properly correlated.

1.15.1 Containers for Rearing

To rear specimens successfully, rearing cages must simulate closely the natural conditions in which the immatures were found. A screen cylinder capped on both the top and bottom can be an effective rearing container (Fig. 1.47A). Almost any container will serve as a temporary cage for living insects or mites. A paper bag is a simple temporary cage that is very handy on field trips. Plant material or a soil sample containing insects or mites is placed in the paper bag, which is then sealed. A paper bag also can be placed over the top of a plant on which insects or mites are found. The bottom edge of the bag is tied tightly around the exposed stems. Stems then are cut and placed in a jar of water. Paper bags are not transparent and must be removed to observe the specimens or to determine when the foliage needs to be changed. Clear plastic bags are better suited to such viewing. However, they are not recommended for more than short-term use because they are airtight and specimens may be damaged by drowning in condensed water inside the bag.

A glass jar whose lid is replaced by a piece of organdy cloth or gauze held in place by a rubber band forms a simple temporary cage (Fig. 1.47B). A few such jars in a collecting kit are useful for holding live insects. For aquatic species, a watertight lid on the jars is advisable. Aquatic insects often die when transported over a considerable distance in water that sloshes in the container. Fewer specimens will die if the jar is packed with wet moss or leaves. After arrival at your destination, release the insects into a more permanent rearing container.

Certain aquatic insects may be reared readily indoors in an aquarium or a glass jar. The main goal is to duplicate their natural habitat. If the specimen was collected from a rapidly flowing stream, it probably will die indoors unless the water is aerated. Other insects do well in stagnant water. Aquatic vegetation usually should be provided in the aquarium, even for predaceous specimens, such as dragonfly nymphs, which often cling to underwater stems. Keep sufficient space between the surface of the water and the aquarium cover to allow the adult insect to emerge. The amount of space will vary according to the insect being reared. For example, a dragonfly needs considerable space and a stick, rock, or other object above the water on which to perch after emerging so that the wings will expand fully.

Aquatic insects can be reared in their natural habitat by confining them in a wire-screen or gauze cage. Part of the cage must be submerged in water and anchored securely. The screen used in aquatic cages should be coarse enough to allow food through yet fine enough to retain the insects being reared.

Most adult insects, both terrestrial and aquatic, are teneral when they first emerge and should not be killed until the body and wings harden and the colors develop fully. Hardening may require a few minutes, hours, or days. Keep even small flies alive for one full day after they emerge. Specimens killed while teneral will shrivel when mounted. Some insects, especially butterflies and moths, will beat their wings against the cage and lose many scales or tear their wings if kept in cages too long after emerging. Providing adequate space in which emerging insects may expand their wings fully and move about slightly is therefore critical in the design of rearing cages.

Some beetles and other boring insects often are abundant in bark and wood. Excellent adult specimens may be obtained if pieces of infested wood are placed in glass or metal containers. However, remember that rearing these insects sometimes requires considerable time. Cages made of wood or cardboard are not suitable for such insects because those found in wood or bark usually are well equipped, both in immature and adult stages, to chew their way through a cage made of such material and thus escape.

Arthropod Collection and Indentification Techniques

Figure 1.48
Flowerpot cage for long-term rearing.

A flowerpot cage is one of the best containers for rearing plant-feeding species over an extended period. The host plant, if its size and habitat permit, is placed in a flowerpot and a cylinder of glass, plastic, or wire screen placed around the plant (Fig. 1.48). Another type of flowerpot cage is made by inserting a cane or stick, taller than the plant, into the soil in the pot. One end of a net or muslin tube is fitted over the edge of the pot and is held in place by a string. The other end of the tube is tied around the top of the stick. An advantage of the flowerpot cage is that the plant is living, and so it is not necessary to add fresh plant material daily.

Plant-feeding mites will not wander far as long as suitable host material is available for them. Because mites are wingless even as adults, they can be confined in an open rearing container by making a barrier around the top edge or upper inner sides of the container with petroleum jelly or talcum powder.

Emergence cages are essentially rearing cages that are used when it is impractical or impossible to bring specimens indoors. Emergence cages may also be considered as traps and are discussed under that heading (see Section 1.6.5). With plant-feeding insects, a sleeve consisting of a muslin tube with open ends is slipped over a branch or plant and tied at one end (Fig. 1.49). The insects are then placed in the tube, and the loose end of the tube is tied. This cloth tube can be modified to allow observation of the insects by replacing the midsection with a "window" of clear plastic or wire screen. If the insects in the tube require duff or debris in which to pupate, the tube should be placed perpendicular to the ground and duff or debris placed in the lower end.

Figure 1.49
Muslin sleeve cage.

1.15.2 Rearing Conditions and Problems

1.15.2.1 Moisture

The moisture requirements of insects and mites are varied. Examination of the habitat from which specimens were collected should provide clues about their moisture requirements in captivity. Many insects in the pupal stage resist desiccation. Species that normally infest stored foods also require very little moisture. In fact, many produce their own metabolic water. Most species found outdoors require higher levels of humidity than generally exist indoors. Additional moisture can be added to indoor rearing cages in several ways. To increase the humidity in a cage, keep a moist pad of cotton on top of the screen cover of the cage or place a moist sponge or a small glass vial filled with water in the cage. The mouth of the vial is plugged with cotton and the vial laid on its side so that the cotton remains moist. Pupae may be held for long periods in moist sawdust, vermiculite, sphagnum, or peat moss. In a flowerpot cage, the water used to keep the plant alive and moisture released through transpiration should provide sufficient moisture for the plant-feeding insects and mites. Spraying the leaves daily also may supplement moisture requirements in rearing cages. Too much moisture may result in water condensation on the sides of the cage, which may trap the specimens and damage or kill them. Excess moisture also enhances the growth of mold and fungus, which is detrimental to the development of most insects and mites. A 2–3% solution of table salt sprayed regularly in the cage will help prevent mold and fungus growth.

1.15.2.2 Temperature

Of all the environmental factors affecting the development and behavior of insects and mites, temperature may be the most critical. Because arthropods are cold-blooded, their body temperature is usually close to the temperature of the surrounding environment, and their metabolism and development are directly affected by increases and decreases in temperature. Each stage of an insect or mite species has a low and a high point at which development ceases. These are called *threshold* temperature levels.

Most species that are collected and brought indoors for rearing can be held at normal room temperature; the optimum temperature for rearing varies from species to species and with different stages of the same species. As with all rearing techniques, every attempt should be made to duplicate optimum natural conditions. Specimens that normally would overwinter outdoors should be kept during the winter in rearing cages placed in an unheated room, porch, or garage. Never place an enclosed rearing cage in direct sunlight; the heat becomes too intense and may kill the specimens.

1.15.2.3 Dormancy and Diapause

Insects and mites cannot control the temperature of their environment. Instead, they make physiological adjustments that allow them to survive temperature extremes. In regions with freezing winters, insects and mites have at least one stage that is resistant to low temperatures. The resistant stage may be egg, larva, nymph, pupa, or adult. When winter arrives, only the resistant stage survives. *Dormancy* is the physiological state of an insect or mite during a period of arrested development. *Diapause* is the prolonged period of arrested development brought about by such adverse conditions as heat, drought, and cold. This condition can be used to advantage in rearing. For example, if rearing cages must be unattended for several days, then many specimens can be refrigerated temporarily to slow their activity and perhaps force diapause. This measure should be used with caution because the degree and duration of cold tolerated by different species vary.

Dormancy: the physiological state of an insect or mite during a period of arrested development.

Diapause: the prolonged period of arrested development brought about by such adverse environmental conditions as heat, drought, and cold.

The reverse situation, that of causing diapause to end, is equally useful. Overwintering pupae that normally would not develop into adults until spring can be forced to terminate diapause early by chilling them for several weeks or months, and then bringing them to room temperature so that normal activity will resume. Often, mantid egg cases are brought indoors accidentally with Christmas greenery. The eggs, already chilled for several months, hatch when kept at room temperature, often to the complete surprise and consternation of the unsuspecting homeowner.

1.15.2.4 Light

Most species of insects and mites can be reared under ordinary lighting conditions. However, artificial manipulation of the light period will control diapause in many species. If the light requirements of the species being reared are known, then the period of light can be adjusted so that the specimens will continue to develop and remain active instead of entering diapause. Light and dark periods can be regulated with a 24-hour timing switch or clock timer. The timer is set to regulate light and dark periods to correspond with the desired lengths of light and darkness. For example, providing 8–12 hours of continuous light during a 24-hour cycle creates short-day conditions that resemble winter; providing 16 hours of continuous light during a 24-hour cycle creates long-day conditions that resemble summer. Remember that many insects and mites are very sensitive to light; sometimes even a slight disturbance of the photoperiod can disrupt their development.

1.15.2.5 Food

The choice of food depends upon the species being reared. Some species are detritovores and will accept a wide assortment of dead or decaying organic matter. Examples include most ants, crickets, and cockroaches. Other insects display food preferences so restricted that only a single species of plant or animal is acceptable. At the time of collection, carefully note the food being consumed by the specimen and provide the same food in the rearing cages.

Carnivorous insects should be given prey similar to that which they normally would consume. This diet can be supplemented when necessary with such insects as mosquito larvae, wax moth larvae, mealworms, maggots, and vinegar flies or other insects that are easily reared in large numbers in captivity. If no live food is available, a carnivorous insect sometimes may be tempted to accept a piece of raw meat dangled from a thread. Once the insect has grasped the meat, the thread can be gently withdrawn. The size of the food offered depends on the size of the insect being fed. If the offering is too large, the feeder may be frightened away. Bloodsucking species can be kept in captivity by allowing them to take

> **Artificial manipulation of the light period will control diapause in many species.**

blood from a rat, mouse, rabbit, or guinea pig. **A human should be used as a blood source only if it is definitely known that the insect or mite being fed is free of diseases that may be transmitted to the person.**

Stored-product insects and mites are easily maintained and bred in containers with flour, grains, tobacco, oatmeal or other cereal foods, and similar products. Unless leaf-feeding insects are kept in flowerpot cages where the host plant is growing, fresh leaves from the host plant usually should be placed in the rearing cage daily and old leaves removed.

1.15.2.6 Artificial Diets

Some species can be maintained on an artificial diet. The development of suitable artificial diets is complex, involving several factors besides the mere nutritional value of the dietary ingredients. Because most species of insects and mites have very specific dietary requirements, information regarding artificial diets is found mainly in reports of studies on specific insects or mites.

1.15.2.7 Special Problems and Precautions in Rearing

Problems may arise in any rearing program. Cannibalism, for instance, is a serious problem in rearing predaceous insects and necessitates rearing specimens in individual containers. Some species resort to cannibalism only if their cages become badly overcrowded. Disease is also a problem and can be caused by introducing an unhealthy specimen into a colony, poor sanitary conditions, lack of food, or overcrowding.

Cages should be cleaned frequently and all dead or clearly unhealthy specimens removed. Exercise care not to injure specimens when transferring them to fresh food or when cleaning the cages. Mites and small insects can be transferred with a camel's-hair brush.

Attacks by parasites and predators also can be devastating to a rearing program. Carefully examine the host material when it is brought indoors and before it is placed in the rearing containers. This reduces the possibility of predators and parasites being introduced accidentally into the cages. Also, place rearing cages where they will be safe from ants, mice, the family cat, and other predators.

SUGGESTED READINGS:
Rearing
- Needham 1937
- Hodgson 1940
- Gerberich 1945
- Chu 1949
- Hopkins 1949
- Harwood & Areekul 1957
- Levin 1957
- Fischer & Jursic 1958
- Lumsden 1958
- Peterson 1964
- Krombein 1967
- U.S. Department of Agriculture, Extension Service 1970
- Ford 1973
- Smith 1974
- Sauer 1976
- Clifford et al. 1977
- Smith et al. 1977
- Barber & Matthews 1979
- Banks et al. 1981
- Edwards & Leppla 1987
- Gray & Ibaraki 1994

Arthropod Collection and Identification Techniques

Suggested Readings:
Molecular Research
- Arensburger et al. 2004
- von Hagen & Kadereit 2001
- Wells et al. 2001
- Wells & Sperling 2001

1.16 Collecting Insects for Molecular Research

Collection of insect or mite specimens for molecular research requires few special techniques. In most cases, arthropods may be collected in much the same manner as discussed for general collections. Clean, contaminant-free specimens must be ensured, however. Keep in mind that some insects may carry foreign debris on their bodies, such as pollen or other potential contaminants, that may confound some studies. Some arthropods also carry parasites (either externally or internally) that may pose confusion when DNA or RNA extractions are done. Some mites commonly conceal themselves under wing covers of insects or other locations and may potentially go unnoticed.

Specific preservation techniques of the sample must follow the protocol determined by the procedure set out in Chapter 2. It is imperative to know the ultimate and specific use of the specimen beforehand, however, because the preservation techniques may affect the molecular technique being performed.

Common methods include preservation in ethanol, freezing on dry ice and then holding in an ultralow freezer, and placement in liquid nitrogen or other specific products, such as RNALater®, that helps preserve the more fragile RNA materials.

Packaging and labeling must also be done with care when specimens are meant for molecular studies. Using a soft pencil to label specimens in liquids is still the best method. Labeling specimens both inside and outside of the package or vial should be standard operating procedure because the extreme environment in which the specimens are held often causes labels, even with the best adhesive, to detach. Voucher specimens are often valuable. These may consist of properly preserved whole insects or parts thereof (e.g., wings of larger insects) to positively link the insect identification to the molecular data.

Purposes of molecular studies of insects vary widely but may also include taxonomic studies. Measuring degree of genetic relationships is valuable and is sometimes much more sensitive than traditional methods.

Chapter 2: Agents For Killing and Preserving

> **Tips on how to preserve various orders of arthropods are provided at the end of Chapter 6.**

Agents for killing insects are discussed in Section 1.3.4. Many instances allow for insects and mites of all kinds to be killed and preserved at once in liquid agents, especially acetone. Before proceeding, however, first determine the advisability of using a liquid killing agent rather than a dry gaseous agent. Some kinds of insects are best kept dry; other kinds of insects should not be allowed to become dry. Many insects, especially larger ones, may be killed by an injection of a small amount of alcohol. Death is nearly instantaneous, which is desirable with moths, large beetles, and large Orthoptera. Generalized directions for the treatment of different orders are provided at the end of Chapter 6.

Preservation of insects in alcohol is a complex subject about which there is some controversy and misunderstanding. If you wish to specialize in an insect group suited to preservation in one or another kind of concentration of alcohol, consult specialists in that group or experiment to find what yields the best results.

Ethyl alcohol mixed with water (70–80% alcohol) is usually the best general killing and preserving agent. For some kinds of insects and mites, other preservatives or higher or lower concentrations of alcohol may be better. Many histological or molecular preservation techniques call for preservation in nearly 100% (absolute) alcohol. However, absolute ethyl alcohol is often difficult to obtain, thus some collectors use isopropanol (isopropyl alcohol), with generally satisfactory results. Isopropanol does not seem to harden specimens as much as ethyl alcohol. It is readily obtained and is satisfactory in an emergency. Although controversy exists concerning the relative merits of ethyl alcohol and isopropanol, the choice of which to use is not so important as what concentration to use. This choice depends on the kind of insect or mite to be preserved.

Parasitic Hymenoptera are killed and preserved in 95% alcohol by some professional entomologists. This high concentration prevents the membranous wings from becoming twisted and folded, setae from matting, and soft body parts from shriveling. This concentration also may be desirable if large numbers of insects are to be killed in a single container, such as in the killing jar of a Malaise trap, because the insect body fluids will dilute the alcohol. On the other hand, soft-bodied insects (such as aphids and thrips, small flies, and mites) become stiff and distorted when preserved in 95% alcohol. These specimens should be preserved in alcohol of a lower concentration. Adult bees should not be collected in alcohol because their usually abundant body setae become badly matted. Adult moths, butterflies, mosquitoes, moth flies, and other groups with scales or long, fine setae on the wings or body may become severely damaged and worthless if collected in alcohol, regardless of the concentration.

Formalin (formaldehyde) solutions are not recommended by some workers because the specimens become excessively hardened and then become difficult to handle. However, formalin is a component of some preservatives.

Larvae of most insects should be collected in alcohol or killed in boiling water to "fix" their proteins and prevent the specimens from turning black. Larvae should be left in hot water for 1–5 minutes and, depending on the size of the specimens, then transferred to 70–80% alcohol. Large specimens or small specimens that have been crowded into one vial should be transferred to fresh alcohol within a day or two to reduce the danger of diluting the alcohol with body fluids. If the alcohol becomes too diluted, the specimens will begin to decompose. Water is not a preservative.

For some groups, preservation is better if certain substances are added to the alcohol solution. Thrips and most mites, for example, are best collected in an alcohol–glycerin–acetic acid (AGA) solution. Larval insects may be collected into a kerosene-acetic acid (KAA) solution. Formulas for these and other solutions are given in Appendix I.

Glycerin is an excellent preservative for many kinds of insects, and it has characteristics that should not be overlooked by collectors. Glycerin is stable at room temperature and does not evaporate as rapidly as alcohol. The periodic "topping off" or replacement of alcohol in collections is expensive and time-consuming. Glycerin-preserved material does not require the attention that alcohol-preserved material does. When specimens are preserved in glycerine, some colors are preserved better and longer, and the specimens remain supple. When specimens are removed from glycerine, they may be inspected for relatively long periods under a microscope without shriveling or decomposing.

Some histological, cytological, or physiological studies require that specimens be preserved in special media (Walker & Boreham 1976). For instance, mosquito eggs may be preserved in a lacto-glycerol medium (Murphy & Gisin 1959).

Larvae and most soft-bodied adult insects and mites can be kept almost indefinitely in liquid preservatives. However, for a permanent collection, mites, aphids, thrips, whiteflies, fleas, and lice usually are mounted on microscope slides (see Section 3.3.9). Larvae are usually kept permanently in alcohol, but some specimens may be mounted by the freeze-drying technique (Dominick 1972) or by inflation (see p. Section 3.1.2). Many insects collected in alcohol are later pinned for placement in a permanent collection. Hard-bodied insects such as beetles can be pinned directly after removal from alcohol, but for them and all softer insects, such as flies and wasps, follow the process described in Section 3.2.4.

Chapter 3: Storage of Specimens

3.1 Temporary Storage

After specimens have been collected, time is often not immediately available to prepare them for permanent storage. There are several ways to keep specimens in good condition until they can be prepared properly. The method chosen depends largely on the length of time that the specimens must be stored temporarily.

3.1.1 Refrigeration

Medium-sized to large specimens may be left in tightly closed bottles for several days in a refrigerator and still remain in good condition for pinning. Smaller specimens, such as parasitic Hymenoptera, may be left overnight in the freezer compartment of a refrigerator without adverse effects. Some moisture must be present in the containers so that the specimens do not become "freeze-dried," but excessive moisture will condense on the inside of the bottle as soon as it becomes chilled. Absorbent paper placed between the jar and the insects will keep them dry. When specimens are removed for further treatment, place them immediately on absorbent paper to prevent moisture from condensing on them. Insects may be placed in alcohol, as described previously, and kept for several years before they are pinned or otherwise treated.

3.1.2 Dry Preservation

It is standard practice to place many kinds of insects in small boxes, paper tubes, triangles, or envelopes for an indefinite period, allowing them to become dry. We do not advise storing soft-bodied insects by such methods because the specimens become badly shriveled and subject to breakage. Diptera should never be dried in this manner because the head, legs, and especially the antennae become broken very easily.

Almost any kind of container may serve for dry storage. However, tightly closed, impervious containers of metal, glass, or plastic should be avoided because mold may develop on specimens if even a small amount of moisture is present. Little can be done to restore a moldy specimen.

Dry-stored specimens must be labeled with complete collection data in or on each container (see Section 3.3). Avoid placing specimens collected at different times or places in the same container. If specimens with different collection data must be layered in the same container, include a separate data slip with each layer.

Arthropod Collection and Indentification Techniques

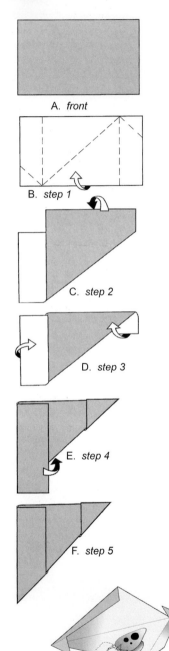

Figure 3.1
Steps in folding paper triangles.

To ensure that specimens do not slip from one layer to another, cut pieces of absorbent tissue, glazed cotton, or cellucotton a little larger than the inside of the container. Place a few layers of this material in the bottom of the container, then add a few insects (do not crowd them), then add more layering material, and so on until the container is filled. If much space is left, add enough cotton to keep the insects from moving about but not enough to produce pressure that will damage them. To prevent parts of the insects from getting caught in the loose fibers, use plain cotton only for the final layer. Insect parts are very difficult to extract from plain cotton without damage.

One method of keeping layered specimens soft and pliable for several months includes the use of chlorocresol in the bottom of the layered container and a damp piece of blotting paper in the top. The container must be impermeable and sealed while stored; plastic sandwich boxes are useful containers for this method. Add about 5 ml of chlorocresol crystals to the bottom, cover with a layer of absorbent tissue, follow with the layers of specimens, then a few layers of tissue, and finally a piece of dampened blotting paper as the top layer. The cover is then put in place and sealed with masking tape. Keep boxes of layered specimens in a refrigerator. See Tindale (1962).

Some insects, such as small beetles, should be glued to triangles (see Section 3.2.5) directly from the layers for permanent preservation. If the specimens are to be pinned or otherwise treated, they must be relaxed as described in Section 3.2.1.

3.1.3 Papering

Pinning specimens when they are fresh is preferable. However, the storage method known as *papering* has long been used successfully for larger specimens of Lepidoptera, Trichoptera, Neuroptera, Odonata, and some other groups. Papering is a traditional way of storing unmounted butterflies and is satisfactory for some moths. Unfortunately, moths often will have their relatively soft bodies flattened, legs or palpi broken, and the vestiture of the body partly rubbed off. To save space in most large collections, file Odonata permanently in clear cellophane envelopes instead of pinning them. Cellophane is preferred over plastic because cellophane "breathes."

Papering consists of placing specimens with the wings folded together dorsally (upper sides together) in folded triangles (Fig. 3.1) or in small rectangular envelopes of glassine paper, which are the translucent envelopes familiar to stamp collectors. Glassine envelopes have become almost universally used in recent years because of the obvious advantages of transparency and ready availability. In many collections, glassine has become partly superseded by plastic. However, many collectors still prefer folded triangles of a softer, more absorbent paper, such as ordinary newsprint, and believe they are superior for preserving specimens. Specimens can become greasy after a time, and the oil is absorbed by paper such as newsprint but not by glassine. Moreover, glassine and plastic are very smooth, and specimens may slide about inside the envelopes during shipping, losing antennae and other brittle parts. Although softer kinds of paper do not retain creases well when folded, this shortcoming may be circumvented by preparing the triangles of such material well before they are needed and pressing them with a weight for a week or so. Triangles are easy to prepare if the paper is folded as shown in Fig. 3.1.

Some Lepidoptera are most easily papered if first placed in a relaxing box for a day or two. The wings, often reversed in field-collected butterflies, may then be folded dorsally without difficulty. Do not pack specimens together tightly before they are dried, or the bodies may be crushed. Do not store fresh specimens immediately in airtight containers or plastic envelopes, or they will mold. Write collection data on the outside of the envelopes before inserting the insects.

3.2 Mounting Specimens

Specimens are mounted so that they may be handled and examined with the greatest convenience and with the least possible damage. Well-mounted specimens enhance the value of a collection; their value for research may depend to a great extent on how well they are prepared. Standardized methods have evolved over the last two centuries in response to the aesthetic sense of collectors and to the need for high-quality research material. Although the style and technique of mounting may vary from one worker to another, the basic procedures outlined here are widely accepted. Methods of preparation are subject to improvement, but in the interest of uniformity it is best to follow currently accepted practices until the superiority of other methods has been proved.

> **The value of specimens may depend on how well they are preserved.**

The utility of a mounted specimen is determined by how well it is preserved, how safe it is from damage, and how much of the specimen can be examined conveniently. Utility supercedes beauty in research collections, whereas beauty should be a major consideration in nontechnical displays.

Preparation of specimens for a permanent collection is discussed here. Ideally, specimens prepared for a permanent collection should be fresh. The body tissues of fresh specimens have not hardened or dried, and thus the specimen is easier to manipulate. Specimens that have been in temporary storage must be specially treated before mounting. Dry specimens usually must be relaxed (Weaver & White 1980), and those preserved in liquid must be processed so that

Arthropod Collection and Indentification Techniques

they will dry with minimal distortion or other damage. Some specimens may be kept permanently in a liquid preservative or in papers or envelopes, as discussed earlier.

3.2.1 Preparing Dry Specimens for Mounting

Dry insects to be pinned must first be relaxed. Moistening the specimen softens the body and appendages so that the body will not break when the pin is inserted into the thorax, and parts of the relaxed specimen may be rearranged or repositioned. Insects, particularly Lepidoptera, whose wings are spread should be relaxed even if they have been killed recently. The muscles of Lepidoptera stiffen in a matter of minutes. The thoracic flight muscles are very strong and make adjustment of the wings difficult. Treatment in a relaxing chamber, however, usually will make this procedure much easier (Fig. 3.2). Eight hours in a relaxing chamber should suffice, but larger specimens may require 24–36 hours. Simply leaving specimens in a cyanide jar for a short time sometimes will relax them, but this method is not reliable. See Lane (1965) for a discussion of mounting insects of medical importance.

Figure 3.2
Insect relaxing chambers.
A. glass relaxing chamber
B. homemade box relaxing chamber

High humidity must be provided in a relaxing chamber for periods from several hours up to about three days. The actual amount of time depends upon the circumstances. Care must be exercised to ensure that the specimens do not become wet during humidification. Growth of mold should be avoided because it will destroy specimens left too long in a relaxing chamber. A chemical mold inhibitor can be added to the relaxing chamber. Parasitic Hymenoptera killed with ethyl acetate remain pliant for several hours and can be relaxed easily if they have been allowed to dry. However, some killing agents (especially chloroform, ether, and carbon tetrachloride) may harden muscles to such an extent that specimens are brittle and seemingly impervious to the humidity of the relaxing chamber. In such instances other techniques must be employed. In Korea, for example, butterflies are injected in the thoracic muscles with very hot water through a fine hypodermic needle before the wings are spread. Occasionally, some specimens cannot be relaxed satisfactorily by any method, they should be mounted soon after capture and killing.

Glass dishes and jars with covers (low, widemouthed jars and casserole dishes are excellent), plastic food storage containers, and earthenware crocks make excellent relaxing chambers. Glass and earthenware containers are not so immediately affected by fluctuations in temperature as are other types and thus may relax insects more evenly. Containers 5–15 cm deep are most convenient; clear plastic sandwich boxes not more that 2.5 cm deep will serve for small specimens if they are not on pins. A layer of damp sand, peat, or crumpled paper toweling is placed in the bottom of the container and covered with a layer of cotton, cellulose wadding, or jeweler's cotton. This layer will not absorb water readily and will prevent direct contact between the insects and the moisture beneath.

Some workers object to the use of cotton because of the tendency for insect legs to become entangled in it and break off. If this is a problem, cover the cotton with a single piece of soft tissue paper. For very small specimens, a lining of tissue paper or some absorbent material with a smooth surface is advantageous. Heavy paper, such as blotting paper or cardboard, may be used in place of cotton, but this should be supported at least 1 cm above the moist bottom layer to avoid direct contact. Wooden or plastic strips or fine-mesh plastic screen also may be used for this purpose.

> **Dry insects must be carefully relaxed prior to pinning.**

In temperate and arid regions, mold probably will not be a problem if insects are relaxed for no longer than two days at normal room temperature. However, relaxing chambers in regular use must be kept clean, with frequent renewal of the contents. If mold is likely to develop, as may happen with large specimens held more than two days, a few crystals of naphthalene, paradichlorobenzene, phenol, or chlorocresol may be sprinkled in the bottom of the relaxing chamber. A little thymol, which is more potent, may be used. Many of these chemicals may damage plastic boxes, so select relaxing chambers with care.

Insects held too long in a killing jar, and those that were originally papered and pinned but not spread, or layered (placed in small boxes between pieces of soft tissue) may be relaxed by placing them in a relaxing chamber. Papered specimens will relax faster if removed from their envelopes. Beetles and other insects whose wings do not require spreading may be held less than 24 hours in a relaxing chamber. Small moths and delicate Neuroptera also should be relaxed sufficiently after 12–24 hours to allow the wings to be spread. Large moths may take 48 hours or longer if the relaxing chamber is kept at room temperature. The process can be hastened and the chance of mold developing greatly reduced if the relaxing jar is subjected to a slight raising and lowering of temperature between 18 and 27°C. The process is greatly accelerated if the relaxing jar is placed in, or floated on, warm water for an hour or more; specimens may be relaxed within 3–6 hours in this way. If the warm-water treatment is overdone, however, the specimens may be spoiled by the absorption of too much moisture. Some colors, especially nonmetallic greens in Lepidoptera, are unstable and may be completely lost by exposure to too much humidity. Such material requires special attention; the specimens should be left in the relaxing chamber for the shortest possible time and ideally should be pinned and spread when fresh. Experience soon enables one to judge the best procedure for the particular kind of material being prepared.

The length of time that insects may be left safely in a relaxing chamber partially depends on the temperature. At 18–24°C, the specimens may be left for about three days; beyond that time they will begin to decompose. If the relaxing chamber is placed in a refrigerator at 3–4°C, the specimens may be kept two weeks, although they may be slightly damaged from excessive condensation over time. If relaxing chambers containing fresh specimens are placed in a deep freeze at −18°C or lower, the specimens will remain in comparatively fresh condition for months, but not indefinitely. At extremely low temperatures the specimens gradually desiccate and eventually will become dried. However, a freezer may be used to keep them fresh for a month or two and is a great convenience.

Even when specimens have been relaxed suitably for spreading, the wings may still seem stiff. In this instance, the wing muscles must be loosened by forcing the wings to move up and down. This may be safely done by pressing the tips of curved forceps firmly against the costal vein very near the base of the wing. The forceps should have the tips ground or honed smooth and not too sharp. Repeat this procedure separately with all four wings, or they will revert gradually toward their original positions. With care, all the wings may be loosened in this way without leaving visible marks.

Occasionally, it is necessary to relax and reposition only a part of an insect, such as a leg that may conceal characters needed for identification. This is accomplished by putting a drop or two or Barber's Fluid (see Appendix I) or ordinary household ammonia directly on the leg. Most household ammonia contains detergent, which helps wet and penetrate insect tissue. After a few moments, perhaps after adding a little more fluid, the body part may be pried carefully with a pin. When the part moves easily, it may be placed in the desired position and held there with a pin fixed in the same substrate that holds the pin on which the specimen is mounted. Leave the positioning pin in place until the fluid dries thoroughly.

A few methods for relaxing insects involve heat. These and other methods are summarized in a reference describing a steam-bath method (Weaver & White 1980).

3.2.2 Degreasing

Many dried insect specimens become greasy with time. This is particularly evident in some beetles, butterflies, and moths. The grease is derived from internal fatty tissue. Traditional methods of degreasing involve soaking specimens in benzene, ethyl ether, or carbon tetrachloride for a few days (Peterson 1964). Periodically the fluid is replaced as it changes color. Vapor degreasing is a relatively new technique (McKillop & Preston 1981) that employs trichloroethane as a solvent heated to vaporization in a metal tank. The apparatus is elaborate and may be available at some universities or museums. The toxicity of degreasing solvents should be kept in mind. Some curators prefer hexane because it is very clean, it does not affect pigments, and the odor is less annoying.

3.2.3 Preparing Liquid-Preserved Specimens

Sabrosky (1966) discussed mounting specimens that have been preserved in alcohol. Most specimens preserved in fluid must be removed from the fluid in a way that minimizes distortion or matting of setae as the specimens dry. Only specimens with hard exoskeletons, such as beetles and some bugs (Pentatomidae, Cydnidae), may be mounted without special treatment when removed from the preserving fluid and mounted directly on pins or points. The following methods have been used routinely for removing specimens from the usual fluid preservatives. Some specimens are left in better condition than if they had been pinned while fresh. This is especially true of small Diptera.

The following equipment is needed.

1. Two screw-top jars about 5 cm in diameter with a cork cemented with epoxy on the top of each lid. A label on the outside should clearly identify the contents. One jar should contain about one-third Cellosolve (2-ethoxyethanol, ethylene glycol ethyl ether) and the other jar one-third xylene.

2. A small dish, such as a watchglass.

3. Absorbent tissue from which to twist small "pencils" for absorbing xylene.

4. Insect pins or double mounts (see Section 3.2.5) for mounting the specimens.

5. Adhesive in a jar with a rod in its stopper.

6. Narrow-pointed forceps.

7. A few small cards of blotting paper.

When specimens are ready for preparation, remove as many from the preservative as can be pinned or placed on triangles or card points in an hour. Experience will suggest how many specimens should be removed from the preservative. Place them on blotting paper, then drop them from the blotter into the jar of Cellosolve, and place a label with the collection data on the pin stuck into the cork cemented to the lid of the jar. This may be done at the end of the day and the specimens left in the Cellosolve overnight. Otherwise, leave the specimens in Cellosolve for about three hours or longer for large specimens.

> **WARNING!**
>
> Xylene is now considered to be carcinogenic. A new and widely used chemical, Histo-Clear, is a promising substitute.

They may even be left over a weekend. The same jar of Cellosolve may be used several times, up to about 10 times if the insects are small. This part of the treatment removes water and other substances from the specimens. However, Cellosolve does not evaporate readily, so it must be removed subsequently with another solvent that will evaporate readily.

The next step is to use forceps to remove the specimens carefully from the Cellosolve. Place them again briefly on blotting paper and then place them in the jar containing xylene. The identifying label on the pin in the cork must also be transferred. Small specimens should be left in the xylene for about one hour, larger specimens for up to four hours. Specimens left too long in xylene become extremely brittle and difficult to put on a pin or triangle without losing legs, antennae, or the head. As with the Cellosolve, the xylene also may be used many times until it becomes so contaminated with Cellosolve that specimens dry slowly when removed. Specimens wet with Cellosolve or xylene are somewhat pliable and appendages may be repositioned slightly.

After one hour in xylene, remove the smallest specimens first. Remove each with small forceps and place it in a dish. The forceps will pick up a small amount of xylene and the specimen will be left lying in it. While it is in the dish, position the specimen correctly for mounting. The wings, if present, will float out flat, sometimes requiring a little adjustment with a pin or the tip of the forceps. When the specimen is positioned correctly, take a corner of absorbent tissue and touch it to the specimen to remove the excess xylene. Larger specimens may be pinned directly in the usual manner. Just before the xylene fully dries from the surface of a small specimen, the tip of a triangle or a tiny pin called a *minuten* (already attached to its carrying insect pin) should be touched to adhesive and then be touched to the specimen, picking it up.

> **Specific directions for labeling and pinning insects, including double mounting and using minuten pins, follow later in this section.**

If a minuten is used, it may be inserted into the thorax of the specimen. A little final adjustment of position may then be made, and the specimen is ready for its label and place in the collection. Having touched the tip of the minuten to the adhesive will leave a small amount of adhesive around the place where the minuten has pierced the specimen and will keep it from working loose when fully dried. Specimens placed on standard insect pins also should have a small amount of adhesive placed around the site where the pin protrudes from the lower side of the specimen. Specimens pinned after having been in fluid preservatives do not cling as firmly to the pin as those pinned fresh.

This treatment will leave surface pile, setae, and bristles in a loose, unmatted, natural condition. Small insects that shrivel considerably after having been pinned fresh will usually dry in better condition if pinned or placed on triangles after this treatment.

3.2.4 Direct Pinning

This section pertains entirely to insects because mites should never be mounted on pins. Direct pinning refers to the insertion of a standard insect pin directly through the body of an insect. Only insect pins should be used; ordinary straight pins are too short and thick and also have other disadvantages. Standard insect pins are 38 mm long and range in thickness generally from size #000 to #7. An integral "upset" head is made by mechanically squeezing out the end of the pin, or a small piece of metal or nylon is pressed onto the pin. A well-made upset head is considered essential by most entomologists; other kinds of heads sometimes come off, leaving a sharp point that easily can pierce a finger. Pins of #2 diameter are most useful (0.46 mm in diameter). In the past, most entomologists avoided the use of very slender pins of size 000 to 1, preferring to use double mounts (see Section 3.2.5), but now that soft polyethylene or plastic foam is commonly used for pinning bottoms in trays and boxes, these smaller sizes are not so impractical. Pins of larger diameter (numbers 3–7) may be needed for large insects.

Standard insect pins are currently made of spring steel (which is called "black") or stainless steel with a blued or a lacquered (japanned) finish. The black pins may corrode or rust with even slight exposure to moisture or to the body contents of the insects. Stainless steel pins are more expensive than black pins, but their rustproof nature makes them desirable for use in permanent collections. However, their points are somewhat more easily turned than those of black pins when piercing an insect with a hard cuticle, and their shafts are not as rigid. For that reason, pierce an especially hard cuticle with a strong steel pin before inserting a stainless steel pin. Lacquered pins have a surface on which the insect may be less likely to become loose than it might on a bare pin.

Insect pins made of German silver or brass were once common. However, the action of the insect body contents produce a greenish verdigris on the pin inside the insect and eventually corrodes through the pin.

For handling a large number of pinned specimens, pinning or dental forceps (Fig. 3.3) may be helpful. Their curved tips permit grasping the pin below the data labels and enable the curator to set the pin firmly into the pinning-bottom material of the box or tray without bending the pin. The forceps are also invaluable when removing pins strongly corroded into cork pinning bottoms. The pin is grasped firmly above the cork and turned a little before it is lifted. Unfortunately, pinning forceps are impractical in dealing with large-winged specimens, such as many Lepidoptera.

Figure 3.3
Pinning forceps.

Figure 3.4
Proper pin positions in common insect orders; black dots indicate pin placement.

Insects should be pinned vertically through the body with a pin of appropriate thickness. Exercise care so that the pin does not damage the legs as it passes through the body. Many insects are pinned to the right of the midline so that all the characters of at least one side will be visible. Do not attempt to pin specimens unless they are relaxed or recently killed. Inserting a pin into a dry specimen may cause it to shatter. When pinning relaxed specimens or specimens taken from Cellosolve and xylene, a little glue may be needed where the pin emerges from the specimen to prevent the dry specimen from working loose and rotating on the pin. Application of adhesive is unnecessary when mounting recently killed insects.

Standard pin placements for some of the more common groups of insects are illustrated in Fig. 3.4.

1. **Orthoptera.** Pin the specimen through the back of the thorax to the right of the midline. For display purposes, one pair of wings may be spread. Some orthopterists prefer to leave the wings folded because of limited space in most large collections (see Beatty & Beatty 1963).

2. **Large Heteroptera.** Pin the specimen through the triangular scutellum to the right of the midline. Do not spread the wings.

3. **Large Hymenoptera and Diptera.** Pin the specimen through the back of the thorax or slightly behind the base of the forewing and to the right of the midline. Characters on the body must not be obscured. Legs should be pushed down and away from the thorax; wings should be turned upward or sideways from the body. The wings of most Diptera and many parasitic Hymenoptera will flip upward if the specimen is laid on its back before pinning and pressure is applied simultaneously to the base of each wing with a pair of blunt forceps. Wings should be straightened if possible so that venation is clearly visible. Folded or crumpled wings sometimes can

be straightened by gentle brushing with a camel's-hair brush dipped in 0% alcohol. Peterson's XA mixture (equal parts of xylene and ethyl alcohol) is recommended for treating Hymenoptera wings.

4. **Large Coleoptera.** Pin the specimen through the right forewing (wing cover or elytron) near the base. Do not spread the wings of beetles.

5. **Large Lepidoptera and Odonata.** Pin the specimen through the middle of the thorax at its thickest point or just behind the base of the forewings. Spread the wings as described in Section 3.2.7.2.

Figure 3.5 Pin heights.
A. correctly pinned specimen
B. specimen mounted too low
C. specimen not level

The proper height of an insect on a pin will depend somewhat on the size of the specimen. As a guideline, enough of the pin should be exposed above the thorax to grasp the pin without the fingers touching the specimen. A specimen mounted too high on a pin probably will be damaged in handling. If a specimen is pinned too low on a pin, the specimen's legs may be broken when the pin is inserted into a tray or box. Also, insufficient space may be left for labels if a specimen is mounted too low on a pin. Fig. 3.5 illustrates correct and incorrect examples of pinning.

After the pin is inserted and before the specimen is dry, the legs, wings, and antennae should be arranged so that all parts are visible. With most insects, the legs and antennae simply can be arranged in the desired position and allowed to dry. Occasionally, the appendages must be held in place with insect pins until the specimen dries. Long-legged species or specimens with drooping abdomens can present problems. Temporarily mounting insects on a piece of styrofoam board or other soft material is recommended. Legs and abdomens may be thusly supported until dry with a series of strategically placed support pins or a piece of stiff paper pushed up on the pin from beneath. When the specimen is dry, supports can be removed. A spreading board or a block (Section 3.2.7) may be necessary for moths, butterflies, and other insects that should be mounted with their wings spread.

Some entomologists glue small insects directly to the side of a standard insect pin. This procedure is not recommended because often too much of the insect is obscured by the glue or pin and the adhesive does not bond well to the pin. A double mount is recommended for small insects.

3.2.5 Double Mounts

Double mounts are intended for insects that are too small to be pinned directly on standard pins yet should be preserved dry. This term refers to insects mounted

Figure 3.6
A properly double-mounted insect on a minuten pin.

on a minuten pin or card point (triangle), which in turn is mounted or attached to a standard insect pin. Minutens are available from supply houses in 10- and 15-mm lengths and in two or three thicknesses. They are generally constructed of stainless steel and are finely pointed at one end and headless on the other end. Double mounts are assembled by inserting the minuten into a small cube of soft, pithy material, such as fine cork, balsa wood, or fine-textured plastic. Some entomologists prefer polyporus, a pure white material obtained from a bracket fungus. Unfortunately, polyporus is expensive and difficult to obtain, especially in the United States. Many entomologists prefer silicone rubber, which is obtained from plastics suppliers. Silicone rubber is made into plaques by pouring the polymerized material, a thick creamy liquid, into a flat-bottomed plastic container to a depth of about 2.5 mm and allowing it to solidify over several hours. The rubber is then lifted easily from the mold and cut with a sharp knife or razor blade into strips and finally into cubes. With most materials, the minuten must be inserted point first. With silicone rubber, however, the dull end of the minuten is inserted first, until it strikes the surface on which the cube is lying so that it will be held firmly. Minutens should be handled with forceps; they are so thin that either end can easily pierce a finger.

Sometimes it is preferable to mount an insect on a minuten before inserting the minuten into the mounting cube. For convenience, prepare a series of minuten mounts beforehand, already attached to standard no. 3 pins. To mount extremely small insects on minutens, pick up a droplet of adhesive with the prepared minuten and touch it to the specimen between the base of the legs. When mounting an insect on a minuten, the pin needs to extend just barely through the insect. If the insect is lying on a glass surface when it is pierced with the minuten, a little extra pressure will curl the point of the minuten back into the insect and ensure that the specimen will not come off the minuten.

Figure 3.7 Proper mounting of card points.

Many entomologists prefer to mount insects on a minuten in a vertical position in a short strip of pith or cork, with the minuten parallel to the main pin. The insect lies in an excellent position for examination under a microscope and is less liable to be damaged in handling than it would be otherwise. See Peterson et al. (1961).

Figure 3.8 Point punch.

Card points are slender, small triangles of stiff paper. They are pinned through the broad end (Fig. 3.7) with a no. 2 or 3 insect pin, and the insect is then glued to the point. Card points may be cut with scissors from a strip of paper; they should be no more than 12 mm long and 3 mm wide. A special punch for card points (Fig. 3.8), available from entomological supply houses, will make better, more uniform points. Many European and North American entomologists who specialize in collecting parasitic Hymenoptera prefer to use small rectangular cards for mounting specimens

Part 1: Basic Tools and General Techniques

1–10 mm long. The cards are available in various sizes and may be purchased through entomological equipment dealers. If the collector desires a specific size of card, a special card punch can be manufactured. The specimen is mounted directly onto the card, with one side of the thorax contacting the adhesive.

Card points and rectangular cards should be made only from high-quality, acid-free paper. The paper should be as good as or better than that used for data labels (Section 3.3.1). Do not use "index cards" as point or rectangular-card stock. If specimens are in good condition and are well prepared, they may be kept in museum collections for a long time, perhaps for centuries. Most of the paper in common use does not have that kind of life expectancy. Most paper typically becomes yellow and brittle with age. Paper made especially to last, such as that used for herbarium sheets in botanical collections, is highly recommended.

> **Paper made especially to last, such as that used for herbarium sheets in botanical collections, is highly recommended.**

The choice of adhesive for mounting insects on card points and rectangles is equally important. Bonding characteristics, solubility, and availability are three factors that should be considered in the selection of an adhesive. Unfortunately, the aging properties of many glues are not known and we cannot make unequivocal recommendations. The adhesive should bond to the specimen so that the specimen will not fall from the card with time. The adhesive should be soluble in a solvent that is inexpensive, nontoxic, and available. The solvent permits specimens to be remounted when necessary; expensive, toxic, or exotic solvents make remounting difficult. Ordinary white glue or carpenter's glue is readily available in the United States. It is a superior adhesive and it is soluble in water. Clear acetate cement and fingernail polish are also commonly utilized for double mounting insects, but they are less satisfactory. Fingernail polish does not bond well to some specimens and requires a special solvent to thin the polish or to remount specimens. White shellac is a possibility but is less satisfactory because it is usually too thin and flows over a specimen when applied.

For many years some professional entomologists have used a very viscous polyvinyl acetate for double mounting and similar uses. The acetate can be obtained in granular, bead, or pellet form. A small quantity is placed in a bottle containing a glass stirring rod and covered with absolute ethyl alcohol. The acetate will dissolve in a day or two into a thick solution. If it is too thick, the acetate will "string out," and more ethyl alcohol should be added. Use the rod to stir the ethyl alcohol into the acetate. If the solution is too thin, the bottle should be left open to allow some of the ethyl alcohol to evaporate. After a period of use, the solution will thicken, and then more ethyl alcohol must be added. Specimens adhere very well to a pin or a point with this solution, and they may be removed with 95% ethyl alcohol.

Some entomologists use shellac gel (Martin 1977) as an adhesive for double mounts. We do not recommended this material because its preparation involves boiling white shellac solution, a procedure with some danger from fire. It is hoped that a safer method of preparation may eventually be found for this useful medium.

Figure 3.9
Double mount using card point method.

Adhesives should not be permitted to get so thick that they form "strings." If this happens, add a little solvent to the adhesive until it attains the proper consistency. Nor should an adhesive be so thin that it flows over a specimen. Only a small amount of adhesive should be used to glue the specimen to the card point or rectangle. Excessive glue may obscure certain sutures or sclerites necessary for identification, just as the card point may conceal certain ventral structures if allowed to extend beyond the midline of the insect.

For most insects, the card point is attached to the *right* side of the specimen, with the left side and midventral area clear. For better adhesion with some insects, the tip of the card point may be bent downward at a slight angle to fit against the side of the specimen (Fig. 3.9). The bend increases the surface area of the point that makes contact with the specimen. Only a very small part of the point should be bent, and this is accomplished with fine-pointed forceps. With a little practice, it will be easy to judge how much of the point to bend and at what angle to fit the particular insect being mounted.

One method to ensure that the specimen is oriented properly on the point is to place it on its back with its head toward you; then with the pin held upside down, touch a bit of adhesive to the bent point and apply it to the right side of the insect. If the top of the point can be slipped between the body of the insect and an adjacent leg, a stronger mount will result. The card point should be attached to the side of the thorax, not to the wing, abdomen, or head. Some insects, such as small flies and wasps, are best mounted on unbent points.

Opinions differ on when to use direct pinning and when to use a double mount. Perhaps the decision is best determined through experience. A general rule of thumb: Do not use a double mount if you can mount a small insect on a no. 1 or 0 pin without damaging the specimen. Insects too heavy to be held on the point by adhesive yet too small to be pinned with standard pins may be attached to card points via the following method. Puncture the right side of the insect at the place where the card point usually would be placed, and insert into this puncture the tip of an unbent card point with a little glue on it. For puncturing specimens, use a needle ground and polished to make a small, sharp scalpel. Some specimens, such as moths, should never be glued to points; other specimens should never be pinned with minutens. The following suggestions will serve as a guide.

Part 1: Basic Tools and General Techniques

1. *Small moths, caddisflies, and neuropteroids.*
   Mount the specimen on a minuten inserted through the center of the thorax with the abdomen positioned toward the insect pin. The mount must be sufficiently low so that the head of the pin can be grasped easily with fingers or pinning forceps. Do not glue moths to points. Ideally, such specimens should be spread in the conventional manner despite their small size.

2. *Mosquitoes and small flies (recently killed).*
   Pin the specimen with a minuten through the thorax with the left side of the specimen positioned toward the main pin. Note that the minuten is vertical. This is more advantageous than if it were horizontal because the specimen is less liable to come into contact with fingers or pinning forceps. Placing a small amount of glue on the tip of the minuten before piercing the specimen will help hold soft-bodied insects. See Schlee (1966).

3. *Small wasps and flies (not recently killed).*
   Mount the specimen on an unbent card point, with the point inserted between the coxae on the right side of the insect (keeping clear of the midline), or glue the tip of the point to the mesopleuron. Alternatively, microhymenoptera can be mounted directly on rectangular cards.

4. *Small beetles, bugs, leafhoppers, and most other small insects.*
   Glue the card point with its tip bent down to the right side of the specimen.

3.2.6 Pinning Blocks

Pinning blocks (Fig. 3.10) allow for insects and labels to be easily mounted at uniform heights on the pin. Specimens should be mounted on the pin such that the pin can be grasped between the thumb and index finger without touching the specimen and yet keeping the specimen above the bottom of the box or tray.

Figure 3.10
3-step wooden pinning block.

After piercing the insect through the appropriate spot on the specimen, insert the pin into the deepest hole in the pinning block until the pin goes no further. This pushes each specimen to the same height on each pin. Repeat this process for positioning label heights on pins in subsequent holes (Fig. 3.11). Pinning blocks improve the general appearance of a collection and help preserve specimens from breakage.

Double mounts should conform to the same rule as for direct pinning. Do not place a double mount too high or too low on the pin. Double-mount cubes or points may be adjusted at any time; a directly pinned insect is virtually impossible to move without damage after it has dried. If points become loose on the main pin, place a little adhesive at the connection. Consult Borgmeier (1964) for additional details.

Figure 3.11
Plastic pinning block illustrating hole depths.

3.2.7 Mounting and Spreading Boards and Blocks

The appearance of an insect collection can be enhanced if specimens are pinned in a lifelike position. Most entomologists find that a mounting board made of styrofoam, cork, or cardboard is convenient. Pins on which fresh specimens have been mounted should be inserted into the mounting board deep enough that the specimen rests slightly above the flat surface. The legs, wings, abdomen, and antennae can then be conformed into the desired position by strategically placing holding pins. Leave the specimen in place until dry (sometimes 1–2 weeks). Once dry, the specimen will hold this shape indefinitely.

In advanced collecting, mounting boards allow an entomologist to prominently display certain specific characters used in identification.

Insects to be preserved with their wings spread uniformly are generally set and dried in this position on spreading boards or blocks. Spreading boards are more commonly used than spreading blocks. These pinning aids vary greatly in design, but the same basic principle is inherent in all of them: a smooth surface on which the wings are spread and positioned horizontally; a central, longitudinal groove for the body of the insect; and a layer of soft material into which the pin bearing the insect is inserted to hold the specimen at the proper height. An active collector will need several spreading boards because the insects must dry for a considerable time (about 2 weeks for most specimens, even very small ones) before being removed from the boards. Spreading boards may be purchased from biological supply houses or may easily be made as described here if the proper materials can be obtained. When purchasing spreading boards, avoid (1) too hard or too soft a material for the pinning medium under the central groove, (2) too hard an upper pinning surface, and (3) top pieces without the same thickness at the center (an especially common fault in beveled boards). This last defect may be corrected by sanding down the higher side; evenness is especially critical when working with small specimens.

Figure 3.12
Spreading board.

3.2.7.1 Construction of Spreading Boards

A spreading board of simple design (Fig. 3.12) requires the following materials.

1. Two top pieces, 9 mm by 4.8 cm by 38 cm, preferably of seasoned basswood (a fine-grained, durable wood from trees of the genus *Tilia*). Holes made in basswood by insect pins tend to close after they are removed. If the surface

of the board is lightly sanded after use, especially when working with small specimens, its smooth, even quality can be maintained through many years. Basswood is sometimes known as "whitewood." Wood from trees of the genus *Liriodendron* is also sold under this name. If basswood cannot be obtained, well-seasoned white pine selected for softness and "pinnability" is serviceable. A third choice is 20-cm (8-inch) beveled redwood siding. Beveled top pieces are desirable because the beveling, sloping inward, compensates for the tendency of the wings of spread specimens to droop slightly after the specimens are removed from the board. The 20-cm siding is actually 19.1 cm wide, and a 0.9-m piece of it will provide two pairs of top pieces. One pair cut slightly more than 4.8 cm wide from the wide side of the board and planed to exact width will make a pair 11 mm thick at the narrow side, and another pair cut from the same side of the board will provide a second pair about 8 mm thick at the narrow side. Redwood is stiff and fine-grained; it splits and splinters easily.

2. Two end pieces of any good, fine-grained wood, 2 cm square by 10 cm.

3. One strip of entomological or gasket cork or foam, 6 mm by 3 cm by 34 cm.

4. One base of plywood or any fine-grained wood, 6 mm by 10 cm by 38 cm.

These materials are for a spreading board with a central groove 6 mm wide. Boards with grooves of several widths will be needed. For the larger Lepidoptera (macrolepidoptera, or "macros"), the most useful widths are 3, 6, and 9 mm. For very large moths, a width of 17 mm is required; the board will also have to be as much as 15 cm in total width, with a groove depth of 16 mm. For small moths (microlepidoptera, or "micros"), special boards with groove widths of 1.5–2 mm will be needed, with the groove shallow enough for minutens, the width and thickness of the top pieces (1) must be altered accordingly. The pinning medium (3) could be of polyethylene foam; but to give specimens firm support, the entire depth between the top pieces and the base would have to be filled with the material. A dense, finely textured plastic foam known as Plastazote is better than polyethylene for entomological applications. A strip of model maker's balsa wood, selected for pinning softness, may also be satisfactory.

The end pieces (2) should be glued with epoxy or another good adhesive and nailed to the top pieces (1), with the proper groove width maintained. Then the pinning strip (3) should be firmly glued to the underside of the top pieces (1), on the same side on which the end pieces (2) were fastened, and should cover the central groove. Finally, after the adhesive has set, the base (4) should be attached. It is affixed to the end pieces (2) with two flat-headed wood screws (no. 5, 19 mm) countersunk into the base piece and screwed into each end piece. The base may be removed easily later if replacement of the pinning strip is necessary.

3.2.7.2 Using the Spreading Boards

Before spreading specimens, the spreading boards and the following materials should be at hand.

1. Pins (called setting pins) of size 00 or 000 for bringing wings into position. Setting pins used by some lepidopterists are made by inserting a minuten into the end of a round matchstick and securing it with a drop of glue.

2. Strips of glassine or tracing paper (the translucent, smooth paper used for tracing, not what a draftsman calls tracing paper). Cellophane, plastic film, or waxed paper should not be used. Their disadvantages include expanding with moisture, becoming electrostatically charged, and containing a substance that pulls scales off the wings. The strips of tracing paper should be long enough to extend from the base of the wings to a little beyond the end of the wings of the specimens being spread. Strips about 25 mm long are convenient for spreading most Lepidoptera. Short strips are appropriate when spreading specimens that have been relaxed from a dried condition; strips long enough to cover several specimens in a row on the board are commonly utilized for recently captured insects. The strips are often applied with a narrow fold alongside the body of the specimen with the fold upward; this provides a rounded edge that reduces the likelihood that a sharp edge will displace a row of scales. This fold may be made by holding the strip on a spreading board with 2–5 mm of it overhanging the edge of the board, running a finger along the overhang to bend it down, and then firmly folding it back.

3. Glass-headed pins at least 2.5 cm long for holding the strips in place. Ordinary no. 2 or 3 insect pins with nylon heads may also be used, but some lepidopterists find them hard on the fingers.

With this equipment ready, the collector is prepared to mount and spread the specimens. The specimens must be properly relaxed (see Section 3.2.1), even the recently collected ones, before any attempt is made to spread the wings. Insert an insect pin of appropriate size through the middle of the thorax, leaving at least 7 mm of pin above the specimen. The pin should pass through the body as nearly vertically as possible to avoid having the wings higher on one side than on the other. Pin the specimen into the central groove of the spreading board so that the wings are exactly level with the surface of the board. Carefully draw each wing forward, with the point of a setting pin inserted near the base and immediately behind the strong veins that lie near the front of the wings (Fig. 3.13). If care is taken not to tear the wing, the fine setting pins should leave holes so small that they are barely visible. The hind margin of the forewings should

Part 1: Basic Tools and General Techniques

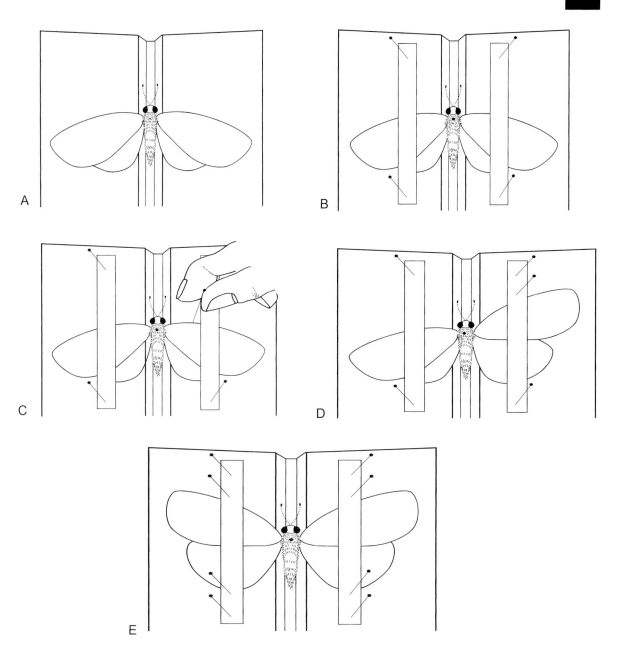

Figure 3.13
Steps to follow when using the spreading board.

be at right angles to the groove in the board. Bring the hindwing into a natural position, with its base slightly under the forewing. The setting pins will hold the wings in position until they can be secured with the paper strips.

The strip is placed near the body of the insect, with its fold upward and toward the insect. A glass-headed pin is inserted in the middle of the folded part of the strip just outside the margin of the forewing. The pin may be tilted slightly away from the wing to keep the strip down against the wing. The tip is then carefully stretched backward and another pin placed just behind the hindwing. A third pin in the notch where the forewings and hindwings meet is usually enough. None of the three glass-headed pins on each side of the specimen should pass through the wings. After the paper strips are in place, the setting pins may be removed. Twisting the setting pins a little as they are removed will prevent a possible bent tip from hooking onto a wing vein and pulling the whole wing out of place.

Fresh specimens may be arranged closely on the board in series of five or even more before the paper strips are applied to cover all the wings. Relaxed specimens should be treated individually because they dry so quickly that antennae may break or the wings curl if the spreading is not completed promptly.

To prevent the abdomen from drooping as the specimen dries, support it with a pin on each side, crossing beneath. Pins may also be used to arrange and hold the antennae and legs in position until they dry. The appearance of many insects may be improved by gently blowing on them before spreading, to remove extraneous loose scales and to straighten the setae or, with small moths, the fringes of the wings. When working with small insects, a large magnifying lens mounted on an adjustable stand may be very helpful.

Specimens relaxed from a dried condition present additional problems. The wings may be stiff and require loosening (see Section 3.2.1). If the wings of a relaxed specimen are turned upward and do not lie on the surface of the spreading board, the paper strips may need to be pinned over the wings to hold them down before they are positioned. Because the wings of relaxed specimens are still relatively stiff, skillful manipulation is needed to spread the wings without tearing or leaving excessively large pinholes. If the wings do not move readily under gentle pressure, do not force and possibly break them. Return the specimen to the relaxing chamber.

3.2.7.3 Construction of Spreading Blocks

The spreading block, a modification of the spreading board, is designed to accommodate only one specimen at a time. Blocks often are preferred by

Part 1: Basic Tools and General Techniques

specialists in microlepidoptera but also can be used for other insects. The design is simple (Fig. 3.14), consisting of a wooden cube about 2 cm on a side for most insects, with a groove across the middle of one face. The width and depth of the groove vary to suit the size of the insect to be spread, usually 1.5–2 mm in width and deep enough to accommodate a strip of fine cork, polyporus, or similar pithy material into which the pin or minutens are lodged to hold the insects being spread. The groove should be cut parallel with the grain of the wood, and the top surface of the block should be sanded very smooth. Before the pinning strip is wedged or cemented into the bottom of the groove, a hole about 1 mm in diameter should be drilled squarely in its middle. The pin can extend into this hole when the insect becomes level with the spreading surface. A few gashes, made by pressing the blade of a thin knife in the upper corners of the block near each end of the groove, should be made to catch the thread that will hold the wings of the insect.

The insect to be spread, mounted either on a standard insect pin or on a minuten, is pinned into the groove as with a spreading board, and the wings are manipulated by gentle blowing and using a setting pin. A piece of fine silk or nylon thread is then caught in one of the knife gashes and brought over the wings, and, if necessary, once around the block and again over the wing at another point, and finally caught again in the knife gash. A small piece of tracing paper may be placed on the wings before passing the thread over them, but if special scale tufts are found on the wings it is better to omit the paper and leave the tufts in a natural position.

Specimens on spreading boards or blocks should be placed in a warm, well-ventilated place to dry for at least two weeks. Even very small moths require that long to dry completely and become stiff. If they are placed in a low-temperature oven, such as is used for drying plant specimens, two days may suffice. Specimens relaxed from a dry condition, as already noted, dry quickest, but even they should be left for several days. Fresh specimens, even large ones, may be dried in two days or less with heat. Where humidity is low and there is ample sunshine, the spreading boards or blocks may be placed in cardboard cartons painted black and left out in full sunlight for about two days. Occasionally, specimens may become greasy, but otherwise no harm results. The spreading boards or blocks must be kept where they are safe from mice, bats, dermestid beetles, lizards, psocids (booklice and barklice), and ants, especially in tropical climates. One preventive measure that is sometimes advisable involves placing the boards or blocks on bricks set in pans of water. If they are hung from the ceiling, then a mosquito net around them may be necessary.

Again, we must emphasize that the collection data are as valuable as the specimen. Always keep temporary data labels with specimens on spreading

SUGGESTED READING:

Spreading boards and blocks
- Edwards (1963)
- Lewis (1965)
- Tagestad (1974, 1977)

Figure 3.14
Spreading block for small insects.

Arthropod Collection and Indentification Techniques

Figure 3.15
Riker Mounts.

boards or associated with them in some way to ensure that there is no confusion or loss of data when they are removed from the boards. Specimens are scientifically worthless without accurate collection information.

Spreading is a highly individualistic skill, subject to wide variation. Nearly everyone, with practice, evolves his or her own technique, so that two workers may appear to follow different procedures and yet produce equally good results. There is no single standardized technique with respect to the fine points of spreading.

3.2.7.4 Riker Mounts

Sometimes it is desirable to prepare specimens for exhibition in such a way that they may be handled for close examination without risk of damage. Riker mounts have long been used for this purpose. They may be purchased from entomological supply houses, but similar cases may be easily constructed. The Riker mount (Fig. 3.15) is a flat cardboard box about 3 cm deep, filled with cotton and with a pane of glass or plastic set into the cover. Unpinned specimens are placed upside down on the glass of the cover, spread into position with some cotton held in place by small weights, and allowed to dry thoroughly (about two weeks). Then the weights are removed, enough cotton is added to hold the specimens firmly in place, a little fumigant is added to kill any pests or their eggs that might have been laid in the box, and the bottom part of the box is put in place. When the box is closed, it should be sealed completely to prevent access by pests. Plant material or associated insect frass may also be dried in place with the specimens to make valuable reference collections.

Riker mounts are practical only for relatively large insects, such as butterflies, larger moths, beetles, and dragonflies, that are suitable for such display. Although Riker-mounted specimens are useful for classroom instruction and general display, they are not used for storage of insects in a scientific collection, where specimens must be available for examination from all angles under magnification. Riker mounts should be inspected periodically for pests and kept out of sunlight, which will cause fading of colors and general deterioration. It is sometimes desirable to put pinned specimens into Riker mounts. To do so, remove the pin or cut it off flush with the surface of the insect (Holbrook 1927).

3.2.7.5 Inflation of Larvae

A common practice in the nineteenth and early twentieth centuries was to preserve larvae, mainly caterpillars, by inflation. That practice has largely been abandoned in favor of alcoholic preservation or freeze-drying. These latter methods permit more thorough examination of all parts of the specimens, even internal organs, which must be removed before inflation. Some of the colors of the larger larvae are better preserved in inflated specimens than in alcohol, but color photography has made preservation of the larval colors less essential. However, the technique is still potentially useful and, if done well, is not to be discouraged. For instructions on how to inflate larvae, consult Banks (1909, pp. 69–70), Hammond (1960), and Martin (1977).

3.2.7.6 Artificial Drying

The disadvantages of the conventional methods of drying are numerous.

1. Time is involved in transferring the specimens or the liquids.
2. Often the appendages must be teased away from the body.
3. The body sometimes shrivels.
4. Specimens become hard or brittle.
5. Compounds such as xylene and acetone are toxic.

Modern techniques for artificially drying specimens preserved in fluid involve freeze-drying and critical-point drying.

Many soft-bodied insects and other arthropods may be preserved by freeze-drying in a very lifelike manner and with no loss of color. Freeze-drying is a great improvement over traditional methods of preservation in fluids or by inflation, especially with lepidopterous larvae, in which important characters of color and pattern are largely lost in all the traditional methods. Unfortunately, the cost of freeze-drying equipment is high, amounting to several hundred dollars (Blum & Woodring 1963; Roe & Clifford 1976).

Briefly, the procedure consists of killing the insect by first freezing it in a natural position and then dehydrating it under vacuum in a desiccator jar kept inside a freezer at -4 to -7°C. With a vacuum of 0.1 μm at -7°C, a medium-sized caterpillar will lose about 90% of its moisture and about 75% of its weight in 48 hours. The frozen condition prevents distortion while drying. The time required to complete drying is variable, at least a few days with small specimens and more than a week with larger specimens. When dry, specimens can be brought to room temperature and pressure and permanently stored in a collection. Like all well-dried insect specimens, they are rather brittle and must be handled carefully. Freeze-drying has also yielded excellent specimens of plant galls formed by insects. Freeze-drying can distort or destroy ultrastructure by differential thermal expansion or the formation of ice crystals from unbound water. Thus the technique is not appropriate for tissues.

Critical-point drying (CPD) (Gordh & Hall 1979) is a procedure that was developed to prepare specimens for study under the scanning electron microscope. Like freeze-drying, CPD is relatively expensive and sophisticated because special equipment and chemicals are required. Nevertheless, the results are ideally suited for specimens collected into or preserved in fluid. The technique involves passing specimens through an intermediate fluid (acetone, ethyl alcohol, Freon) and then into a transitional fluid (CO_2, Freon, nitrous oxide), and then subjecting them to a critical temperature and pressure.

The CPD procedures are simple and relatively rapid. Specimens in alcohol are placed in a small mesh-screen basket with a lid. Specimens then are dehydrated through a series of increasing concentrations of ethyl alcohol (ETOH), ending in a solution of 100% ETOH. The specimens within the basket are passed through two washes of 100% ETOH. Specimens that have been preserved in 70% ETOH can be taken through the alcohol series without rehydration. Leave the basket of specimens in concentrations of alcohol for 20–30 minutes. Recently collected specimens preserved in alcohol or small Lepidoptera larvae may require 1–2 hours in each concentration. Specimens can be left for longer periods in concentrations of alcohol above 50%. After removal from the last wash of 100% ETOH, the specimens in their basket are placed within the chamber of the critical-point drier and are processed according to drier instructions. Several critical-point driers are manufactured. Liquefied CO_2 (research grade) is the ideal transitional fluid because it is easy to use, comparatively inexpensive, and less noxious than others.

Critical-point drying has distinct advantages over conventional methods of specimen preparation. Many specimens can be handled simultaneously, which saves considerable time. Pigmentary colors remain lifelike. Specimens do not collapse or shrivel. Appendages can be manipulated and are more supple than in air-dried specimens. Critical-point drying is most effective for small, soft-bodied insects, such as parasitic Hymenoptera and Nematocera. As freeze-drying and critical-point drying become more popular, we may see excellently prepared dry specimens of immature insects in entomological collections alongside pinned adults.

Figure 3.16
Observation block.

3.2.7.7 Embedding

Preservation of various kinds of biological specimens in blocks of polymerized transparent plastics was popularized some years ago and is still of some interest. The process is rather complicated and laborious, but if carefully done it will yield useful preparations, especially for exhibits and teaching. Directions for embedding insect specimens may be found in the following references and in directions furnished by suppliers of the materials: Foote (1948), Fessenden (1949), and Hocking (1953).

3.2.8 Mounting Specimens for Microscopic Examination

Insects are mounted on pins to allow them to be moved, sorted, or viewed under a microscope when necessary. Many of the external morphological features of insects, which are important for identification, can be seen under low-power microscopy, such as with a dissecting scope. Pinned specimens can be handled most easily and with the least amount of damage if they are transferred from the permanent storage box to a small observation base (Fig. 3.16) constructed of cork and a small amount of modeling clay or Plasticine® mounted on a small wooden or cardboard base. The base should be small enough to sit directly on the stage of the dissecting microscope, allowing enough room for it to be turned as needed. The pinned insect can then be inserted into either the clay or the cork at nearly any angle such that observations can be made of any side of the insect. Commercially available observation stages (Fig. 3.17) can be purchased that offer more flexibility under a microscope.

Figure 3.17
Commercially available observaton stage.

Arthropod Collection and Indentification Techniques

> The techniques and materials used to prepare specimens for high-power microscopic examination vary considerably and are influenced by the kind of insect or mite as well as the researcher's preferences and objectives.

Many insects and mites must be examined at relatively high magnification with a dissecting microscope (80–250X) or a compound microscope (320–970X). This is particularly true of small mites, thrips, whiteflies, aphids, scale insects, fleas, lice, and some parasitic Hymenoptera. The microscope is also necessary to clearly see minute details of larger insects, such as the genitalia. Small specimens (ca. 1-4 mm long) and parts of specimens must be specially prepared and placed temporarily or permanently on microscope slides. We call this process *whole-mount preparation*. Whole-mounts are distinct from slide preparation of tissues for histological purposes, although many of the techniques are similar. Slide mounting is sometimes not advisable for large, thick, or complex structures that must be examined from several angles. Complex or large structures may be examined in a liquid (such as glycerin) and preserved in microvials. Whichever course is adopted, the preliminary treatment of specimens is the same.

The techniques and materials used to prepare specimens for high-power microscopic examination vary considerably and are influenced by the kind of insect or mite as well as the researcher's preferences and objectives. The information given here will provide the reader with basic concepts and principles involved in preparing specimens for such study. For more specific instructions, consult the references listed in **Suggested Reading** at the end of Section 3.2.8.6.3. For reagents and media formulas mentioned here, see Appendix I.

Microscopy methods vary considerably according to the condition of the specimen, the objectives of the study, and the mounting medium employed. However, certain features are common to all processes. The following procedures are given for whole-mounting specimens to be examined microscopically and do not apply to histological preparations.

3.2.8.1 Maceration

Maceration is the chemical removal of muscles and other soft tissues while leaving the sclerotized or chitinized parts of insects that are needed for identification. Maceration eliminates external secretions, foreign matter, some organs, muscles, and fat body without damage to chitinous parts. This is accomplished by immersing the specimen in a 5–10% solution of sodium hydroxide (NaOH) or potassium hydroxide (KOH). Other macerating agents are lactic acid and lactophenol. These chemicals are strongly caustic and must be handled carefully to avoid damage to the skin and eyes. If spattered on the skin, immediately wash off the chemical with water.

KOH may be used for maceration, but this chemical must be used cautiously because delicate specimens may be easily and quickly damaged or completely destroyed in it. Often, NaOH performs as well as KOH or even better. Delicate specimens should be cleared in NaOH; other specimens should be cleared in KOH. Fine ducts lost with KOH remain even after lengthy treatment with NaOH, which will damage only teneral or newly emerged specimens.

NaOH supplied by chemical firms in pellet form is most convenient; three pellets in about 10 ml of water may be used for a day. The solution, however, becomes useless if left overnight. Even if NaOH boils dry on a hotplate set a little above its boiling point, a specimen in it will seldom be damaged. Under similar conditions, a specimen may be completely dissolved in KOH. Adding water to a specimen boiled dry in NaOH solution may restore the specimen.

The length of time the specimen is subjected to maceration is determined by the size and physical characteristics of the specimen, the concentration of the chemical agent, and the temperature at which the treatment occurs. Overnight treatment at room temperature is generally adequate for most specimens; more refractory specimens may require longer periods at higher temperatures. The cuticle of relatively large, strongly sclerotized insects sometimes must be punctured with a fine needle to allow the agent to penetrate the body. Puncture the specimen along the intersegmental membrane between the sclerites but not through them. Heating accelerates the action of the macerating agent, but care must be taken to avoid damage by excessive action, especially if the specimen is teneral. Do not heat the genitalia of microlepidoptera because heat distorts the sclerotized parts. Immersion of the genitalia in cold 10–20% KOH solution overnight is recommended. Boiling for one minute in a 10% solution of NaOH will clear most other small genitalia. For directions on how to macerate insect genitalia, see Section 3.2.8.1.

3.2.8.2 Washing

When the specimen is adequately macerated, remove the caustic macerating agent from the specimen by washing it in distilled water, when available, or tap water. Remove the specimen from the macerating container and place the specimen in a small dish, a watchglass, or a small plastic bottle cover. If the specimen is placed in water for several minutes for manipulation, dissection, or examination, it will then be ready for further treatment. Adding a drop of acetic acid (white vinegar) will guarantee that no caustic substance remains.

3.2.8.3 Dehydration

Specimen dehydration is usually necessary before the specimen is mounted on a microscope slide. Dehydration is essential if the mounting medium has a resin base. Dehydration is accomplished by immersing the specimen in a graded series of ethyl alcohol (ETOH). After maceration, wash the specimen in distilled water

Arthropod Collection and Indentification Techniques

and place the specimen sequentially in 50% ETOH, 70% ETOH, 80% ETOH, 90% ETOH, 95% ETOH, and two washes of absolute ethyl alcohol. Each wash extracts water from the tissues such that the final treatment with absolute ethyl alcohol leaves the specimen dehydrated. The amount of time in each wash varies with the size and nature of the specimen. Some specimens must gradually dehydrate over a period of several hours to avoid distortion.

3.2.8.4 Staining

Specimen staining is sometimes necessary with insects and mites because their immersion in the mounting medium may make colorless and transparent tissues virtually invisible if the medium has a refractive index close to that of the tissues of the specimen (Stein et al. 1968). If a phase-contrast microscope is available, then staining (even with colorless specimens) is not necessary. Staining specimens can be a complicated matter, and histological texts often devote considerable attention to the process. Consult a good histological text for kinds of stains and protocols for specific results (Humason 1979). Bleaching, usually accomplished with hydrogen peroxide, may also be required in very dark-colored specimens.

Several kinds of stains are available from biological supply houses. Acid fuchsin is generally used for aphids, lice, and scale insects. Thrips and fleas should never be stained; most acarologists do not stain mites if they are to be mounted in Hoyer's Medium. A simple stain for the exoskeleton of insects is made by dissolving a few Mercurochrome crystals in water. Specimens may be immersed in the stain solution for a few minutes, depending on the degree of staining needed, and then briefly rinsed in water. Consult Gier (1949) and Carayon (1969) for information pertaining to cuticular stains.

3.2.8.5 Bleaching

If specimens are too dark to reveal sufficient detail after maceration, they may be bleached in a mixture of one part strong ammonia solution to six parts hydrogen peroxide solution. The length of time the specimen is left in the ammonia–peroxide solution depends on the amount of bleaching needed. Specimens should be examined frequently, and care must be exercised because specimens are irrevocably damaged by excessive treatment with bleach.

3.2.8.6 Mounting Medium

Selection of a mounting medium is very important. Mountants are substances in which a specimen is placed for observations, usually beneath a coverslip for microscopical observation. Mountants are classified as temporary or permanent. Mountants may be solid (plastic), semisolid (gelatin), liquid (glycerin), or gas (air). Mountants miscible in water include Hoyer's Medium, lactic acid, polyvinyl

> **Small specimens (ca. 1–4 mm long) and parts of specimens must be specially prepared and placed temporarily or permanently on microscope slides. We call this process wholemount preparation.**

alcohol, and glycerin jellies. Mountants miscible in alcohol include resins, gum mastic, Venice Turpentine, and Gum Sandarac. Mountants miscible in aromatic hydrocarbons include natural resins (Canada balsam) synthetic resins (Permount, Euparol, Histoclad, Piccolyte, Hyrax), and mineral oils (Nujol).

3.2.8.6.1 Mounting

Specific mounting techniques depend on the purpose of the preparation. Some workers feel that preparations may be needed only temporarily in routine work and may be discarded after examination. We recommend permanent microscope slide mounts whenever possible or maintaining the preparation permanently in glycerin in a microvial. If the preparation is kept in a microvial, see Section 3.2.8.7. If the specimen is mounted on a microscope slide, then further treatment depends on the mounting medium.

3.2.8.6.2 Temporary Mounting

A temporary mount can be made with lactic acid, glycerin, or other medium on a 2.5- by 7.5-cm cavity slide. The specimen is placed near the edge of the cavity and wedged into position by manipulating a coverslip over the cavity and the specimen. A fine needle will help bring the specimen into a desired position before the coverslip is centered over the cavity. After the specimen is positioned and the coverslip is centered, a commercial ringing compound, nail polish, or quick-drying cement is used to seal around the edge of the coverslip.

Temporary slide mounts may be kept for a year or more without damage to the specimens. However, these mounts occupy more space in a collection than other slide mounts and are more expensive than conventional slides. Thus, the specimens usually are placed in vials of alcohol for storage. Temporary mounts are advantageous, in that the specimen can be turned easily and viewed from many angles. However, because of the thickness of the mounts, a vertical illuminator operated through a microscope or some alternate method of direct lighting is generally needed. Genitalia and other body parts may be examined. Drawings can be made with the aid of an ocular grid in the microscope while the parts are lying in water in the dish in which they were dissected and extended. Water gives contrast to the structure, which may be difficult to see in glycerin. The water should be "dead," or boiled to drive out gases that may form bubbles in or on the object. The object may be held in place with a minuten bent at a 90° angle and laid over the object or by piercing it at a convenient place.

Hoyer's Medium and polyvinyl alcohol (PVA) are aqueous mounting media. Slides made with aqueous mounting media are not permanent. Their impermanence creates several problems that must be evaluated by preparators before these mounting media are selected. Advocates of Hoyer's Medium note that the U.S. National Collection at the Smithsonian Institution contains some 40-year-old slides of mites mounted in Hoyer's Medium that are still in good condition. However, 40-year-old mounts are still young in terms of slide preparations, and such mounts may eventually deteriorate.

In spite of its popularity, this medium contains chloral hydrate, a controlled substance that cannot be obtained without a permit. Further, chloral hydrate progressively clears specimens. This means that over a period of years, the specimen becomes transparent and sometimes invisible. Remember that Hoyer's Medium is a water-based medium and thus impermanent. Some workers ring Hoyer's mounts with Glyptol or fingernail polish. These ringing agents fail in time, the mountant dries, and the specimen must be removed, or it is ruined.

Microscope slide preparations made with Hoyer's Medium or PVA, particularly of large or thick specimens, tend to crystallize with age and eventually require remounting. Remounting is a relatively simple yet time-consuming process that exposes specimens potentially to damage or destruction each time they are remounted. To remount a specimen, soak the slide in water until the cover glass can be removed; then lift the specimen carefully, and transfer it to a new slide. Specimens permanently mounted require no additional curatorial time, and specimens are not periodically exposed to damage or destruction. Considering the amount of time and effort devoted to whole-mount preparations, a permanent mounting medium is preferred.

Some preparators find specimens easier to arrange if the Hoyer's Medium is diluted with water. However, in the process the mounts may collapse as the excess water evaporates. If Hoyer's Medium is selected, then we strongly recommended that the medium be prepared exactly as directed (see Appendix I) and used undiluted. Aqueous media are affected by ambient moisture; mounts made in very humid conditions may not dry satisfactorily. These problems notwithstanding, Hoyer's Medium is preferred by some acarologists because its refractive index is excellent for use with mites, and specimens can be mounted directly from the collecting fluid without clearing or fixing. The specimens are cleared after mounting by heating the slides briefly on a hotplate set at 65°C until the medium barely begins to bubble. Do not allow Hoyer's Medium to boil, or the specimens may be damaged beyond use. Such mounts can be prepared quickly for immediate study but should be placed in an oven for curing.

3.2.8.6.3 Permanent Mounting Media

The standard medium for permanence is Canada balsam. Specimens must be dehydrated through a graded series of ethyl alcohol concentrations before mounting them in Canada balsam. Some preparators are reluctant to use Canada balsam because it has some drawbacks. It yellows with age, which sometimes complicates photography. Inexperienced workers find it difficult to manipulate delicate specimens in balsam. Mites require special treatment, primarily because their cuticle differs from that of insects. These difficulties notwithstanding, Canada balsam is selected as a mountant for many groups of insects, including scale insects, thrips, and parasitic Hymenoptera.

Euparal is a satisfactory mounting medium for most insects (other than scale insects and thrips). This synthetic material has been used for many decades, with favorable results. When it was unobtainable, especially during World Wars I and II, an inferior compound was used. Euparal may be obtained from medical or entomological supply houses and other sources, all of which import it from Germany. Euparal's formula is a proprietary secret. Euparal does not require dehydration of specimens before mounting. Good preparations may be obtained with specimens taken directly from 80% ETOH and with specimens immersed in 95% ETOH for only a minute. The medium is water-white and remains so indefinitely. For remounting, in case of breakage, specimens may be removed by soaking in absolute ethyl alcohol.

Euparal has a decided advantage over other media, in that small air bubbles trapped in slide preparations are absorbed by the medium during drying. The process sometimes requires several days. Euparal shrinks considerably in drying, which is also a disadvantage. In moderately thick preparations, this results in shrinkage away from the edges of the coverslip. Shrinkage may be countered by adding additional Euparal until there is no further shrinkage. Alternatively, a large coverslip, 2.2 cm in diameter, will pull down around the edges when drying instead of allowing the medium to draw inward. Euparal is relatively fast-drying. Allow the slide to remain overnight in an oven set at about 35°C or in the open at room temperature for a few days.

To place specimens in the medium, put one or more drops of the medium in the center of a clean 2.5- by 7.5-cm glass slide. Determining the precise amount of medium to use requires some experience. Sufficient mountant is needed to run under the entire coverslip. Because Euparal shrinks, a little more mountant is required than with other media. An excess of any medium around the edge of the coverslip is undesirable. Place the cleared and washed (also stained or bleached, if necessary) specimen in the medium on the slide. Make certain that the specimen is completely immersed and that air bubbles are absent. Arrange

SUGGESTED READING:
Mounting Specimens
- Hood 1947
- Mitchell & Cook 1952
- Baker 1958
- Murphy & Gisin 1959
- Hazeltine 1962
- McClung 1964
- Richards 1964
- Strenzke 1966
- Singer 1967
- Wirth & Marston 1968
- Willey 1971
- Freeman 1972
- Grimstone 1972
- Burrells 1978

the specimen in the desired position with a fine needle. If the specimen is thick, then place a few pieces of broken coverslip or plastic sheet around it to prevent crushing when the coverslip is applied. With some preparations (such as ovipositors of tephritid flies), a considerable amount of pressure during drying is desirable to obtain maximally flattened and comparable preparations. Then gently lower a coverslip onto the specimen with forceps, holding the coverslip at a slight angle so that it touches the medium first at one side to prevent air entrapment. Apply gentle pressure with the forceps to fix the position of the specimen. Prepare specimens in more than one position (e.g., dorsal side up and dorsal side down). Do not mount parts of more than one specimen on one slide, because all individuals in the series may not be taxonomically identical.

Coverslips are an important consideration in slide preparations and can be purchased from many biological supply houses. Coverslips are manufactured in many sizes and grades. Outline shapes include rectangles, squares, and circles. The size of coverslip selected should be influenced by the size of the specimen. The coverslip should not be so large that the specimen occupies a minute portion of the area covered. The coverslip should not be so small that all of the specimen is not adequately covered. The specimen should be centered under the coverslip and in the center of the slide. This may be accomplished by taping a microscope slide on the bottom and left edges of a card stock (Fig. 3.18). Place a microscope slide on the taped bottom slide and mark the two top corners. Remove the slide and draw diagonal lines from each dot to the top corners of the taped slide; where the two lines intersect, center a round coverslip. Trace the outline of the coverslip, which serves as a target for positioning the specimen within the mountant. With use of this template, all specimens in a slide collection are placed in the same area of the slide. This placement consistency is helpful when examining many specimens under the microscope and very helpful when the microscope is fitted with a mechanical stage.

3.2.8.6.4 Ringing

Canada balsam, Euparal, and other permanent mounts do not require ringing. However, special compounds should be applied around the edge of the coverslip and the adjacent area of the slide to seal temporary mounts, such as aqueous media and media that do not harden as they dry. Transparent fingernail polish is used by some workers, but this is not effective for more than a year. Glyptol is a product utilized as a ringing compound by some taxonomists interested in preserving specimens in a temporary mountant. Gyptol lasts for several years, but its life expectancy for slide ringing is not known. A second application of Glyptol after the first has dried is likely more effective than one application.

Part 1: Basic Tools and General Techniques

3.2.8.6.5 Curing

Allow slides to dry completely while they remain in a horizontal position. Do not handle slides before the mountant is dry because movement, gravity, and touching of the coverslip can cause the specimen to move in the mountant. In the case of small specimens, setae can be sheared from the body or appendage. Avoid storing slides in a slide box, mailing them, or allowing people to use them until the mountant has dried completely. If an oven is available, set the temperature at 40–45°C and leave the slides in it for at least 24 hours and preferably for 3–4 weeks. Avoid excessive heat because this can cause bubbles to form and ruin the mount.

The final stage in preparing permanent mounts is thorough drying or hardening of the mounting medium. Special microscope slide warming trays and ovens may be purchased through biological supply houses. However, special warming equipment is not necessary, and the process may be done in any clean environment or in an oven under gentle heat. Avoid excessive heat (more than 40°C), which can cause bubbles in the mountant or movement of the specimen under the coverslip. The microscope slides should be carefully labeled, preferably before drying. If more than one mount is being made at a time, some recognition mark or code must be applied to reagent containers and anything associated with the specimen so that the final mount may be correctly labeled.

3.2.8.6.6 Labeling

Collection data should accompany specimens at all times during preparation. As with pinned specimens, the information on a microscope slide label is as important as the specimen.

3.2.8.7 Preparation and Storage of Genitalia

The structures at the apex of the insect abdomen are called the *postabdomen*, *terminalia*, or *genitalia*. The last term is more restrictive and refers morphologically only to certain organs of the eighth and ninth abdominal segments. These structures are very important in identification. Many insects cannot be identified to species without critical examination of these parts. In some insects, these parts are seen easily without special preparation; in other insects, special positioning of the genitalia when the insects are pinned is sufficient. In many insect species, these structures are withdrawn, folded, or convoluted. For these species, the abdomen or a large part of it must be removed and the genitalia specially prepared as follows.

1. Carefully remove the abdomen by grasping it with forceps near the thorax. Bending the abdomen slightly upward and then downward will usually

Figure 3.18
Steps in making a slide template.

Arthropod Collection and Indentification Techniques

break it free of the specimen. Perform this operation over a small dissecting dish containing water or 70% ETOH into which the part can fall. If the specimen is in a fluid and soft, then the abdomen may be severed with fine scissors.

2. Place the severed abdomen in a small beaker or crucible containing three pellets of sodium hydroxide (NaOH) in about 10 ml of water. Then set the container on a hotplate at a temperature a little above that needed to boil the solution. Place a cover loosely over the container to prevent the specimen from being thrown out and lost if the solution "bumps" when heating. Use a copper-mesh screen between the hotplate and crucible to eliminate "bumping." Allow the solution with the specimen in it to boil for a minute. Great care must be taken to avoid getting NaOH on any part of the skin. If that should happen, wash it off immediately with plenty of water.

3. Remove the specimen with forceps and return it to the dissecting dish. Clean the tips of the forceps with water to remove caustic material. Examine the abdomen to see that muscles and most internal organs have been dissolved. If they have not completely dissolved, then return the specimen to the solution and heat it a little longer.

4. When the specimen is well macerated, use a pair of no. 1 stainless steel insect pins (glued in wooden handles with a drop of epoxy) to pull the genitalia into an extended position. Remove unwanted parts or debris. If much unwanted material is present, transfer the specimen to a clean dish of 70% ETOH. Water or dilute alcohol is better than glycerin in which to examine small, colorless specimens because fine structures are more clearly visible. Tap water may be used, but it should be boiled to remove dissolved gases that may collect on and in the specimen. These gases may be very difficult to remove. The specimen then may be examined and identified; it may be placed in an aqueous solution of a few grains of dry Mercurochrome for staining (see Section 3.2.8.4). The specimen may be held at various angles with a bent minuten for study or sketching. If the specimen is preserved for permanent reference, then it may be stored in a microvial or mounted on a microscope slide.

When several specimens must be identified simultaneously, they may be placed in a "spot," "well," or "depression" plate. The plate is a white or black ceramic dish, generally with 12 wells on the surface. Place one specimen in each well. With more than one specimen per well, be absolutely certain that the abdomens are correctly associated with the specimens. To ensure this, place the abdomens in the wells in the same order or configuration as the specimens are arranged in their holding container, and mark one side of the plate to indicate its orientation.

Part 1: Basic Tools and General Techniques

Add water and one pellet of NaOH to each well. After the NaOH has dissolved, place one abdomen or part of it in each depression and warm the plate gently under an incandescent bulb for about one hour. After some of the water has evaporated, replace it with fresh distilled water. Also at this time, examine the abdomens and expel any large air bubbles trapped within that might prevent penetration of the caustic. Reposition the plate under the bulb. A thin stream of macerated tissues soon will issue from the abdomens into the fresh solution.

After an hour or so, depending on the degree of maceration desired, transfer the abdomens for a few minutes to the wells of a second plate that are filled with 70–80% ETOH. Add a small amount of acetic acid to neutralize the caustic. While in these wells, the abdomens may be gently manipulated to remove any remaining tissues. Wash and dry the first plate, place two drops of glycerin in each well, top with 70–80% alcohol, and transfer the abdomens to this plate. Exercise care to keep them in the proper order.

The abdomens may now be examined or left in a clean place for several days if necessary. Avoid the accumulation of dust by covering the container. The glycerin will not evaporate. If the genitalia are to be permanently preserved, place the parts in a microvial (Fig. 3.19). Mount the specimen on a microscope slide only if the specimen is relatively flat and all important characters can be seen in the final position. For example, the ovipositors of fruit flies (Tephritidae) are flat and may be fully extended; the ovipositor and sheath, including spermathecae, can be mounted on a slide with all necessary characters well displayed. The postabdomens of male tephritids are not suited to such treatment because they are nearly as thick as wide and must be examined in profile, ventral, and posterior view.

5. If the specimen is mounted on a microscope slide, then place the specimen in a small dish of 95% ETOH briefly (one minute is usually sufficient). Place a few drops of Euparal in the center of a clean microscope slide. Remove the specimen from the ETOH and immediately place it in the Euparal. Position the specimen and break any large bubbles present before carefully lowering the coverslip. If insufficient Euparal is present to run to the entire circumference of the coverslip, add a little mountant at the edge of the coverslip. This process can be messy. However, the process can be neat if forceps are used to collect the Euparal from the bottle. With pressure from the thumb and forefinger, nearly close the tips of the forceps and immerse them in the Euparal. This action collects fluid between the tips and shafts of the forceps. Next, touch the tips of the forceps to the microscope slide at the edge of the coverslip. When contact is established between the slide, coverslip, and forceps, release the closing pressure on the forceps and allow

Figure 3.19
Specimen mounted with microvial containing genitalia.

the tips to expand. Capillary action will conduct the Euparal beneath the coverslip and increase the volume of mountant beneath the coverslip. Repeat the process until the mountant fills the area beneath the coverslip.

Label the slide and allow the mountant to dry overnight in a warm oven or for a few days in a clean, dust-free environment (see Section 3.2.8.6.5). Small bubbles will disappear, and the specimen will become a little more transparent.

The specimen or its parts may be kept in a microvial if the preparation is not suitable for mounting on a slide. The best microvials are made of transparent plastic with neoprene stoppers. Microvials with an inner lip are particularly desirable. Earlier workers used glass microvials with cork stoppers, but the tannin in the cork is injurious to the specimen. Tannin in solution also affects the glycerin in which the specimen is kept and the pin on which the preparation is held. Whatever kind of microvial is used, before placing the specimen in the vial, add just enough glycerin to the bottom to cover the specimen completely. A throwaway injection syringe is excellent for this purpose. It may be kept filled with enough glycerin for many preparations. A small container of squeezable plastic with a fine tubular nozzle is made for model makers to dispense plastic cement. This is an excellent glycerin dispenser. After placing the specimen in the vial, add the stopper. A dull-pointed pin inserted between the stopper and vial allows air to escape as the stopper is pressed into the vial. When the stopper is completely seated in the vial, the pinpoint is extracted. The insect pin holding the specimen and label is then thrust through the stopper, thereby preserving the specimen and its parts together. The parts may be removed from the microvial and reexamined in water or ETOH at any time and then replaced.

With the aid of an ocular grid in the microscope, the genitalia may be examined and sketched when lying in a small dish of water. (Water provides more contrast than glycerin to delicate structures that may be difficult to see.) Lay a bent minuten over the object to hold it in place, or pierce the object at a convenient place. The specimen can be held in place by a dab of petroleum jelly, but the jelly must be dissolved before the specimen is replaced in a microvial or mounted on a slide.

NOTE: Dissecting, staining, and mounting Lepidoptera genitalia are highly specialized procedures that are not included here. See Clarke (1941) for a description of mounting genitalia of Lepidoptera on slides.

3.2.8.8 Mounting Wings

Wings of many kinds of insects can be mounted on microslides for detailed study or photography. Wings covered with scales, such as wings of Lepidoptera and mosquitoes, must first have the scales removed or at least bleached for study of the venation.

Wings are bleached by immersion in an ordinary laundry bleach (sodium hypochlorite) solution. Wet the wings with ethyl alcohol to activate the bleach. Immersion in the bleach 1–3 minutes is usually sufficient. When the veins become visible, remove the specimen or part from the bleach and wash it in water. Sometimes the scales must be removed under water by brushing the wings carefully with a fine brush or with the tip of a small feather. The descaled wing may then be stained, if desired, in eosin-Y or in an aqueous solution of Mercurochrome crystals for a few to several hours and then washed again. The wing is then ready for mounting as described here, or it may be dried on a slide, placed under a coverslip, and the coverslip ringed with a ringing compound to hold it in place.

Wings not needing descaling may be removed from a fresh specimen. Wings on a dried specimen should be treated with a drop of household ammonia (containing detergent) placed at the base of the wing and allowed to stand in a closed receptacle for about an hour. The wing may be removed with fine-pointed forceps by piercing the body cuticle surrounding the wing base and then pulling the wing loose. This method yields the complete wing, including the basal sclerites. Then wet the wing with 70% ETOH and place it in water for about 10 minutes to soften it. If the wing is from a dried specimen, place the wing in water and then heat the water carefully until it just begins to boil. This will aid in removing air from the larger veins. While the wing is in the water, carefully remove dirt that may be present with a fine brush, but avoid removing setae. Also remove unwanted parts of thoracic cuticle and muscles at the base of the wing.

Next, place the wing in 95% ETOH for about a half a minute and add a few drops of Euparal to a slide. Remove the wing from the alcohol and immediately place it in the Euparal on the slide. Position the wing as desired and make sure that its basal part is well stretched. Alternatively, especially with very delicate wings, arrange the wet wing on the bare slide first, and then pour the mounting medium on top. Carefully apply a coverslip, touching it to one side of the Euparal first at a slight angle from horizontal to avoid entrapping bubbles. Press the coverslip down on the wing carefully to expand it as much as possible and to force bubbles out of the basal veins and elsewhere. Then cure the slide in a warm oven overnight or in open, clean air for a few days.

3.2.8.9 Mounting Larvae of Diptera, Coleoptera, Lepidoptera, and Other Groups

The study of the immature stages of many insects is important for identification, especially of economically important species. Special techniques are often needed to identify immature insects because their cuticle is soft. Critical-point drying and freeze-drying leave the immature stages of most insects well preserved. Unfortunately, this technology is not always available. A method is given here for preparing dipterous larvae that may also be used for the immature stages of other groups. Dipterous larvae, especially those of the higher Diptera, have mouthparts (a cephalopharyngeal skeleton), anterior and posterior spiracular structures, anal plates, cuticular spicules, and other features that are taxonomically important. These parts usually must be examined at high magnification and require special treatment. The larvae of Diptera, Coleoptera, Lepidoptera, and many other groups are best killed in boiling water because it leaves them in prime condition for study.

For cursory examination of the internal cephalopharyngeal skeleton, place the larva in a dissecting dish with a few drops of water. Pierce the cuticle in a few places near the anterior end of the larva and apply a few drops of pure liquid phenol. Do not get phenol on your skin; if you do, wash with water immediately. In a short time the tissues will become transparent. The larva may be returned to 75% ETOH after examination, and the tissues will again become opaque.

For more detailed study and permanent larval preparation, place the specimen in water in a dissecting dish and cut the cuticle with fine dissecting scissors along one side. Start near the anterior end, pass below the spiracle, and continue almost to the posterior end. Then place the larva in an NaOH solution and boil as described in Section 3.2.8.1. When the larva is well macerated, remove the body contents, almost separate the posterior spiracular area from the remainder of the integument, and pull the cephalopharyngeal skeleton a short way out of the body. Place the integument in 95% ETOH while adding a few drops of Euparal on a slide. Then put the integument in the Euparal, opened outward so that the cephalopharyngeal skeleton with the mouth hooks lies away from the integument and the posterior spiracular area lies with both spiracles upward. Apply the coverslip and carefully press it into place. This should give a clear view under high magnification of the cephalopharyngeal skeleton in lateral view, the anterior spiracles, all structures of dorsum and venter of one side, anal plates, and posterior spiracles. The last, often somewhat domed or conical protuberances, may be distorted; the sunray setae and relationships of one spiracle to the other should be easily observed.

As with the genitalia, the larval integument is sometimes best preserved in glycerin in a microvial. Other parts of the insect body, such as antennae, legs, and palpi, may be mounted on slides in Euparal in the same manner as described for the genitalia, wings, and larvae.

References by Clarke (1941), Hardwick (1950), and Zimmerman (1978) concern specialized procedures for making slide mounts of lepidopterous genitalia. Richards (1964) and Wirth and Marston (1968) describe methods using Canada balsam for mounting aphids, scale insects, and other small insects. For methods to use with mites (Acari), see references at the end of Section 3.2.8.6.3. Further procedures are also given in Appendix II.

3.3 Labeling

To have scientific value, specimens must be accompanied by a label or labels giving, at the very minimum, information about where and when the specimen was collected, who collected it, and the host or food plant. During preparation and mounting, specimens should bear temporary labels with this information. Any time a sample is subdivided, the label must be copied so that every specimen continues to be accompanied by the data. Many collectors keep a field notebook to record more detailed information, such as general ecological aspects of the area, abundance and behavior of the specimens, and other observations noted at the time of collection. Byers (1959) describes a rapid method for making temporary labels in the field. A system of code numbers may be used to associate field notes with the specimens collected. However, the code number should appear in addition to the basic data on the label with the specimen. *A code number by itself is never a valid permanent label for a specimen.*

Samples taken for forensic evidence must be sealed and fully labeled. Chain of custody must also be documented, including the name of the person collecting the specimen, dates of collection, and similar information, each time the samples are transferred from one person to the next, verified by signatures.

3.3.1 Paper

The selection of an appropriate paper for labels is very important. Ordinary paper contains acids that weaken the cellulose fibers and subsequently break the paper down over time. Most paper manufactured today is produced from wood pulp that is bleached. Chlorine residue from the bleaching process increases acidity. Other sources of acidity include rosin and alum that are used to prevent ink from running. Paper manufactured from rags is superior because cotton fibers in the paper have superior strength and hold up over time and in solution (alcohol). The best paper has 100% rag content. The paper used for making labels should be heavy so that the labels remain flat and do not rotate loosely on the pin. The

The best paper has 100% rag content.

surface of the paper should be smooth enough to write on with a fine pen. Lined ledger paper (100% rag and 36-pound weight) is best. Smooth-calendered, two-ply Bristol board is also good; it is usually available at art supply stores. Also desirable is a heavy, high-rag-content paper that is used for professional-grade herbarium sheets; it may be obtained from biological supply houses. Labels made from poor-quality paper become yellow and brittle with age, tend to curl, disintegrate in liquid preservatives, and are generally unsatisfactory.

3.3.2 Pens

Handwritten labels are made with "rapidograph," or technical drawing pens. Several brands (Staedtler, Koh-i-nor, Rotring) are available from art stores, craft shops, and bookstores. Technical drawing pens come in several sizes that produce lines of different widths. For most label information, 0.25 mm (no. 000) to 0.30 mm (no. 00) are suitable sizes for achieving small, legible writing. Proper care and maintenance of technical writing pens is necessary for optimal performance.

3.3.3 Ink

The ink should be a superior grade of India ink that is permanent and will not "run" if the labels are placed in jars or vials of liquid preservative. Be sure the ink is completely dry before placing the label in the liquid. When possible, use a waterproofing spray (artist's fixative) on the labels after they are dry. India ink is not always available or convenient to use when collecting in the field. However, labels written with a firm hand and with a moderately soft lead pencil are satisfactory. Labels created using a ballpoint pen or hard-lead pencil soon fade and become illegible, especially when placed in liquids.

3.3.4 Lettered and Printed Labels

Labels may be lettered carefully by hand with a fine-pointed pen, such as a crow quill or Hunt's 104 drawing penpoint. Personal computers and printers also can provide a rapid, inexpensive, and professional label for museum-quality specimens (Inouye 1991). Printed labels, with four-point type, are preferable and are advisable if more than a few labels of one kind are needed. They may be printed in advance, with blank spaces left for the date, or the full data may be printed. Alternatively, computer-generated labels with the proper font and size printed on sufficiently heavy, good-quality paper can be printed by the sheet. Then permanent labels can be cut from the sheets. Photo-offset methods can also produce satisfactory labels from typewritten copy, but the proper kind of paper must be specified. Print on only one side of a label. A satisfactory alternate approach is to order printed labels from a commercial firm.

> **The maximum size of a label should be about 7 by 18 mm.**

3.3.5 Size of Labels

The size of the label is important. Labels must be neither too large nor too small. In determining the size of labels, a relationship must be established between the size of the insect on a pin and the amount of data a label will hold. Most insects are small, and the collection data can occupy more label space than the planar surface area occupied by the individual insects. Try to make labels of a certain maximum size, and use multiple labels for each specimen to accommodate all of the information that must be presented. The maximum size of a label should be about 7 by 18 mm. In four-point type this translates into five lines of 5 pica length, or about 13 capital letters. Commercially printed labels can be much smaller. Use several labels if need be for each specimen to accommodate all of the information that must be presented. Large beetles and butterflies need larger labels, but avoid excessively large "barndoor" labels because they do not hold well on a pin. Even with very small insects, do not skimp on the amount of data just to make a small label. One advantage of moderately large labels for small insects is that if a pin with such a label is accidentally dropped, the label will often keep the insect from being damaged. On the other hand, large labels may damage nearby insects in a box when the pin holding the label is removed from the box. If capital and lowercase letters are used, then spaces between words are not necessary. If there is any chance of ambiguity, then use full spellings. With only one line of data, the label should be wide enough so that when the pin is inserted, all data are legible.

3.3.6 Label Data

The information provided on the data label is as important as the specimen on the pin. The information must be concise, accurate, and unambiguous. Indispensable data must answer the questions of where and when the specimen was collected and include the name of the collector. This kind of data, usually known as locality, date, and collector data, should be given as follows.

3.3.6.1 Locality

The collection locality should be given in such a manner that it can be found on any good map. If the place is not an officially named locality, then it should be given in terms of approximate direction and distance from such a locality. Alternatively, the coordinates of latitude and longitude may be given. The Smithsonian Institution (U.S. National Museum) recommends that for localities in the United States and Canada, the name of the state or province be spelled in capital letters, such as ALBERTA, UTAH, KANSAS, INDIANA, and NO. DAKOTA. This method should also be used for foreign countries, such as NETHERLANDS, CHINA, EQUADOR, and HONDURAS. The next subordinate region should be cited in capitals and lowercase letters, such as counties and parishes in the United States and Canada and provinces elsewhere. Here are a few examples, with virgule (/) indicating the end of a line: ARIZONA:Cochise Co./15kmNEPearce (=15 km northeast of Pearce); NEWFOUNDLAND:Hermitage Dist./12kmWStAlbans; EGYPT: Cairo/SuezRoad 38kmW/Suez; EGYPT:Mud.-Al-Tahrir/22km/SWAbulMatamir; or EGYPT:Mud.-Al-Tahrir/30°05E,30°15N. Current two-letter abbreviations for American states and zip codes should not be used because they are not self-explanatory and may not be permanent.

3.3.6.2 Date

Avoid ambiguity in providing the date of collection. Cite the day, month, and year in that order. Use the international convention of writing day and year in Arabic numerals and the month in Roman numerals without a line over and under the numerals. (Americans tend to put the month before the day; Europeans tend to put the day before the month. A Roman numeral indicating the month avoids confusion involving dates such as 4-8-78 and 8-4-78). Place a period or hyphen between each number: 4.VIII.78 (= August 4, 1978), or 16.V.86 (= May 16, 1986). If a few consecutive days have been spent collecting in one locality but not more than a week, then the extreme days may be cited. For example, 1-6.IX.1989 means that specimens were collected from the 1st through the 6th of September 1989. If three consecutive nights of light trapping were at one place, the median day may be cited, such as 19.VIII.1980 for trapping on the nights of the 18th to 20th of August 1980. For reared specimens, the dates of collection of the immature stages and of adult emergence should be cited. "Pupa 10-XI-1981, emer. 14-VIII-1982" indicates that the pupa was collected on 10 November, 1981 and that the adult emerged on 14 August, 1982.

3.3.6.3 Collector

The name of the collector is regarded as invaluable information on data labels. Spell out the last name of the collector or collectors, using initials for given names. If the last name is a common one, such as Smith, Jones, or Williams, always include middle-name initial. If a collecting party consists of more than three collectors, then use the leader's name followed by et al.: T. J. Gibb et al.

3.3.6.4 Other Data

Many kinds of information may be important but not relevant or available for all insects. For instance, cite the hosts of parasitic insects and the host plant for phytophagous insects when this information is known. Details of the habitat (elevation, ecological type, and conditions of collection) are important and are usually put on a label in addition to the primary data. Such data may include "*swept from Salsola kali,*" "*McPhail trap in orange grove,*" "*at light,*" "*3,200 m,*" "*sandy beach,*" and "*under bark dead Populus deltoides.*" Do not use vernacular names of hosts unless the host is common and widespread, such as orange

Part 1: Basic Tools and General Techniques

or horse. If the specific name of a host is not known, at least give the genus. "*Vaccinium sp.*" is better than no name or "*huckleberry.*" Even the family name of the host is helpful if no more specific name is available. The presumed nature of the association between insect and plant should be clearly indicated, for example, "*Resting on flowers of Vaccinium sp.*" The word "*ex*" (Latin for "*out of*") should mean that the insect was observed feeding on or in or was bred from the mentioned plant, for example, "Ex seed *Abutilon theophrasti.*"

Forensic evidence also should include additional data, such as agency of collector, case number, victim or subject's name, a brief description of the item, and name and signature of the person sealing the evidence.

As mentioned earlier, it is a good idea for collectors to keep a notebook in which details of locality, habitat, and other important data are recorded. Each individual locality may be assigned a notebook or code number with which the collecting jars and vials are marked until the specimens can be prepared. However, citation of such a number on permanent labels is virtually worthless. Even when large files of such data are kept, they are seldom available many years later when a researcher needs to know what a cited notebook number means. Cryptic codes should be avoided. If information is maintained on a computer, then the files should be copied onto separate storage disks and stored in a safe place away from the computer.

Figure 3.19
Proper labeling of double mount.

3.3.7 Placing the Labels

For double-mounted insects, insert the pin through the center of the right side of the label (Fig. 3.19), with the long axis of the label oriented in the same direction as the card point. Take care that the pin is not inserted through the writing on the label. For specimens mounted by direct pinning, the label is centered under the specimen, with the long axis of the label coinciding with the long axis of the specimen. The left margin of the label is toward the head of the insect. An exception to this is when specimens have the wings spread, such as Lepidoptera. In these cases, the label always should be aligned transversely, at right angles to the axis of the body, with the upper margin toward the head. Labels may be moved up the pin to the desired height by using a staging or pinning block (Fig. 3.19). The middle step of the block will give about the right height if only one label is used. With more than one label, space the labels on the pin beneath the specimen so that the information on the labels can be read without moving any of them.

3.3.8 Labeling Vials

Specimens preserved in fluid should be accompanied by one rather large label that includes all collection data. Do not fold the label because small specimens

Vial labels should be written with a moderately soft lead pencil or with India ink.

Arthropod Collection and Indentification Techniques

may be damaged or lost when the label is removed. Multiple labels or small labels that float in the vial may also damage specimens. Further, when two labels lie face to face, they cannot be read.

Always place labels inside the vial. Labels attached to vials may become defaced, destroyed, or detached, regardless of the method or substance used to affix them to the vial. The label should be written with a pencil or in ink. If a pencil is used, then select a moderately soft lead pencil because hard-lead pencil writing becomes illegible in liquid. If a pen is chosen, then select India ink and ensure that the ink is well dried so that it will not dissolve or run when immersed in the liquid. *Never* use a ballpoint or felt-tip pen for labeling vials.

3.3.9 Labeling Microscope Slides

Preserving specimens on microscope slides is time-consuming and tedious. The labels for slide-mounted specimens should be given equal care and consideration. Labels made expressly for this purpose can be obtained from biological supply houses. Labels with pressure-sensitive adhesive are now available that seem durable with time. Modern adhesives are superior to the older glues, which often failed with time or were consumed by cockroaches or psocids. Labels should *not* be affixed to slides with transparent tape.

Figure 3.20
A properly labeled, microscope-slide-mounted insect.

All microscope slide labels should be square (Fig. 3.20). Specimens should be centered on the microscope slide and a label placed on either side of the specimen. Never put labels on the underside of a microscope slide. The label on the left side of the specimen should contain taxonomic information. The label on the right side of the specimen should include collection information. The sequence of information should also be standardized. Put as much data on the label as possible, including the kind of mounting medium in which the specimen is preserved. This kind of information is important when specimen remounting is required (Waltz & McCafferty 1984).

3.3.10 Identification Labels

When specimens are sent to an expert for identification, they should be accompanied by permanent collection labels giving all essential data. If associated field notes are available, copies of these should accompany the specimens. When the identification has been made, the scientific name of the specimen and the name of the identifier should be printed on a label associated with the specimen. On pinned specimens, this information is always printed on a separate label placed below the collection label or labels on the same pin. When a series of specimens consists of the same species, the identification label is often placed only on the first specimen in the series, with the understanding that all other specimens to the right in that row and in following rows belong to the same species. The series ends with another specimen bearing an identification label. Identifications for specimens preserved in alcohol or on slides may be written on the same label as the collection data or on a separate label, depending on the preference of the collector or person making the identification.

Figure 3.21
Vial rack.

3.4 Care of the Collection

If care is taken and a few basic precautions are followed, a collection of insects or mites can be maintained indefinitely. The information given here is general; institutions and individuals will want to adapt materials and procedures to fit their own needs and resources.

3.4.1 Housing the Collection

The serious collector should consider using standard equipment for housing a collection because this ensures uniformity of container size and style when additions are necessary. Homemade boxes, drawers, and cabinets may serve a useful purpose, but they cannot be easily incorporated into museum collections. Standard equipment is available from many entomological and biological supply houses.

Material preserved in liquid usually needs little attention beyond the occasional replacement of preservative and stoppers. Small vials may be stored in racks so that the stoppers are not in contact with the liquid (Fig. 3.21). The use of storage racks for vials expedites rearrangement and examination of the material. Alcohol evaporates in closed containers, and vials should be examined periodically to ensure that the specimens do not become dry. Specimens that have become dry in vials, because of the evaporation of preservative, become ruined. If the vials cannot be inspected frequently, then vials containing larvae or large insects should have their stoppers replaced by cotton plugs. Several cotton-plugged vials can be placed upside down in a large jar filled with preservative. Use of cotton plugs is not recommended for very small or delicate specimens because they may become entangled in the cotton fibers. Jars with screw tops or clamping lids, as are used in home canning, are ideal for this purpose. Jars specifically designed for museum use can be obtained from biological supply houses. Stoppers of neoprene or other synthetic materials generally are superior to cork stoppers, but good-quality cork

Figure 3.22
Storage containers for microscope slides.
A. plastic slide storage box
B. wooden slide storage box

Arthropod Collection and Indentification Techniques

Figure 3.23
Storage containers for pinned insects.
A. Schmitt box
B. Cornell drawer

stoppers are usually preferred to plastic screw tops, which are easily broken. Some rubber or rubber-like stoppers in contact with 80% ETOH over long periods of time become bleached and swollen. When specimens stored in such containers are critical-point dried, an undesirable residue forms on the specimens. Many of the newer flanged plastic stoppers are excellent.

Microscope slides are usually stored in wooden or plastic boxes obtainable from biological supply houses (Fig. 3.22). The inner sides of the boxes are slotted to hold the slides vertically and to separate them from one another. Slide boxes are available in sizes made to hold from 50 to 100 or more slides. If slides are stored vertically, then they must be thoroughly dried before storage or the coverslips may move. Some workers store the slide boxes so that the slides rest horizontally. This is especially desirable if the slides are made with Hoyer's Medium, which may become soft under very humid conditions. Several slide-filing systems are available from suppliers, but whatever system is used, care should be taken to ensure that additional similar equipment will be available in the future for expansion of the collection.

Small plastic slide boxes, usually made to hold five slides, are convenient for keeping slides in a unit-tray system along with pinned specimens. This is particularly advantageous when genitalia are mounted on slides, because the preparations are readily apparent to visiting researchers examining the pinned specimens.

Pinned specimens should be kept in any standard, commercially available insect drawer (Fig. 3.23). These standards are named U.S. National Museum, California Academy of Sciences, Cornell, or Schmitt sizes. Larger North American collections typically use the unit-tray system, with various sizes of unit trays made to fit into a drawer. The pinning bottoms of the unit trays and boxes are now generally made of polyethylene foam. The older standard was pressed cork, but that was extremely variable in quality and usually contained enough tannin to corrode pins and eventually to cement the pins firmly into the pinning material. Polyethylene foam is available in large sheets that are cut to the desired size and cemented into boxes or unit trays.

A serviceable substitute for polyethylene is 6-mm-thick balsa wood boards, obtainable from model maker supply houses. These boards should be individually selected for softness because they are frequently excessively hard. Another good substitute, especially for temporary storage of pinned specimens, is double-thickness corrugated board, which is often used to separate layers or rows of cans in cartons. Single-thickness corrugated board will not hold an insect pin firmly, and the harder board used for making cartons is not suitable.

Any box used to store insect specimens must be nearly airtight to keep out museum pests. The most commonly encountered pests include dermestid beetles, psocids (booklice), and silverfish. In a period of a few months, these insects (and certain other pests) can devour specimens, chew labels, or otherwise ruin a collection made over many years. These pests find their way into the best boxes or insect drawers. Constant vigilance is necessary to prevent them from destroying valuable specimens and priceless scientific information.

3.4.2 Protecting Specimens from Pests and Mold

To kill pests that are actively damaging a collection, apply a liquid fumigant, which acts more rapidly than solid fumigants. Examples of liquid fumigants include carbon tetrachloride, chloroform, ethyl acetate, ethylene dichloride, methyl bromide, and sulfuryl fluoride. These may be registered for restricted use in museums by certified exterminators only. Exercise extreme care when working with any of these compounds. Liquid fumigants volatilize rapidly, may be flammable, and are toxic to humans. Work outdoors if possible, and use some kind of fumigation chamber. One day in the chamber usually is sufficient to kill the pests.

Individuals interested in pest damage, identification, and control in museums should consult Edwards et al. (1981) and Zycherman and Schrock (1988).

The periodic fumigation of all insect storage boxes is necessary. The best-made insect drawers provide for chemical fumigants. In the past, para-dichlorobenzene (PDB) was used in museums as a fumigant. However, this material has demonstrated chronic health effects and is under EPA investigation for possible carginogenic effects on humans. PDB should be avoided. Naphthalene is widely used as a fumigant, but in a technical sense it is regarded as a repellent.

Solid fumigants (repellents) should not be placed loose in a box of pinned specimens. If crystals or flakes of napthalene are used, a small quantity should be placed in a cloth bag or in a pillbox whose top is perforated with tiny holes. This container is pinned firmly into one corner of the box of specimens. Mothballs may be pinned in a box by attaching the mothball to the head of an ordinary pin. This is done by heating the pin and forcing its head into the mothball. When moving boxes, be careful that the mothballs and fumigant containers do not come loose and damage the specimens. To keep pests out of Riker mounts and other display cases, sprinkle naphthalene flakes on the cotton when the mount is prepared.

Another useful method to kill pests in a collection is to cut strips of dichlorvos (738 Vapona strips, No-pest Strips, Vaponite, Nuvan) into small pieces, and secure them in the insect drawers. Dichlorvos is registered for nonrestricted use, but only pest strips should be used. This method gives a fairly rapid kill while avoiding the hazards of using flammable liquids. Under conditions of high humidity, Vapona strips corrode metals and dissolve polystyrene plastic.

Mold is another serious problem with insect collections, especially in moist, warm climates. Mold is a kind of fungus that readily attacks and grows on insect specimens. Once a specimen has become moldy, it is extremely difficult to restore it. If only a few filaments or hyphae of mold are present on a specimen, they may be removed carefully with forceps or with a fine brush and the specimen dried well in a warm oven and then returned to the collection. Only keeping the collection in a dry place will prevent mold. In humid climates it is sometimes necessary to keep insect and other kinds of collections in rooms with artificial dehydration. Some microscope-slide mounting media are also subject to molding.

3.5 Packaging and Shipping Specimens

In mailing insects and mites, there is always a risk of damaging or losing specimens. By following the recommendations given here, the risk can be reduced (Sabrosky 1971; Stein 1976; Hoffard 1980).

3.5.1 Packing Materials

Often specimens must be mailed to colleagues or museums. Specimens are at high risk of damage when they are in the mail service. Thus, mailing specimens is a very important process that requires considerable thought and attention. Mailing cartons should be constructed of strong corrugated board or other stiff material; wood is advisable for overseas shipments. Do not use cardboard boxes that have been sent through the mail several times. They tend to be damaged. Screw-top mailing tubes are ideal for small items, particularly slide-mounted specimens and specimens in alcohol. All containers must be sufficiently large to include ample packing material to minimize the effects of jarring (Fig. 3.24). A rule of thumb suggests a minimum of 5 cm on all sides between the specimens and the inner surface of the mailing carton. The packing material is intended to absorb shock. Containers holding specimens should not be packed so tightly that this objective is negated. Suitable packing materials include excelsior, shaved wood, crumpled newspapers, styrofoam chips, or plastic bits. Clear plastic sheet material with a regular pattern of bubbles (bubble wrap or blister pack) is superior. This material is very lightweight and has excellent shock-deadening properties.

Figure 3.24
Packaging specimens for shipment.

3.5.2 Pinned Specimens

Pinned specimens should always be placed as described here in a small box with a pinning bottom. The box should be well wrapped and placed in a larger carton with at least 5 cm of lightly packed packing material between it and the carton on all sides.

1. Select a sturdy pinning box with a good pinning bottom. The pinning bottom should not be very hard, because hard substrates tend to transmit shock. The pinning substrate should be at least 6 mm thick and cemented securely to the bottom of the box. The box should have a tight-fitting lid or a flap-type lid held in place with a strip of masking tape. Do not mail specimens in an open-top museum tray.

2. Pin the specimens firmly into the pinning bottom. Leave sufficient space between pins so that the specimens can be removed easily. Specimens pinned too closely together sometimes are damaged by adjacent specimens or their labels when removed. Place bracing pins on both sides of heavy or long-bodied specimens to prevent them from rotating (cartwheeling) on their pins and damaging adjacent specimens. Bracing pins may be put on the sides of data labels or card-mounted specimens to prevent cartwheeling. Microvials should have an additional pin at the end of the vial to keep it from coming off its stopper. Pin a loose ball of cotton in the corner of the box. This captures antennae and legs that might fall from specimens in transit. If loose appendages are not captured, they can damage specimens. Vials other than microvials should be wrapped in a box separate from pinned specimens.

3. The box in which the specimens are pinned should be shallow enough so that the heads of the pins almost touch the lid. If the box is too deep, a piece of firm cardboard should be cut to fit into the box and lie on top of the pins. If there are only a few specimens in the box, a few extra pins should be added near all of the corners to keep the cardboard level. Attach a tab made of a piece of adhesive tape folded double, with the ends left free to attach to the top of the inserted cardboard. The insert may be lifted out by the tab. The space between the insert and the lid of the box should be filled with packing material to keep the insert pressed lightly against the tops of the pins when the lid is in place. This prevents the pins from working loose and damaging specimens in transit. Use cotton batting, sponge, or bubble wrap between the lid and insert. Do not use excelsior or any loose material because this can work its way beneath the insert and damage specimens.

4. If only one or two specimens are being shipped, they may be placed in a straight-sided plastic vial with a press-on or screw-on top. The vial should be

of sufficient diameter to hold the labels in a normal position. A cork stopper cut to such a length that its larger end is a little greater in diameter than that of the inside of the vial is pressed tightly into the bottom of the vial. This will provide a good pinning bottom into which one or two pinned specimens may be firmly pressed. Attach the cover of the vial, wrap the vial in enough packing material to hold it firmly in a mailing tube, and attach the cover of the mailing tube, and it is ready to ship.

We recommend fumigating boxes before they are shipped. This eliminates the transmission of museum pests (psocids, silverfish, dermestid beetles) to other collectors or institutions. Do not leave loose fumigant in the box with the specimen or any fumigant balls on pins in containers. They are especially prone to work loose and damage specimens. Boxes that are received should also be fumigated to prevent the acquisition of unwanted museum pests.

3.5.3 Specimens in Vials

The following procedures are recommended.

1. Fill each vial completely with liquid preservative. Air bubbles in vials promote movement of specimens and labels. Stopper the vial tightly by holding a pin or piece of wire between the inner surface of the vial lip and the stopper. Press the stopper into place and permit air or excess fluid to escape; then remove the pin or wire. Cork stoppers should not be used because they crumble and often leak. If they are used, make certain they do not have defects that will allow leakage while specimens are in transit. Screw-top vials should be firmly closed and sealed with a turn and a half of plastic adhesive tape around the lower edge of the cap and part of the vial. Some workers seal vials with paraffin. In our experience this is unnecessary because the paraffin often breaks loose and will not prevent leakage.

2. Wrap each vial with cotton, tissue, foam padding, paper toweling, or similar material. Allow no piece of glass to come into contact with another piece of glass. Several individually wrapped vials may be bound or held with tape or rubberbands as a unit, or they may be placed in a small cardboard box with enough packing material to ensure that they are not shaken around. Placing all of the vials in a self-sealing plastic bag prevents fluid from damaging pinned material in the event that a vial breaks.

3.5.4 Loading Cartons

After pinned specimens, specimens in vials, or both have been prepared properly, they should be placed in a strong carton large enough to hold at least 5 cm of packing material around all sides, including the top and bottom. Use ample packing material to prevent the contents from shifting within the carton, but do not pack the material tightly. Packing material should be resilient, to absorb shocks and prevent damage to the contents being shipped. One or a few vials may be shipped in a mailing tube, as previously described. When shipping more than one box or packet of vials, tie or wrap them together as a unit before placing them in the larger carton. Individual boxes or vials may easily be overlooked and lost when unpacked. For this reason, a packing list should be included in the carton. Vials of specimens in fluid are much heavier than boxes of pinned specimens. Cartons containing many vials may be packed somewhat tighter in the carton than cartons containing only pinned specimens. Pinned and liquid-preserved specimens may be shipped in the same carton. When many vials of liquid-preserved specimens are involved, ship them separately.

3.5.5 Shipping Microscope Slides

Make certain that slides are thoroughly dried and cured. Slides may be shipped in holders made expressly for that purpose (Fig. 3.25). Slide mailers are available from biological supply houses. Even in these holders, wrap a little soft tissue around each end of each slide so that the coverslip does not come into contact with anything. The slides may also be shipped in standard storage boxes. Place soft tissue around each end of each slide and between the slides and the box lid to prevent movement. The box should then be wrapped to hold the lid firmly. The slide box may be treated as described here for pinned and liquid-preserved specimens and placed in a carton with them. If slide holders are not available, a few slides, each wrapped with tissue, may be tied together at each end with tape, rubber bands, or string and wrapped in strong paper. Place the wrapped slides in a mailing tube or carton with packing material.

3.5.6 Shipping Live Specimens

Most insects and mites intended for a collection or submitted to experts for identification should not be shipped alive. To protect American agriculture, federal law prohibits the importation and movement of live pests, pathogens, vectors, and articles that might harbor these organisms unless the shipments are authorized by the U.S. Department of Agriculture. Shipments of live material must comply with all federal, state, and local regulations (see sample form in Appendix IV). Shipments of live insect material without valid permits may be seized and destroyed by Animal and Plant Health Inspection Service (APHIS) quarantine inspectors. In addition to meeting federal laws, the shipment of some species must be approved by state officials. To obtain federal authorization, write well in advance of shipment to: Animal and Plant Health Inspection Service,

Figure 3.25
Microscope slide shipper.

USDA, Federal Building, Hyattsville, MD 20782. Each request is evaluated according to its individual merits.

Immature stages (egg, larva, nymph, pupa) shipped for rearing elsewhere should be placed in tightly closed containers without vent holes. These stages require a minimum of air and will not suffocate. Pupae should be packed loosely in moist (not wet) moss or similar material. Larvae should be packed with adequate food material to last until their arrival. Most beetle larvae and some caterpillars (especially cutworms) should be isolated because they are cannibalistic. To prevent excessive accumulation of frass (fecal material) and moisture, do not overload containers. Plant material held without ventilation becomes moldy, especially when kept in plastic bags. For this reason, pieces of the host plant bearing such insects as scale insects (Coccoidea) should be partially or completely dried before being placed in a container. If this is not possible, they should be packed in a container (such as a paper bag) that will permit drying to continue after closure. Live Heteroptera and other small, active insects are killed easily by excessive moisture in the container. Therefore, provide several tiny vent holes or place a fine-mesh screen over one end of the container when shipping such insects.

Some containers designed to hold living insects are strong and can be shipped without additional packing. Generally the containers should be enclosed in a second carton, with enough packing material to prevent damage to the inner carton. In all cases, affix a permit for shipping live insects in a conspicuous place on the outside of the shipping container. Some express mail service companies will not accept living insects for shipment. Check with company agents before taking the material in for shipment. If specimens are sent overseas, be certain that importation permits of the destination country are obtained, and use airfreight to send the parcel. Check with international carriers for special requirements.

There are no restrictions on mailing dead specimens within the United States or to other countries. If the specimens are sent out of the country, U.S. Customs requires that the contents and value of the package be listed. The statement "Dead Insects for Scientific Study, of No Commercial Value" will suffice. All packages should be marked "FRAGILE," and a complete return address should be included on the outside of each container.

SUGGESTED READING:
Collections
- VanCleave & Ross (1947)
  Artificial Drying
- Woodring & Blum (1963)
- Harris (1964)
- Flaschka & Floyd (1969)
- Dominick (1972)
- Berte (1979)
- Gordh & Hall (1979)
- Hower (1979)
- Kissinger (1982)
  Lettered and Printed Labels
- Ellis et al. (1985)
- Clark & Gregg (1986)
  Care of the Collection
- Darling & Plowright (1990)

To protect American agriculture, federal law prohibits the importation and movement of live pests, pathogens, vectors, and articles that might harbor these organisms unless the shipments are authorized by the U.S. Department of Agriculture.

Part 2: Classification of Insects and Mites

Arthropod Collection and Indentification Techniques

PART 2:
CLASSIFICATION OF INSECTS AND MITES

Introduction 100

Chapter 4: Classification of Insects and Mites 103
    4.1 Key to Classes of Arthropoda   106
    4.2 Class Arachnida   110
        4.2.1   Key to Orders of Arachnida   110
    4.3 Subclass Acari   114
        4.3.1   Key to Some Primary Groups of the Subclass Acari   115
    4.4 Classes Diplopoda, Chilopoda, Pauropoda, and Symphyla   117
    4.5 Class Crustacea   117
    4.6 Class Hexopoda (Insecta)   117

Chapter 5: Synopsis of Insect Orders 121
    5.1 Subclass Class Entognatha: Primitive Wingless Hexapods   121
    5.2. Subclass Ectognatha: Primitive Wingless Hexapods   121
    5.3 Subclass Pterygota (Insecta): Winged and Secondarily Wingless Insects   121
        5.3.1   Exopterygota   122
        5.3.2   Endopterygota   124
    5.4 Key to Orders of Hexapoda (Insects)   125

Part 2 - Introduction

Chapter 6: Descriptions of Hexapod Orders 157
- 6.1 Protura 157
- 6.2 Diplura 157
- 6.3 Collembola 158
- 6.4 Microcoryphia 158
- 6.5 Thysanura 159
- 6.6 Ephemeroptera 159
- 6.7 Odonata 160
- 6.8 Orthoptera 162
- 6.9 Blattodea 163
- 6.10 Mantodea 164
- 6.11 Phasmatodea 165
- 6.12 Gryllblattodea 166
- 6.13 Dermaptera 166
- 6.14 Isoptera 167
- 6.15 Embiidina 169
- 6.16 Plecoptera 170
- 6.17 Psocoptera 172
- 6.18 Zoraptera 172
- 6.19 Mallophaga 173
- 6.20 Anoplura 174
- 6.21 Thysanoptera 175
- 6.22 Hemiptera 176
- 6.23 Homoptera 177
- 6.24 Coleoptera 179
- 6.25 Strepsiptera 181
- 6.26 Mecoptera 182
- 6.27 Neuroptera 183
- 6.28 Tricoptera 184
- 6.29 Lepidoptera 185
- 6.30 Diptera 189
- 6.31 Siphonaptera 192
- 6.32 Hymenoptera 193

Summary 198

Introduction

The number of insects and mites known today staggers the imagination. More than 1.5 million species of insects have been described, and estimates of the number of undescribed species range in the millions. There are more than 50,000 described species of mites, and many times more than that number remain undescribed. Associated with this enormity of numbers we find a biological diversity not seen among other animal groups. Insects and mites can be found in almost every conceivable environment. Mites are found living on plants, molds, and stored foods, in the nares of sea snakes, in the cloacae of turtles, beneath the elytra of beetles, and in the quills of birds. Insects can be collected from such diverse and seemingly inhospitable habitats as in petroleum pools or saltwater marshes, within the canopy of tropical trees, or deep inside solid wood, at high altitudes in the atmosphere or deep inside caves and from every continent, including Antartica. The habits of insects and mites vary from parasitic to free-living and from beneficial to highly destructive. Many insects and mites have developed intimate associations with humans. Some are extremely valuable and provide pollination, food, pest control and other benefits to humans. Other destructive forms can cause enormous losses to agricultural crops, food, clothing, and other materials of value to humans and can injure people or transmit devastating diseases.

Because of the potential destructiveness of some species of insects and mites, outbreaks of pests must be detected rapidly and their population sizes must be estimated accurately. Correct identification of a newly detected pest or newly collected or discovered insect or mite is of paramount importance, because its scientific name is the key to all known information about the insect, its habits, its behavior, and its potential threat or benefit to human welfare. Control recommendations, where necessary and prudent, hinge directly on an accurate identification of the pest. Land grant universities also include cooperative extension specialists who can serve as professional resources for arthropod information and identification. A list of state land grant universities, together with street and e-mail addresses and telephone numbers, is provided in Appendix III. Requests for specific identification can also be submitted through the Systematic Entomology Laboratory in Beltsville, MD. Directions for submitting samples are provided in Appendix IV.

Insects and mites can and should be observed in their natural environments, but most of them, especially the many small ones, must be collected and properly preserved before they can be identified. Correct identification seldom is easy. To identify specimens correctly, they must be properly labeled and preserved in the best condition possible. The identification of a particular insect or mite usually requires examination of minute details of its anatomy with the aid of a hand lens or microscope. If the microscopic details of a specimen's anatomy are concealed, damaged, or missing because of improper handling or preservation, then identification is sometimes impossible, and information about the species to which it belongs cannot be made available. Therefore, adequate preservation and proper labeling of specimens are essential to their identification.

The reasons for identifying insects and mites are many and varied. Insects and mites may be collected to obtain general study material (as for a school course) or for personal interests (as in a hobby). Professional entomologists may survey for the presence or abundance (relative or absolute numbers) of pests. Scientists may acquire material for biological, physiological, ecological, and other types of studies. The pest control industry often requires specimens to be collected and retained to document past history as well as to justify need for pesticide applications. It is essential that forensic entomologists and crime scene investigators collect, preserve, and retain specimens that may become important evidence in legal cases. The purpose for making an insect or mite collection may demand various special collecting methods.

What to collect depends on the purpose for which the collection is intended. When important pest insects and mites are needed for identification, specimens should be collected in large numbers. A sample of 20 specimens should be considered the minimum; larger numbers are desirable. Remember, excess specimens can be discarded or exchanged, but additional specimens cannot always be collected when needed. All available life stages should be collected. Frequently, insects and mites cannot be identified accurately from immature stages, and they must be reared to the adult stage for correct identification. Photographers should collect the specimens they photograph when positive identification is desired. Minute or critical diagnostic characters often are not discernible in photographs. If specimens are destined for display cases, it may be important to collect and identify a sample of the host plant for the display as well. Collections that include gall-producing insects and mites must also display the gall. Reference collections often are made more valuable if they contain samples of associated insect damage or frass, which can help identify similar problems later (Kosztarab 1966). For example, the architecture of ground nests for bees, wasps, mole crickets, and termites can be preserved with wax or fiberglass resins (Chapman et al. 1990) (Brandenburg & Villani 1995).

Biology students in high school and college are often required to collect and identify specimens from as many orders or groups as possible. Other beginning enthusiasts attempt to collect every specimen they find. The experience and knowledge gained in making a general collection are of value in helping the collector decide on a specialty. However, with so many different kinds of insects from which to choose – over 80,000 described species in North America alone – most students eventually concentrate on one or two of the major insect or mite

Arthropod Collection and Indentification Techniques

SUGGESTED READING:
- Seber 1973
- Lee et al. 1982

groups. Specimens other than those in a chosen group may be collected for exchange with other collectors or for deposition in an institutional collection.

Part 2 of this text presents basic identification techniques and descriptions of the different arthropod groups. References listed in the text and in **SUGGESTED READINGS** are intended for those who desire to more specific information about the topic.

Chapter 4: Classification of Insects and Mites

Classification is essential in any type of biological work. Identification is an element of classification. Identification to the species or genus level is necessary to determine the best collecting methods and techniques for preservation and how best to control pest organisms. Identification to the family or order level is necessary for the most elementary study of insects and mites. The keys and diagnostic information presented here are designed to enable a novice collector to recognize the most common orders. Students must remember that the number of insect and mite species is immense. We cannot provide a complete classification here, and none is available in any other single reference. Keys to aid in diagnosis to family levels may be found in general entomology text books such as Borror et al. (1989), Gullan and Cranston (2004). A glossary of basic entomological terms is often helpful in understanding complex terminology associated with identification and classification. A basic glossary is provided at the end of this work. Nichols and Schuh (1989) is a more extensive English-language glossary that may be helpful in entomological work.

Classification involves the arrangement of organisms in groups based on sets of "characters" common to each group. Traditionally, the "characters" have been based on anatomical study. However, another method of classification is molecular taxonomy. In molecular taxonomy, proteins and genes are used to determine evolutionary relationships. The percentage of DNA that is similar between species is taken into consideration. Signature sequences, such as those derived from rRNA, also serve to determine separation between different arthropods. Once a species has been carefully examined, specific genes or proteins can be used as molecular clocks, which determine the divergence from a common ancestor.

Kingdom
    Phylum
        Subphylum
            Class
                Subclass
                    Order
                        Suborder
                            Superfamily
                                Family
                                  Subfamily
                                      Tribe
                                          Genus
                                            Subgenus
                                                Species
                                                    Subspecies

Categories commonly recognized, arranged in order of rank from highest to lowest.

Arthropod Collection and Indentification Techniques

Groups are arranged in a taxonomic hierarchy that proceeds stepwise from highest to lowest level: kingdom, phylum, class, order, family, genus, and species. In many cases, there are also intermediate superdivisions and subdivisions as well as the less used groups of tribe, cohort, and phalanx. This stepped arrangement is known as a hierarchical system, was developed by the Swedish naturalist and taxonomist Carl Linnaeus (1707-1778). The kingdom Animalia is divided into several phyla. The phylum Arthropoda contains the insects, mites, and all other animals with an external skeleton and segmented appendages. This phylum is divided into classes, as shown in the following key. Additional keys are given to orders and in a few cases to suborders. For further identification, the reader is directed to the vast taxonomic literature.

The keys given here are in a format often found in North American publications. Each couplet is composed of two statements, labeled **a** and **b**. Start at couplet **1a**. Read the two choices and select the one that best describes your specimen. At the end of each couplet is a named taxon or a number directing you to another numbered couplet, from which you must again choose. Follow the numbers, comparing alternatives in each couplet, until a name is reached. A similar procedure may be followed with other keys in the literature to which you may wish to refer. The use of keys is far more satisfactory than comparison with pictures or even with identified specimens because the most important, diagnostic characteristics are specifically stated. The characters used in taxonomic keys often involve seemingly minute differences that might not be suspected by an inexperienced person.

We urge students to refer to an elementary textbook on entomology and to become familiar with the basic elements of insect anatomy. Even such words as "face" and "leg" do not mean quite the same when referring to an insect as they do in reference to a human. It is also necessary to distinguish between an immature and an adult. This kind of information is found in general textbooks.

Scientific names are used throughout the world in many kinds of publications. Each species has one scientific name, and all of the biological information regarding a species is tied to the scientific name. For instance, the scientific name *Homo sapiens* refers to humans and not other primates. In essence, the scientific name is a shorthand method of conveying a considerable amount of information. The scientific name of a species is composed of two words: the genus (written with an initial capital: *Homo*) and the species (a specific epithet or trivial name, written in lowercase: *sapiens*). This combination, known as a *binomen*, and must be written in a typeface different from the text in which the name appears. Scientific names are usually italicized and are often followed by the name of the person (the author) who described the species (e.g., the honey bee: *Apis mellifera* Linnaeus). When italics are not possible, scientific names are underlined, which conveys the same meaning as italics. The author's name is never italicized or underlined. Each genus name may be used only once in the animal kingdom; the same species may be used in different genera, but not more than once in a genus. The application of scientific names is regulated by the International Commission on Zoological Nomenclature, which publishes the *International Code of Zoological Nomenclature*. A similar code for plant names differs in some respects from the Zoological Code.

Relatively few people are qualified to identify organisms, including insects, to species, and therefore scientific names cannot always be applied with confidence. Humans have applied common names to insects in all languages, but sometimes these names do not translate. Also, one common name may be applied to several species. Further, one species may have several common names, even in different regions of one country. This imprecision in the application of names creates problems. In an attempt to solve part of this problem, within the United States certain insects and mites of recognized pest status have common names selected by a committee of the Entomological Society of America. The standardized list is published in *Common Names of Insects and Related Organisms*. This document was last revised in 1997 (Bosik 1997). Authors' names are not used with common names.

Common insect names ending in words applied to most insects of an order (such as *horse fly* and *house fly* of the order Diptera) are written as two words. Names, such as *butterfly* and *caddisfly*, for insects belonging to other orders (that is, not true flies) are written as one word.

The names of groups above the rank of genus are single words, capitalized but not printed in a different typeface from that of text in which they appear. Only the names of families and subfamilies are regulated by the International Code, which specifies that family names must end in "-idae" and subfamily names in "-inae." These endings are added to the stem of the genus name that is considered typical of the higher taxon, and the generic name is called the *basonym* of the group name. All group names are plural.

Names above the rank of family are not regulated by the International Code and are therefore subject to considerable difference in usage. They are sometimes formed, as are those of families and subfamilies, on a basonym, but many are formed on a word descriptive of a diagnostic character (for example, the name of the order Diptera means "two-winged" in classical Greek). For these reasons, alternative names of some classes and orders are cited here.

The name of the person who first proposed a scientific name is often appended to the name as a reference; sometimes the year of the proposal is also added, after a comma. These references, when added to a binomen, are placed in parentheses if the species name was originally combined with a different genus name. This name and year reference is not a necessary part of the binomen, but it is of value to taxonomists because it sometimes helps verify that different researchers are discussing the same taxon.

4.1 Key to Classes of Arthropoda

This key includes all living classes of Arthropoda except Pentastomida (or Linguatulida) and Tardigrada. Pentastomids and tardigrades are numerically small groups that are characterized by small to minute body size and lack antennae. These groups are rare and are not likely to be confused with other Arthropoda. In fact, some authorities exclude them from the Arthropoda.

1a. Antennae absent (in immature stages and some adult insects, the antennae are considerably reduced or absent; such forms will not key well here)...2
1b. Antennae present 4

2a. Body usually with seven pairs of appendages, including five pairs of legs; abdomen rudimentary (sea spiders; not further considered here) .... **PYCNOGONIDA**

2b. Body with six pairs of appendages (rarely fewer), including four pairs of legs (rarely five); abdomen usually well developed but sometimes fused with cephalothorax..3

3a. Abdomen with booklike gills on ventral surface; large animals up to 50 cm long, with hard, expanded shell and long, spinelike tail (horseshoe crabs; not further considered here) Class MEROSTOMATA, order or subclass... **XIPHOSURA**

Part 2: Classification of Insects and Mites

3b. Abdomen without booklike gills; smaller forms rarely over 7 cm long, body not as above (spiders, scorpions, mites, etc.; see Section 4.3.1).. **ARACHNIDA**

4a. With two pairs of antennae (one pair may be rudimentary in sowbugs); head and thorax fused to form cephalothorax; breathing by gills (crabs, lobsters, shrimps, sowbugs, etc.; see Section 4.5)...**CRUSTACEA**

Arthropod Collection and Indentification Techniques

4b. With one pair of antennae; head and thorax separate; breathing by tracheae 5

5a. Body with head, thorax, and abdomen; thorax with three pairs of legs at some stage in life cycle; abdomen sometimes with appendages that resemble thoracic legs; wings present in most species (all insects; see Section 4.6) ....**HEXAPODA (INSECTA)**

5b. Body more or less wormlike; most body segments behind head with legs; body with nine or more pairs of legs; wings absent (myriapodan classes) 6

6a. Most body segments each with two pairs of legs; slow-moving animals (millipedes; see Section 4.4)............ ....**DIPLOPODA**

6b. Most segments of body with at most one pair of legs; speed variable 7

7a. Body rather large, more or less flattened, with 15 or more pairs of legs; often reddish brown, rapidly moving animals (centipedes; see Section 4.4)..**CHILOPODA**

Part 2: Classification of Insects and Mites

7b. Body small to minute (not over 8 mm long), usually cylindrical, with 9–12 pairs of legs, whitish or pale-colored 8

8a. Antennae branched; nine pairs of legs; body minute (1–1.5 mm long); found in leaf litter etc. (pauropods; not further considered here) .... **PAUROPODA**

8b. Antennae not branched; 10–12 pairs of legs; body to 8 mm long, cylindrical, centipede-like; found in moist habitats (symphylans; see Section 4.4) .... **SYMPHYLA**

Arthropod Collection and Indentification Techniques

4.2 Class Arachnida

The Arachnida are the second largest class of arthropods in terms of number of species and species important to agriculture. Most arachnids, including all spiders, are predaceous. Some arachnids are parasitic on animals, including humans. Many mites (subclass Acari) are important pests of plants. All arachnids should be collected and preserved in 70–80% alcohol. Mites should be treated specially (see Section 3.2.8).

2.1 Key to Orders of Arachnida

1a. Abdomen not segmented, or if segmented (rarely) then with distinct sclerites (as in Asiatic family Liphistiidae); spinning organs on ventral side of abdomen 2

Ventral view of spider.

1b. Abdomen distinctly segmented; without silk-spinning organs on abdomen 3

2a. Abdomen joined to cephalothorax by narrow, short stalk; abdomen usually soft and weakly sclerotized, abdomen with spinning organs (spiders; not further considered here)......................... **ARANEIDA**

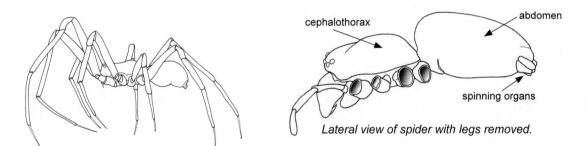

Lateral view of spider with legs removed.

Part 2: Classification of Insects and Mites

2b. Abdomen broadly joined to cephalothorax; abdomen comparatively tough or comparatively strongly sclerotized; abdomen without spinning organs (some mites spin silk from palpi or mouthparts (mites and ticks; see Section 4.2)...**Subclass ACARI**

Ventral view of mite.

Dorsal view of tick.

3a. Abdomen with posterior segments forming a long, tail-like projection 4
3b. Abdomen without a tail-like projection..6

4a. Tail-like projection of abdomen six-segmented, ending in bulbous, clawlike sting; abdomen broadly joined to nonsegmented cephalothorax; venter of second abdominal segment with a pair of comblike organs, the pectines (scorpions; widespread in warm, dry areas; not further considered here)...............**SCORPIONIDA**

Ventral view of scorpion showing pectines.

pectine

4b. Tail-like projection of abdomen very slender, many-segmented but not ending in sting; abdomen narrowed to base, without ventral comblike organs 5

5a. Anteriormost leglike appendages (pedipalpi) slender; body minute, not more than 2 mm long; found in warm areas (not further considered here) .... **PALPIGRADI (MICROTHELYPHONIDA)**

Lateral view of Palpigradi.

5b. Anteriormost leglike appendages (pedipalpi) very stout, contrasting with very long first pair of legs; body 2–65 mm long (whip scorpions, vinegaroons; not further considered here) .... **UROPYGI** (THELYPHONIDA; PEDIPALPIDA in part)

Dorsal view of Uropygi.

6a. Abdomen constricted at base; front legs very long with long tarsi; pedipalpi clawed at apex; tropical animals 4–45 mm long (tail absent, whip scorpions; not further considered here) .... **AMBLYPYGI** (SCHIZOMIDA; PEDIPALPIDA in part)

Dorsal view of Amblypygi.

6b. Abdomen broadly joined to cephalothorax; front tarsi not lengthened; other characters variable 7

7a. Pedipalpi with large, pincerlike claws (pseudoscorpions; not further considered here) .... **PSEUDOSCORPIONIDA** (CHELONETHIDA)

Dorsal view of Pseudoscorpionida.

pedipalpus

7b. Pedipalpi without pincerlike claws 8

8a. Head distinct from three-segmented thorax; chelicerae large and powerful, with pincers moving up and down; pale-colored, nocturnal, large animals (up to 7 cm long); occur in Florida and the southwest (wind scorpions; not further considered here) .... **SOLPUGIDA**

chelicera

Dorsolateral view of Solpugida.

8b. Cephalothorax present, not divided into head and three segments; chelicerae usually smaller, pincers not moving up and down; other characters variable 9

9a. Abdomen apparently four-segmented with lateral and dorsal sclerites, and small, several-segmented endpiece; eyes absent; heavy-bodied animals 5–10 mm long, with moderately long legs; tropical to Texas (ricinuleids; not further considered here) .... **RICINULEI**

Dorsal view of Ricinulei.

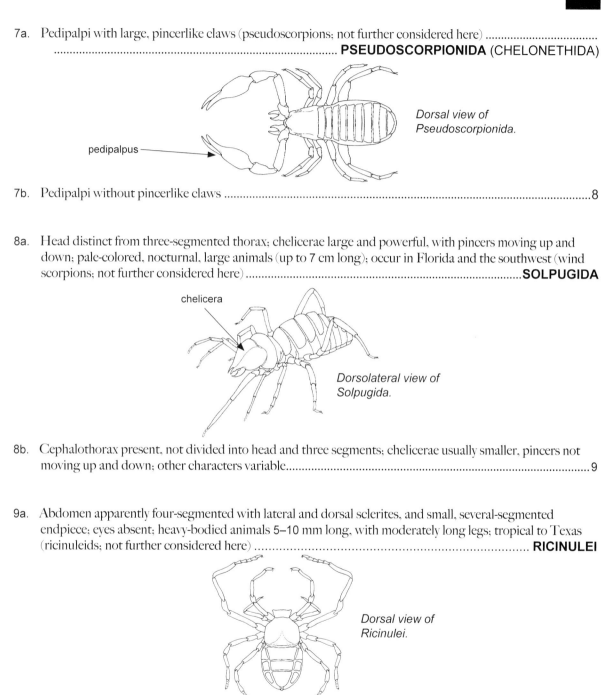

9b. Abdomen usually appearing seven-segmented from above, without separate lateral sclerites; body 5–10 mm long, with legs very long; distribution widespread (daddy longlegs, harvestmen; not further considered here) ....**OPILIONES (PHALANGIDA)**

Lateral view of Opiliones.

4.3 Subclass Acari

The acarines are so varied in form that their anatomical terminology has developed along considerably different lines than that of the insects. Most acarines are very small and require special techniques for their preservation and examination. The small size, morphological plasticity, and complex terminology make acarines considerably more difficult to identify than most insects. Only recently has their higher classification attained a measure of stability.

Past acarologists considered the Acarina as an order equivalent to the other orders of the class Arachnida (such as Araneida and Scorpionida). Current students of the mites regard the Acarina as the subclass Acari without reference to the status of other orders. The classification of the subclass Acari is variously constructed. Wolley (1988) divides the subclass Acari into the orders Astimata, Gamasida, Opilioacarda, Ixodida, Holothyrida, Oribatida (Sarcoptiformes, Oribatei, Cruyptostigmata, and Oribatoidea), and Actinedida (Trombidiformes and Prostigmata). Evans (1992) scheme lists the orders Notostigmata, Holothyrida, Ixodida, Mesostigmata, Prostigmata, Astigmata, and Oribatida. The group names used here are left without rank designation until problems with higher classification are resolved. Specific procedures for mounting mites may also be found in Appendix II.

A few definitions for the more unfamiliar words may be helpful in the following key. A *sejugal furrow* is a line of demarcation that separates the podosoma and opisthosoma; a *hypostome* is the anteroventral region of the gnathosoma (foremost part of the mouthparts); *Haller's organ* is a sensory organ found on tarsus I of ticks.

Part 2: Classification of Insects and Mites

4.3.1 Key to Some Primary Groups of the Subclass Acari

1a. Without visible stigmata (breathing pores) posterior of coxae II; coxae not free, often fused with ventral body wall, forming coxosternal regions delimited by epimera, but sternum lacking; sejugal furrow or interval present, causing legs III to be farther from legs II than the latter are from legs I; number of legs sometimes reduced .... **Order ACARIFORMES, including PROSTIGMATA, ASTIGMATA, and CRYPTOSTIGMATA**

1b. With one to four pairs of dorsolateral or ventrolateral stigmata posterior of coxa II; coxae free, distinct; sternum nearly always present (lacking in Ixodida); sejugal furrow or interval lacking; distance between legs II and III not greater than between I and II and III and IV, all of which are present .... **Order PARASITIFORMES** 2

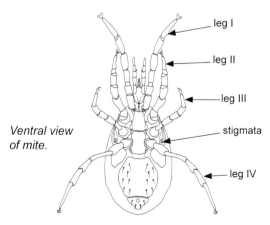

2a. Pedipalpal tarsus without claws; hypostome modified into piercing organ with backward-directed teeth; stigmata present behind coxa IV or lateral above coxal intervals II to III, each surrounded by stigmal plate; sternum absent; Haller's organ present on upper side of tarsus I (large, bloodsucking acarines called ticks; length usually well over 2 mm) .... **Suborder IXODIDA**

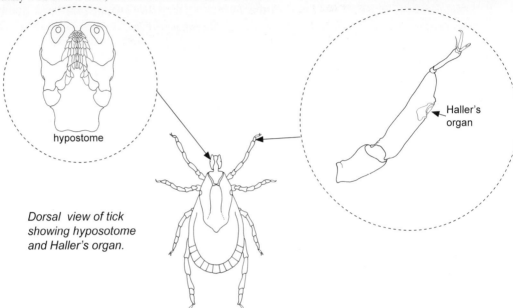

Dorsal view of tick showing hyposotome and Haller's organ.

2b. Pedipalpal tarsus with terminal, subterminal, or basal claw (simple or tined); hypostome serving only as floor of gnathosoma, without teeth; Haller's organ lacking (usually smaller mites) .... **Suborders MESOSTIGMATA, HOLOTHYRINA, and OPILIOACARIDA**

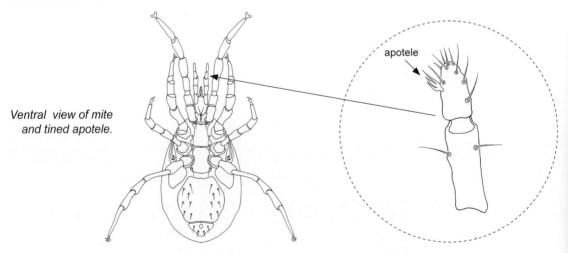

Ventral view of mite and tined apotele.

4.4 Classes Diplopoda, Chilopoda, Pauropoda, and Symphyla

These four classes were once combined as one class, the Myriapoda, and are still known as the myriapodan classes. All millipedes (Diplopoda) are phytophagous, but most species do not seriously injure plants. The centipedes (Chilopoda) are predaceous and possess a pair of strong, poison-supplied fangs called *toxicognaths* used to hold and kill their prey. Toxicognaths are modified legs, and the venom they inject can be painful. Centipedes are fast moving, in contrast to the slow-moving millipedes. The minute Pauropoda are of little direct agricultural importance. One member of the Symphyla, *Scutigerella immaculata* (Newport), is sometimes a pest in greenhouses; the species is whitish, about 8 mm long, and may become very abundant.

All members of the myriapodan classes may be killed and preserved in 70–80% alcohol. Millipedes have the discouraging habit of curling up when dying. The only simple solution to this problem is to remove them from the killing fluid before they stiffen, straighten them out, and then return them to the alcohol in a narrow, straight-sided vial until they become rigid. Millipedes killed in vapors of ethyl acetate remain relaxed and easily straightened before placing the specimens in alcohol.

4.5 Class Crustacea

The Crustacea are a large class divided into eight subclasses and about 30 orders. Most species of Crustacea are aquatic and are found especially in salt water. Many crustaceans, including crayfish, lobsters, and prawns, are important as human food. Sowbugs (including pillbugs) belong to the subclass Malacostraca and the order Isopoda. Sowbugs are terrestrial isopods that are the only Crustacea of agricultural importance. Sowbugs are found under stones, logs, and debris on the ground. They feed on vegetable matter and may become pests of tender plants. Sowbugs are distinguished by a depressed body, seven pairs of walking legs, and a shield-shaped head; the other body segments extend to the side. The first pair of antennae (antennules) are vestigial; the conspicuous antennae represent the second pair. Eggs are carried by the female in a brood sac on the underside of the body. Breathing involves paired gills on the lower hind part of the body. Because the gills must remain moist, sowbugs cannot withstand drying. Pillbugs are capable of rolling themselves into a tight ball for protective purposes. Crustaceans may be best killed and preserved in 70–80% ETOH.

Sowbugs

4.6 Class Hexapoda (Insecta)

All Arthropoda (including insects) are characterized by hardened or sclerotized parts of the integument. After the arthropod emerges from the egg stage, the animal must molt or shed the hardened cuticle to grow and increase in size. These hardened parts (called *sclerites*) restrict growth. The process of molting must take place several times during the insect's period of growth (the nymphal or larval stage). The number of molts varies among species of insects. In some instances the number of molts is species-specific. In other instances the number of molts may be influenced by the sex of the individual, nutrition, or environmental conditions.

The term *instar* refers to the immature insect between molts. The term *stadium* refers to the amount of time between molts. During the molt, each instar of the growth stage increases in size over the preceding instar. During the molt, the animal also may change its form. The amount of change in form can range from insignificant to profound. The transformation in shape or form is called *metamorphosis*. The primitive hexapods include the Entognatha and the division Exopterygota of the Pterygota (see Chapter 5 Synopsis of Insect Orders). These groups undergo gradual metamorphosis in which successive instars differ relatively little in form. Some aquatic exopterygote immatures are called *naiads* and may be entirely different from the adult because the two stages have different life habits. The immature stages of terrestrial insects with gradual (simple or incomplete) metamorphosis are called *nymphs*. The term *larva* is applied to the feeding stages of insects with complete metamorphosis, although some authors (McCafferty 1981; Merritt et al. 1984) use *larva* to refer to the immature stages of aquatic insects. Stehr (1987) reviews the use of *larva*, *nymph*, and *naiad*.

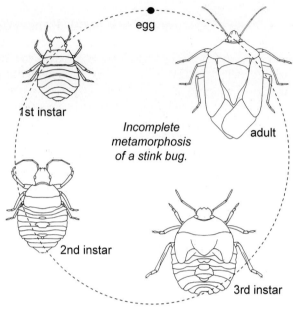

Incomplete metamorphosis of a stink bug.

The wings of insects with "incomplete" metamorphosis develop externally in the later instars but do not become functional until the adult stage. An exception to this rule involves the Ephemeroptera, which experience a molt after the wings become functional. The technical term for adult insects is *imago*; the first imago (adult) of the Ephemeroptera is called a *subimago*.

The wings of insects with complete metamorphosis typically develop internally. Insects with complete metamorphosis change radically in form from one stage to the next. A "resting stage" called a *pupa* (pl., pupae) is interpolated between the larva and adult stages. The pupal exoskeleton protects the insect during the virtual reorganization of its entire body. The stages in the development of insects with complete metamorphosis include the egg, several larval instars, sometimes a prepupal stage, the pupa, occasionally a subimago, and the imago or adult.

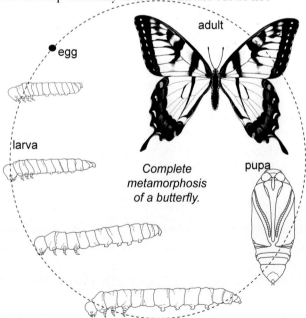

Complete metamorphosis of a butterfly.

Arthropod Collection and Indentification Techniques

NOTES:

Part 2: Classification of Insects and Mites

Some insects are wingless (either primarily or secondarily) in the adult stage and thus are difficult to distinguish from immature stages. Both the immature and the adult insects are treated in the key to the orders. Considerable knowledge of the various groups of insects is needed to determine whether a specimen is an immature or an adult. If functional wings are present, the specimen is certainly an adult (or a subimaginal mayfly). Compound eyes are not found in immature endopterygote insects.

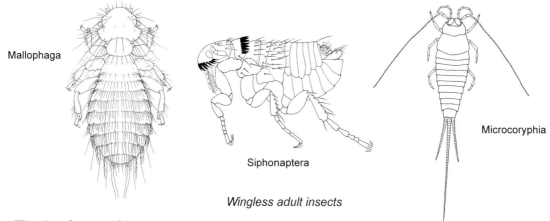

Wingless adult insects

The classification of the insects into orders (ordinal classification) adopted here is a conservative one used in some textbooks. Recent studies on various aspects of insect morphology and physiology have led to the proposal of a considerably different classification, but only time and further study will determine its acceptance or rejection.

A fundamental dichotomy, based on the presence or absence of wings, divides six-legged arthropods (hexapods) into the subclasses Apterygota and Pterygota. The Apterygota of earlier authors included four orders: Protura, Collembola, Diplura, and Thysanura (Richards & Davies 1977). In recent classifications the term Apterygota may be restricted to the Microcoryphia (= Archaeognatha) and Thysanura (Watson & Smith 1991). Other classifications have abandoned the traditional dichotomy of Apterygota-Pterygota and adopted a neutral viewpoint. In the opinion of some phylogeneticists, a more fundamental feature of evolutionary importance involves the ability to extend and retract the mouthparts. Hexapods that maintain their mandibles concealed when not in use are called Entognatha; hexapods whose mandibles remain exposed and not extensible are called Ectognatha. Entognathous Hexapoda include the orders Protura, Collembola, and Diplura; ectognathous Hexapoda include apterygotes of the orders Microcoryphia and Thysanura (Boudreaux 1979, Borror et al. 1989) and the pterygotes.

The following synopsis of the classification includes some synonyms and most of the common English vernacular names for members of the various orders, along with a key to these groups.

Part 2: Classification of Insects and Mites

Chapter 5: Synopsis of Insect Orders

Disagreement exists over the higher classification of insects. Some suborders may be considered orders; some orders may be considered suborders. There is value in understanding the basis for grouping the various orders of insects because each grouping relates to the development or morphology of the insects in question. Here we use the term *hexapod* to refer to wingless arthropods with six legs and three body tagmata. The hexapods can be further classified into two groups, entognatha and ectognatha, depending upon whether the mouthparts are concealed within the head or exposed. In the following list, alternative ordinal names are given in parentheses; common names are given in brackets. The primitive insects with concealed mouthparts (entognatha) include the following orders.

5.1 Subclass Entognatha: Primitive Wingless Hexapods

1. Protura (Myrientomata) [proturans]
2. Diplura (Aptera) [diplurans]
3. Collembola [springtails]

All entognathans should be collected and preserved in a similar manner. They may be collected with a Berlese funnel (Christiansen 1990; Copeland & Imadate 1990; Ferguson 1990), an aspirator, or a small brush dipped in alcohol. To collect the species of Collembola that are often found on the surface of water, use a small dipper (Davidson & Swan 1933). Specimens should be killed and preserved in 80% alcohol. All entognathans are delicate. To minimize risk of breakage, completely fill their vials with fluid. Microscope slide preparations are necessary for critical work with most Entognatha.

5.2 Subclass Ectognatha: Primitive Wingless Hexapods

4. Microcoryphia (Archaeognatha) [bristletails]
5. Thysanura (Zygentoma) [silverfish, firebrats]

The subclass Insecta is restricted to the so-called Pterygota, which display six legs, three body tagmata, and one pair of antennae and wings. The main difference between the hexapods and the insects is that the hexapods are primitively wingless (Apterygota = without wings), while the Insecta possess wings at least sometime in their life (Pterygota = having wings).

5.3 Subclass Pterygota (Insecta): Winged and Secondarily Wingless Insects

Insects that possess wings belong to the Pterygota. However, insects such as lice and fleas do not possess wings and still are classified among the Pterygota. This seeming contradiction is simply explained. Wingless pterygote insects have been shown to be closely related to winged insects, and the ancestors of these wingless pterygotes probably had wings. The wingless condition is common among parasitic insects, insects that live in caves or other subterranean habitats, and insects that live on islands. Some social insects, including termites and ants, display wings only in sexual forms. The workers and soldiers are wingless. Sexuals shed their wings after a nuptial flight.

Arthropod Collection and Indentification Techniques

The subclass Insecta contains two divisions, Exopterygota and Endopterygota, depending upon whether the wings are developed internally or externally. Furthermore, the Exopterygota includes insects with wings that develop externally AND have simple metamorphosis. [Note: In some classification systems Pterygotan orders are further subdivided into Paleoptera and Neoptera, based on the ability to flex wings over the abdomen. Members of the paleopteran orders (Ephemeroptera, Odonata) lack a wing-flexion mechanism and are unable to move wings posteriorly over the abdomen.] Orders of Exopterygotes include the following.

5.3.1 Exopterygota

6. Ephemeroptera (Ephemerida, Plecoptera) [mayflies]
7. Odonata [dragonflies, damselflies]
8. Orthoptera [grasshoppers, locusts, katydids, crickets]

Metamorphosis (Greek, meta = change of; morphosis = form) is another feature of pterygote insects that is important to understand, but difficult to define in a way satisfactory to everyone. Entomologists typically use the term in a restrictive sense to mean the morphological and physiological changes from the immature to the adult insect. Anatomical change is achieved through molting. Metamorphosis may be gradual, as during the transition of structure within a stage, such as the nymphs of a bug. Metamorphosis may be radical, as seen between stages, such as the transition from a caterpillar to a butterfly. Molting serves at least two purposes: growth in size and change in shape.

When viewing all insects and their close relatives, we can see trends in the expression of metamorphosis. For instance, *ametabolous* (Greek, a = without; metabole = change) development involves organisms that lack metamorphosis. Specifically, ametaboly is applied to the Apterygota in which the adult closely resembles the immature stages. Ametaboly is viewed as the primitive condition among hexapods. *Paurometabolous* (Greek, pauros = little; metabole = change) development refers to metamorphosis in which the changes of form between immature and adult are gradual and inconspicuous. Paurometabolous insects are terrestrial. Examples of paurometabolous insects include Dermaptera, Orthoptera, Embiidina, Isoptera, Zoraptera, Corrodentia, Mallophaga, Anoplura, Hemiptera, and most Homoptera. The situation is not straightforward, as noted below.

Hemimetabolous (Greek, hemi = half; metabole = change) development refers to metamorphosis in which form changes between immature and adult are made in one radical move. The anatomical changes between stages are conspicuous. Hemimetabolous development involves aquatic immatures because most hemimetabolous insects are aquatic as immatures. Examples of hemimetabolous insects include Plecoptera, Ephemeroptera, and Odonata. *Holometabolous* (Greek, holos = entire; metabole = change) development involves the interpolation of a pupal stage between immature and adult. This is the most advanced type of metamorphosis found among the Insecta. Examples include Neuroptera, Coleoptera, Strepsiptera, Trichoptera, Lepidoptera, Mecoptera, Siphonaptera, Diptera, and Hymenoptera. Fitting all of the Insecta into one metamorphic category or another is difficult because exceptions occur.

Until recently, the Blattodea and Mantodea were considered suborders of an order called the Dictyoptera. The following orders were included with the Orthoptera for many years:

Part 2: Classification of Insects and Mites

9. Blattodea [cockroaches]
10. Mantodea [mantids]
11. Phasmatodea (Phasmatoptera, Phasmoidea, Phasmida) [leaf insects, walkingsticks]
12. Gryllobattodea (Notoptera) [rock crawlers]
13. Dermaptera (Euplexoptera) [earwigs]
14. Isoptera [termites]
15. Embiidina (Embioptera) [webspinners]
16. Plecoptera [stoneflies]
17. Psocoptera (Corrodentia) [psocids, booklice]
18. Zoraptera [zorapterans]

The following two orders are often called the Phthiraptera. This name is a tongue twister that is derived from Greek (phthir = lice; aptera = wingless) and used to include all lice. This name has an element of redundancy because all lice are wingless. In some classifications, the Phthiraptera are regarded as the order embracing all lice, including Mallophaga and Anoplura. In other classifications, the Mallophaga and Anoplura are regarded as distinct orders. When the Phthiraptera are recognized as an order, the Anoplura are included as a suborder, and the Mallophaga are subdivided into two suborders or divisions: the Amblycera and Ischnocera. Here we continue the classification used by Steyskal et al. (1986) in the interest of continuity and with the proviso that other classifications exist.

19. Mallophaga (Chewing lice) [bird lice]
20. Anoplura (Sucking lice) [body lice, head lice]
21. Thysanoptera [thrips]

The higher classification of the two orders Hemiptera and Homoptera has not been resolved to the satisfaction of all entomologists. Some workers recognize these as distinct orders. Under such a scheme, the Hemiptera include several suborders (Stys & Kerzhner 1975). An alternative classification (Waterhouse 1971) considers the Hemiptera an order that includes the suborders Heteroptera and Homoptera. In Steyskal et al. (1986), Heteroptera were recognized as an order. Here we adopt the view that the Hemiptera and Homoptera should be recognized as distinct orders. No matter what we call them or how we classify them, the two are similar in several respects. In both groups, the mouthparts are adapted for piercing/sucking, and maxillary and labial palpi are not evident. In Hemiptera the mandibles and maxillae have become stylets that are surrounded by a "beak" that arises from the front part of the head. Members of the Homoptera show sucking mouthparts that arise from the posterior-ventral part of the head. Both groups usually have two pairs of wings. Adults of the Hemiptera show the basal part of the forewings thickened and leathery and the apical part membranous. The hindwings are entirely membranous and are covered by the forewings except during flight. When at rest, the wings are held flat over the back, with the apical parts of the forewings overlapping.

22. Hemiptera ("True" bugs) [water bugs, plant bugs, seed bugs, bed bugs, stink bugs]
23. Homoptera [aphids, cicadas, hoppers, leafhoppers, treehoppers, scale insects, whiteflies]

5.3.2 Endopterygota

The next group of orders includes those insects whose wings develop internally. These insects are further classified as the holometabolous insects. Holometabolous insects undergo complete development characterized by egg, larval, pupal, and adult stages. All holometabolous insects display a pupal stage. The term *pupa* (Latin, puppet) originally was used for the chrysalis of a butterfly. Subsequently, entomologists have applied the term to all holometabolous insects. The pupal stage is the last major event in the anatomical maturation of insects through the process of metamorphosis. The pupa represents a phase during which larval anatomical features are destroyed and adult features are constructed. The degree of reconstruction differs among groups. During this process, the pupa is relatively immobile. In holometabolous insects, the pupal stage occurs in time between the larva and the adult.

The division Endopterygota includes insects with complete metamorphosis (= Holometabola) and wings that develop internally. These insects are the most abundant pterygotes in terms of number of species. Orders of endopterygotes include the following:

24. Coleoptera [wood-boring beetles, weevils, leaf beetles, ground beetles, bark beetles]
25. Strepsiptera (sometimes included in Coleoptera) [twisted-winged parasites]
26. Mecoptera [scorpionflies, hangingflies]
27. Neuroptera [snakeflies, antlions, lacewings, mantidflies, dobsonflies, hellgrammites]
28. Trichoptera [caddisflies]
29. Lepidoptera [butterflies, moths]
30. Diptera ("True" flies) [house flies, deer flies, mosquitoes, gnats, midges]
31. Siphonaptera [fleas]
32. Hymenoptera [sawflies, ichneumons, ants, wasps, bees]

Disagreement exists over the higher classification of insects. Some suborders may be considered orders; some orders may be considered suborders. Here we use the term *hexapod* to refer to arthropods with six legs and three body tagmata. The term *insect* is restricted to the so-called Pterygota, which display six legs, three body tagmata, and one pair of antennae and wings. In the following list, alternative ordinal names are given in parentheses; common names are given in brackets.

Part 2: Classification of Insects and Mites

5.4 Key to Orders of Hexapoda (Insecta)

The following key includes suborders sometimes considered as orders. Any comprehensive survey of the larger orders of insects will include exceptions and aberrant forms that do not fit well in a brief key. Treating all aberrant forms is impractical, particularly when including known immature forms. Some orders appear more than once in the key. With specimens that do not key satisfactorily here, the reader should consult other references or an experienced systematic entomologist. This is especially true of pupal forms of Trichoptera, Mecoptera, and some Neuroptera. Interested students should obtain a copy of Borror et al. (1989) for comprehensive keys to the North American insect fauna.

1a. Wings present and well developed 2

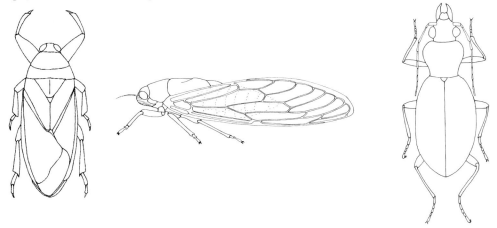

1b. Wings absent or not suitable for flight (wingless adults and immature stages) 34

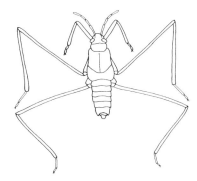

2a. Forewings at least partly horny, leathery, or strongly differing from completely membranous hindwings; hindwings sometimes absent 3

2b. Forewings completely membranous or membranous at base; hindwings variable, usually present and membranous, sometimes modified 12

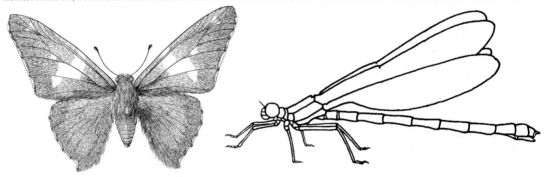

3a. Forewings (wing covers, elytra) uniformly horny or leathery and without apparent veins; hindwings, if present, folded lengthwise and crosswise, concealed beneath forewings when at rest; mouthparts including mandibles 4

3b. Forewings (hemelytra, tegmina) with variable texture and with veins; hindwings not folded crosswise; mouthparts variable 5

Part 2: Classification of Insects and Mites

4a. Apex of abdomen with heavy, forceps-like cerci; wings short and most of abdomen exposed; hindwings delicate, almost circular, radially folded .... **DERMAPTERA,** Section 6.13

4b. Apex of abdomen without heavy, forceps-like cerci; wings usually long or covering most of abdomen; if forewings short, then hindwings elongate or absent .... **COLEOPTERA**, Section 6.24

5a. Mouthparts adapted for sucking, forming a jointed or segmented beak 6

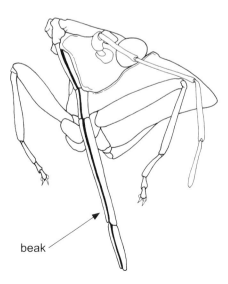

5b. Mouthparts adapted for chewing, forming mandibles that move laterally 7

6a. Forewings hardened or leathery basally and membranous apically, usually held flat on abdomen with apices overlapping; beak arising from anterior part of head and usually capable of movement forward of head..**HEMIPTERA,** Section 6.22

 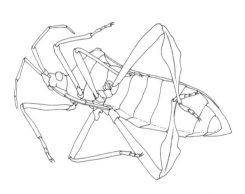

6b. Forewings completely or predominantly membranous, held more or less rooflike over abdomen; beak arising from posterior-ventral part of head and projecting downward and rearward between forelegs..**HOMOPTERA,** Section 6.23

Part 2: Classification of Insects and Mites

7a. Hind wings not folded at repose, similar to forewings; both pairs of wings with thickened, very short basal part separated from remainder of wing by suture; most of wing easily detached at suture; body soft and usually pale-colored; social insects living in colonies (see also couplet 32).... **ISOPTERA**, Section 6.14

7b. Hindwings folded fanwise at repose, broader than forewings; both pairs of wings without basal suture for wing detachment; body hardness and coloration variable.. 8

8a. Minute insects, usually less than 6 mm long; forewings small, clublike; antennae short, with few segments; parasites of other insects...**STREPSIPTERA** males, Section 6.25

8b. Usually large or moderately large insects; forewings usually flat and long; antennae usually lengthened and slender, many-segmented (orthopteran orders)... 9

9a. Hind femora enlarged, modified for jumping... **ORTHOPTERA**, Section 6.8

9b. Hind femora not enlarged, similar in size and shape to femora of other legs or not apparently adapted for jumping... 10

Arthropod Collection and Indentification Techniques

10a. Cerci short, not segmented; body usually elongate and slender, sticklike .... **PHASMIDA,** Section 6.11

10b. Cerci long or short but segmented; body usually not elongate or sticklike 11

11a. Forelegs raptorial with tibial and femoral spines modified for grasping prey; middle and hind legs adapted for walking .... **MANTODEA,** Section 6.10

11b. All legs similar in size and shape, without spines adapted for grasping prey; all legs adapted for walking... .... **BLATTODEA,** Section 6.9

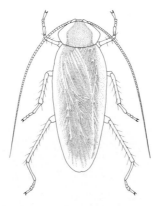

Part 2: Classification of Insects and Mites

12a. Two well-developed wings present: forewings obviously used in flight; hind wings modified, sometimes small and clublike 13
12b. Four wings present: forewings obviously used in flight, hind wings sometimes small but flat or straplike and not clublike 15

13a. Mouthparts forming a piercing-sucking or lapping proboscis, rarely rudimentary or absent; hind wings replaced by clublike halteres; abdomen without tail filaments .... **DIPTERA**, Section 6.30

13b. Mouthparts not functional; hind wings not formed into clublike halteres; abdomen with tail filaments.. 14

14a. Hind wings not halterelike; antennae inconspicuous, with small scape and pedicel, flagellum bristlelike; forewings with numerous crossveins (few mayflies) .... **EPHEMEROPTERA**, Section 6.6

14b. Hind wings reduced to halterlike structures; antennae conspicuous, flagellum not bristlelike; forewings with venation apparently reduced to one forked vein (male scale insects) .... **HOMOPTERA**, Section 6.23

15a. Wings long, narrow, almost veinless, with long marginal fringe; tarsi one- or two-segmented, with apex swollen; mouthparts conical, adapted for piercing and sucking plant tissues (minute insects) ....**THYSANOPTERA**, Section 6.21

15b. Wings relatively broad, veins usually conspicuous and at least one crossvein present, marginal fringe absent or not longer than width of wing; tarsi with more than two segments and apex not swollen; mouthparts variable 16

16a. Wings, legs, and body at least partially covered with elongate, flattened scales (setae) and often with hairlike setae; wings hyaline (transparent) under color pattern formed by scales; mouthparts tonguelike (rarely rudimentary), forming a helically coiled tube; small mandibles present only in a few families of small moths with wingspread not over 12 mm ....**LEPIDOPTERA**, Section 6.29

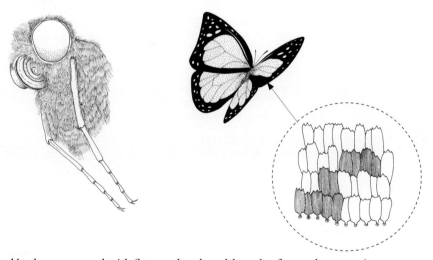

16b. Wings, legs, and body not covered with flattened scales, although a few scales sometimes present; color pattern of wing involving wing membrane or hairlike setae or both; mandibles typically present 17

17a. Hind wings usually larger than forewings, with broad anal area, plaited when wings folded; antennae conspicuous 18

17b. Hind wings usually not larger than forewings, without plaited anal area; antennae often inconspicuous, bristlelike 20

18a. Tarsi three-segmented; cerci well developed, usually long and many-segmented ....
.... **PLECOPTERA**, Section 6.16

18b. Tarsi five-segmented; cerci not well developed 19

19a. Wings with several subcostal crossveins, surface without hairlike setae or scales ....
.... **NEUROPTERA**, Suborder **MEGALOPTERA**, Section 6.27

19b. Wings without subcostal crossveins, surface with hairlike setae or scales . **TRICHOPTERA**, Section 6.28

20a. Antennae short, bristlelike; wings with numerous crossveins forming a network; mouthparts with mandibles near eyes 21
20b. Antennae large or wings with a few crossveins or mouthparts near beak..................................22

21a. Hind wings much smaller than forewings; abdomen with long tail filaments...
....**EPHEMEROPTERA**, Section 6.6

21b. Hind wings about as large as forewings; abdomen without long tail filaments .... **ODONATA**, Section 6.7

Part 2: Classification of Insects and Mites

22a. Head beneath eyes beaklike with mandibles at apex; hind wings not folded; wings usually with color pattern and numerous crossveins; male genitalia usually swollen, turned forward, and with strong pair of forceps.. **MECOPTERA**, Section 6.26

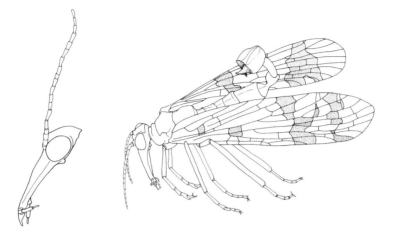

22b. Head beneath eyes not beaklike or formed into a conical tube; wing characters variable; male genitalia without conspicuous forceps 23

23a. Mouthparts sometimes absent, when present consisting of proboscis without chewing mandibles; cerci absent; wings with few crossveins.. 24
23b. Mouthparts with mandibles adapted for chewing; cerci sometimes present; wing venation variable 26

24a. Wings covered with scales that form color pattern; antennae with many segments; mouthparts (when present) consisting of helically coiled haustellum (tongue)......................... **LEPIDOPTERA**, Section 6.29

24b. Wings not covered with scales; antennae with few segments; mouthparts forming a segmented piercing beak 25

25a. Beak arising from anterior part of head .... **HEMIPTERA**, Section 6.22

25b. Beak arising from posterior part of head, extended downward between forelegs ....
....**HOMOPTERA**, Section 6.23

26a. Body and wings covered with whitish powder; wings bordered anteriorly by very narrow cell without row of crossveins; insects less than 5 mm long ....
....**NEUROPTERA, Suborder PLANIPENNIA (Coniopterygidae)**, Section 6.27

26b. Body and wings not covered with whitish powder, other characters differing 27

27a. Tarsi five-segmented 28
27b. Tarsi with four or fewer segments 31

Part 2: Classification of Insects and Mites

28a. Prothorax typically fused with mesothorax; forewings with fewer than 20 cells; hind wings smaller than forewings; abdomen usually constricted at base, forming a petiole; sting or appendicular ovipositor present .... **HYMENOPTERA,** Section 6.32

28b. Prothorax more or less free, sometimes long; forewings with more than 20 cells; forewings and hind wings approximately equal in size; abdomen not constricted to form petiole; sting or appendicular ovipositor not present 29

29a. Prothorax cylindrical, much longer than head; forelegs similar to other legs, not enlarged .... **NEUROPTERA, Suborder RAPHIDIODEA,** Section 6.27

29b. Prothorax not longer than head; if longer, then forelegs enlarged and adapted for grasping prey 30

30a. Costal cell with many crossveins .... **NEUROPTRA, Suborder PLANIPENNIA,** Section 6.27
30b. Costal cell without many crossveins .... **MECOPTERA,** Section 6.26

Arthropod Collection and Indentification Techniques

31a. Wings equal in size or hind wings rarely larger; tarsi three- or four- segmented 32
31b. Hind wings smaller than forewings; tarsi two- or three-segmented..33

32a. Forebasitarsi not swollen; wings dehiscent (see also couplet 7) .... **ISOPTERA**, Section 6.14

32b. Forbasitarsi swollen .... **EMBIIDINA**, Section 6.15

33a. Cerci absent; wings remain attached to body; antennae slender, with 13 or more segments (insects commonly collected) .... **PSOCOPTERA**, Section 6.17

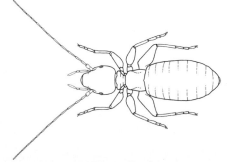

33b. Cerci present, although short, ending in bristle; wings shed; antennae with nine beadlike segments (insects seldom collected) .... **ZORAPTERA**, Section 6.18

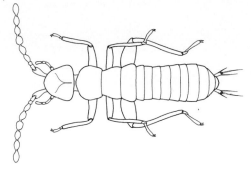

Part 2: Classification of Insects and Mites

34a. Body with more or less distinct head, thorax, and abdomen; legs jointed, enabling animal to move 35
34b. Without distinctly separate body parts, or without legs, or not able to move 78

35a. Parasites of warm-blooded animals 36
35b. Not parasites of warm-blooded animals 40

36a. Adult body strongly compressed laterally; mouth forming a short, sharp, downward-projecting beak; powerful jumping insects; adults found on vertebrate hosts; immatures found in nest of host .... **SIPHONAPTERA**, Section 6.31

36b. Adult body not compressed laterally; mouthparts variable; not powerful jumping insects; habitats variable 37

37a. Mouthparts with mandibles adapted for chewing, directed forward; insects generally oval in outline with more or less triangular head; parasites of birds and mammals .... **MALLOPHAGA**, Section 6.19

37b. Mouthparts forming a beak or otherwise modified; body shape variable; feeding habits diverse 38

Arthropod Collection and Indentification Techniques

38a. Antennae inserted in pits and not visible when viewed from above (also maggot-shaped larvae without antennae) ....**DIPTERA** (a few families), Section 6.30

38b. Antennae present, short but not in pits..39

39a. Beak not jointed; tarsi forming a hook for grasping hairs of host; parasites remaining on host ....
....**ANOPLURA**, Section 6.20

39b. Beak jointed; tarsi not hooked; parasites not remaining on host (bed bugs and related insects) ....
....**HEMIPTERA**, Section 6.22

40a. Aquatic insects, usually breathing by gills; larval and some pupal forms 41
40b. Terrestrial insects, breathing by spiracles or rarely without breathing organs 49

41a. Mouth forming a strong, pointed, downward-curved beak............Immature **HEMIPTERA**, Section 6.22

Part 2: Classification of Insects and Mites

41b. Mouth with mandibles 42

42a. Mandibles extending straight forward, united with maxillae to form piercing jaws ....
.... Some larval **NEUROPTERA**, Section 6.27

42b. Mandibles moving laterally, forming biting jaws 43

43a. Immature insects living within cases formed of sand, pebbles, leaves, twigs, etc.; usually with external tracheae serving as gills .... Some larval **TRICHOPTERA**, Section 6.28

43b. Immature insects not living within cases 44

Arthropod Collection and Indentification Techniques

44a. Abdomen with lateral organs serving as gills (a few larval Trichoptera and Coleoptera also key here)....45

44b. Abdomen without external gills (some larval Trichoptera also key here) 46

45a. Abdomen with two or three long tail filaments ....Immature **EPHEMEROPTERA**, Section 6.6

45b. Abdomen with short end processes (larvae of some Trichoptera key here)...
....Larval **NEUROPTERA**, Suborder **MEGALOPTERA**, Section 6.27

46a. Lower lip (labium) folded backward, extensible, and furnished with a pair of jawlike hooks ....
.... Immature **ODONATA**, Section 6.7

46b. Labium not so constructed..47

Part 2: Classification of Insects and Mites

47a. Abdomen with nonjointed false legs (pseudopods) arranged in pairs on several segments .... Few larval **LEPIDOPTERA**, Section 6.29

47b. Abdomen without pseudopods 48

48a. Thorax in three loosely united divisions; antennae and tail filaments long and slender .... Larval **PLECOPTERA**, Section 6.16

48b. Thoracic divisions without constrictions; antennae and tail filaments short (larvae of some aquatic Diptera and Trichoptera also key here) .... Larval **COLEOPTERA**, Section 6.24

49a. Mouthparts retracted into head and difficult to see; antennae sometimes absent; venter of abdomen with appendages; very delicate, small to minute animals 50

49b. Mouthparts external, conspicuous; antennae always present; venter of abdomen rarely with appendages; body typically larger than a few millimeters 52

Arthropod Collection and Indentification Techniques

50a. Head pear-shaped; antennae absent; abdomen without long cerci, pincers, jumping apparatus, or basal ventral "sucker" .... **PROTURA**, Section 6.1

50b. Head usually not pear-shaped, antennae present; abdomen with long cerci, pincers, or basal ventral "sucker" 51

51a. Abdomen with six or fewer segments, with forked "sucker" at base below and usually with conspicuous jumping apparatus near apex; abdomen lacking conspicuous long cerci or pincers; eyes usually present though often reduced in size .... **COLLEMBOLA**, Section 6.3

51b. Abdomen with more than eight evident segments but lacking a "sucker" at base; abdomen ending in long, many-segmented cerci or strong pincers; eyes and ocelli absent .... **DIPLURA**, Section 6.2

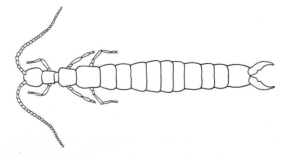

52a. Mouthparts with mandibles adapted for chewing 53
52b. Mouthparts in form of proboscis adapted for sucking 74

Part 2: Classification of Insects and Mites

53a. Body usually covered with scales; abdomen with three prominent tail filaments and at least two pairs of ventral appendages (styli).. **THYSANURA**, Section 6.5

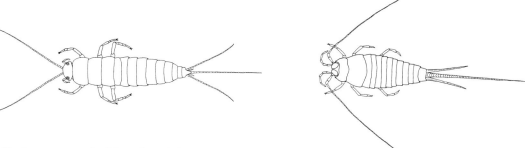

53b. Body not covered with scales; abdomen without three tail filaments or ventral styli 54

54a. Abdomen bearing ventral pairs of false, nonjointed legs (pseudopods) that differ from true legs on thorax; thorax and abdomen not distinctly separated; body caterpillar-like; larval forms.......................................55

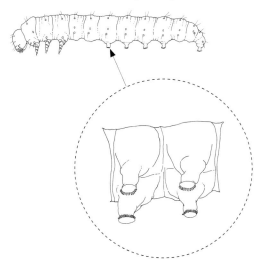

54b. Ventral surface of abdomen without legs or pseudopods; other characters different................................57

Arthropod Collection and Indentification Techniques

55a. Abdomen with five or fewer pairs of pseudopods, none on first, second, or seventh segments; pseudopods tipped with many tiny hooklets and rarely present on second and seventh segments .... Most larval **LEPIDOPTERA**, Section 6.29

55b. Abdomen with 6-10 pairs of pseudopods, not tipped with tiny hooks; one pair of pseudopods on second segment 56

56a. Head with single ocellus (stemmatum) on each side .... Some larval **HYMENOPTERA**, Section 6.32

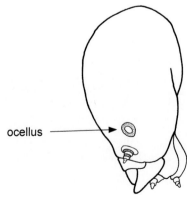

ocellus

Part 2: Classification of Insects and Mites

56b. Head with several ocelli (stemmata) on each side .... Larval **MECOPTERA**, Section 6.26

— occelli

57a. Antennae long and distinct; adult or immature forms 58
57b. Antennae short; larval forms 71

58a. Abdomen ending in strong pincerlike forceps; prothorax free .... **DERMAPTERA**, Section 6.13

58b. Abdomen not ending in forceps; prothorax free or fused to mesothorax 59

59a. Adult abdomen strongly constricted at base, forming a petiole; prothorax fused with mesothorax ....
.... **HYMENOPTERA**, Section 6.32

— petiole

59b. Adult abdomen not strongly constricted at base, broadly joined to thorax 60

Arthropod Collection and Indentification Techniques

60a. Head produced into beak with mandibles at apex .... **MECOPTERA**, Section 6.26

60b. Head not produced into beak 61

61a. Very small insects with soft body; tarsi two- or three-segmented 62
61b. Usually very much larger insects; tarsi usually with more than three segments, or body hard and cerci absent 63

62a. Cerci absent .... **PSOCOPTERA**, Section 6.17

62b. Cerci of single segment, prominent .... **ZORAPTERA**, Section 6.18

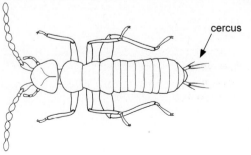

Part 2: Classification of Insects and Mites

63a. Hind femora enlarged and adapted for jumping; wing pads of immatures inverted, hind wing pads overlapping forewing pads..**ORTHOPTERA**, Section 6.8

63b. Hind femora not enlarged or adapted for jumping; wing pads, if present, in normal position................. 64

64a. Prothorax much longer than mesothorax; opposable surfaces of front femora and tibiae with long spines adapted for grasping prey (raptorial insects) .... **MANTODEA**, Section 6.10

64b. Prothorax not greatly lengthened; front legs rarely raptorial... 65

65a. Cerci absent; body often strongly sclerotized; antennae usually with 11 segments ....
....**COLEOPTERA**, Section 6.24

65b. Cerci present; body usually not strongly sclerotized; antennae usually with more than 15 segments 66

66a. Cerci with more than three segments 67
66b. Cerci with one to three segments... 69

Arthropod Collection and Indentification Techniques

67a. Body flattened or dorsoventally compressed, oval in outline; head deflected downward, with mouthparts directed caudad .... **BLATTODEA**, Section 6.9

67b. Body not flattened, elongate in outline; head nearly horizontal, with mouthparts at extreme anterior margin of head 68

68a. Cerci long, five- to eight-segmented; ovipositor long, swordlike; tarsi five-segmented; found in cold habitats remote from human habitation... **GRYLLOBLATTODEA**, Section 6.12

68b. Cerci short, one- to five-segmented; ovipositor absent; tarsi four-segmented; found in tropical and subtropical habitats or in association with humans..**ISOPTERA**, Section 6.14

Part 2: Classification of Insects and Mites

69a. Tarsi five-segmented (three-segmented in Timema [in Pacific coast states], most antennal segments several times longer than wide); body large and sticklike; not communal or social insects .... **PHASMATODEA**, Section 6.11

69b. Tarsi two- or three-segmented; antennal segments beadlike; body usually small, elongate but not sticklike; communal or social insects 70

70a. Front basitarsi swollen and containing silk-spinning gland for producing web in which insects live communally; cerci conspicuous; sexually dimorphic insects but without castes.. **EMBIIDINA**, Section 6.15

70b. Front basitarsi not swollen and not producing silk; cerci inconspicuous; polymorphic social insects with castes .... **ISOPTERA**, Section 6.14

71a. Body cylindrical, caterpillarlike 72
71b. Body more or less compressed, not caterpillar-like 73

72a. Head with six ocelli (stemmata) on each side; antennae inserted in membranous area at base of mandibles... Some larval **LEPIDOPTERA**, Section 6.29

ocelli

72b. Head with more than six ocelli on each side; third pair of legs distinctly larger than first pair ....
.... Larval Boreidae **MECOPTERA**, Section 6.26

73a. Mandibles united with maxillae to form sucking jaws ....
.... Larval **NEUROPTERA**, Suborder **PLANIPENNIA**, Section 6.27

73b. Mandibles nearly always separate from maxillae ....
.... Larval **COLEOPTERA**; **NEUROPTERA**, suborder **RAPHIDIODEA**;
suborder **RAPHIDIODEA**; **STREPSIPTERA**; **DIPTERA**

74a. Body densely covered with scales and hairlike setae; proboscis, if present, coiled under head ....
.... **LEPIDOPTERA**, Section 6.29

74b. Body bare, with scattered hairlike setae or waxy coating; mouthparts not coiled under head 75

Part 2: Classification of Insects and Mites

75a. Last tarsal segment bladderlike, without claws; mouth forming a triangular, nonsegmented beak; very small insects .... **THYSANOPTERA**, Section 6.21

75b. Last tarsal segment not bladderlike, with distinct claws; other characters different 76

76a. Prothorax small, hidden when viewed from dorsal aspect .... **DIPTERA**, Section 6.30

76b. Prothorax evident when viewed from dorsal aspect 77

77a. Beak arising from anterior part of head .... **HEMIPTERA**, Section 6.22

77b. Beak arising from lower posterior part of head .... **HOMOPTERA**, Section 6.23

78a. Legless grubs or maggots; movement by wriggling ....
Larval **DIPTERA** (If aquatic wrigglers, see larvae and pupae of mosquitoes); **HYMENOPTERA**; **LEPIDOPTERA**; **COLEOPTERA**; **SIPHONAPTERA**; **STREPSIPTERA** (in body of wasps or bees with flattened head exposed)

strepsiptera larva

78b. Legless or if legged then each leg with one terminal claw 79

79a. Small animals with little resemblance to most insects; filamentlike mouthparts inserted in plant tissue; usually covered with waxy scale, powder, or cottony tufts .... **HOMOPTERA**, Section 6.23

79b. Body unable to move or able to bend from side to side; mouthparts variable; body enclosed in tight integument, sometimes wholly covering body or sometimes with appendages free, but rarely movable; sometimes enclosed in cocoon (pupae) 80

Part 2: Classification of Insects and Mites

80a. Legs, wings, etc., more or less free from body; biting mouthparts visible 81
80b. Integument enclosing body holding appendages tightly against body; mouthparts evident as proboscis, without mandibles..83

81a. Prothorax small, fused with mesothorax; body sometimes enclosed in cocoon ....
....Pupal **HYMENOPTERA**, Section 6.32

81b. Prothorax larger and not fused with mesothorax; cocoon development variable..82

82a. Wing cases with few or no veins; pupation rarely occurs in silken cocoon..
.... Pupal **COLEOPTERA**, Section 6.24

82b. Wing cases with several branched veins; pupation usually occurs in silken cocoon ....
....Pupal **NEUROPTERA**, Section 6.27

Arthropod Collection and Indentification Techniques

83a. Proboscis usually long, rarely absent; four wing cases, one covering each wing; body often in cocoon........ .... Pupal **LEPIDOPTERA**, Section 6.29

83b. Proboscis usually short; two wing cases, one covering each forewing; body not in silken cocoon, but often tightly enclosed in hardened last larval integument.. Pupal **DIPTERA**, Section 6.30

Part 2: Classification of Insects and Mites

Chapter 6: Descriptions of Hexapod Orders

The following discussion of the orders is arranged according to the preceding synopsis. Early classifications of the hexapods put great emphasis on mouthparts and wings. This emphasis is reflected in the various order names.

6.1 Protura (Fig. 6.1)

Protura are a class of entognathous Hexapoda (sometimes considered a primitive order of minute Insecta). The name Protura is derived from Greek (protos = first; oura = tail). Protura are cosmopolitan in distribution and include about about 180 named species. Characteristically, proturans are small-bodied (less than 2 mm long) and elongate. The head is prognathous, with entognathous piercing mouthparts and well-developed maxillary and labial palpi. The antenna and compound eye are absent, and the thorax is weakly developed. The legs display five segments, and the pretarsus shows a median claw and empodium. The adult abdomen includes 12 segments; sterna 1-3 possess small, eversible styli, but cerci are absent, and the gonopore is positioned between segments 11 and 12. Development of Protura is anamorphic. That is, the animal emerges from the egg with nine abdominal segments and adds a segment with each of the first three molts. The lifestyle of Protura is cryptic. They inhabit leaf litter, soil, or moss or may be found beneath rocks or under bark. Protura are most frequently found in Berlese samples.

Figure 6.1
Proturan

6.2 Diplura (Fig. 6.2)

Diplura are a numerically small, cosmopolitan class of entognathous, epimorphic hexapods whose position in relation to the Insecta is questioned. The name is derived from Greek (diploos = two; oura = tail). Diplura consist of about 700 named species included in four families: Campodeidae, Japygidae, Procampodeidae, and Projapygidae. Diplura usually are small in size, but a few species measure up to 50 mm long. The body is narrow, eyes are absent, and the antennae are moniliform, with each segment containing intrinsic musculature. The legs each include five segments, and the pretarsus displays a pair of lateral claws and occasionally a median claw. The abdomen shows 10 segments; some abdominal segments contain styli and eversible vesicles. Cerci are usually present but variable in development; cerci are annulate in campodeids and forcipate in japygids. The gonostyli and gonopore are positioned between segments 8 and 9. Internal fertilization is not known in Diplura. The male manufactures a spermatophore, which is placed on the substrate. Specimens live beneath logs and rocks; a few records associate Diplura with ants and termites. Campodeids are regarded as phytophagous; japygids are reported as predaceous and probably use their cerci to capture prey.

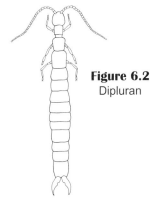

Figure 6.2
Dipluran

6.3 Collembola [Springtails, Fig. 6.3]

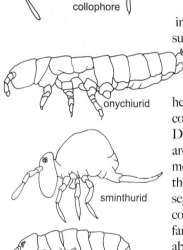

The scientific name of the order is derived from Greek (kolla = glue; embolon = peg) and refers to a tubular appendage (collophore) on the first abdominal sternum that secretes a sticky substance. The common name *springtail* refers to a forked structure called the *furcula* on the fourth abdominal sternum that is held in a place by a clasplike structure called a *retinaculum* on the third abdominal sternum. When the insect jumps, the forked structure is released with sufficient force against the surface of the ground to propel the animal into the air. A few Collembola lack this device.

Collembola are cosmopolitan, small-to-minute, soft-bodied, wingless hexapods. About 2,000 species have been described. The oldest fossil collembolan, *Rhyniella praecursor* (Hirst & Maulik), was extracted from Devonian beds of Rhynie chert in England. The compound eyes of Collembola are reduced to a few facets, and the antenna contains four to six segments. The mouthparts are reduced and concealed within the head (entognathous). The thoracic segmentation is not always clearly defined, and the legs contain four segments (a coxa, a trochanter, a femur, and a fused tibiotarsus). The abdomen consists of six segments. Metamorphosis is simple. Three commonly collected families are the Poduridae, Entomobryidae, and Sminthuridae. Some species are abundant on the surface of snow and have been given the name *snow fleas*. Most species feed in moist habitats such as decaying vegetation and rotting wood, under bark, in leaf debris, on the surface of ponds and streams, in soil, and on algae, fungi, and pollen. Some species feed on dead or moribund invertebrates in the soil. Cannibalism and predation have been documented in some groups. Parthenogenesis has been recorded in a few species. Collembola do not engage in internal fertilization. Males produce and deposit a spermatophore on the substrate. The female straddles the spermatophore, and sperm move into her genital aperture.

Figure 6.3 Collembola

6.4 Microcoryphia [bristletails, Archaeognatha, Fig. 6.4]

Figure 6.4 Microcoryphia

Microcoryphia represent a primitive order of Ectognatha whose members were assigned to the Thysanura. The order name is derived from Greek (micro = small; corypha = head) and refers to the small head. The common name *bristletail* refers to the long cerci and appendix dorsalis. Two modern families are known (Meinertellidae and Machilidae), and about 450 species have been described. The earliest fossil record of Microcoryphia comes from Triassic deposits in Russia (Triassomachilidae). The bristletail body is moderate-sized and laterally compressed, with the thorax arched. The compound eyes are well developed, and ocelli are present. The mandible is monocondylic. The coxae of some legs display styli, and the tarsi of all legs are three-segmented. All species lack wings, and their ancestors never had wings. The abdomen is two-segmented, with styli on some segments; the appendix dorsalis is longer than the cerci. Bristletails are found under bark and stones and in grass or leaf litter. They feed on algae, moss, and other plant material. Bristletails are typically nocturnal and capable of jumping by rapid flexion of the abdomen.

6.5 Thysanura [silverfish, firebrats, Fig. 6.5]

In older classifications, the name Thysanura was applied to include the Microcoryphia (Archaeognatha in part). The name Thysanura is derived from Greek (thysanos = fringe; oura = tail) and refers to an older common name, bristletails, which is more appropriately applied to the Microcoryphia. Thysanura are called *silverfish* because as the insects move quickly, the scales of their body shimmer like those of a silver fish. Thysanura represent an order of small to moderate-sized, fusiform, dorsoventrally compressed insects. The body is usually covered with scales, which are flattened setae. The mouthparts are adapted for chewing; the mandible displays two points of articulation (dicondylic) with the head, and the compound eye is small or absent; ocelli are sometimes present. The thoracic segments are similar in size and shape; coxal styli are absent. The abdomen is two-segmented, and some sterna have styli; the appendix dorsalis and cerci are present and similar in size. Metamorphosis is simple. The most common family is the Lepismatidae. Some species, such as *Lepisma saccharinum* Linnaeus and *Thermobia domestica* (Packard), are found in domestic situations feeding on bookbindings, curtains, wallpaper paste, paper, clothing, and similar articles. Most thysanurans occur outdoors under bark and stones, in grass and leaf litter, or in rotting wood or other debris.

silverfish

Figure 6.5
Thysanura

6.6 Ephemeroptera [mayflies, Fig. 6.6]

The order name is derived from Greek and refers to the short life (ephemeros) as a winged form (pteron). Mayflies are a small order of Paleoptera consisting of soft-bodied, elongate insects with at least one pair of membranous wings and two or three long, slender tails. Adults do not feed, and their mandibulate mouthparts are not strongly sclerotized. Mouthparts of the nymphal instars are adapted for chewing. The nymphs are aquatic, with gills along the sides of their abdomen. Metamorphosis is simple. Mayflies possess a unique developmental stage called a *subimago*, which is the initial winged form. This is not the adult stage; the subimago undergoes one molt before the

adult mayfly

Figure 6.6a
Ephemeroptera

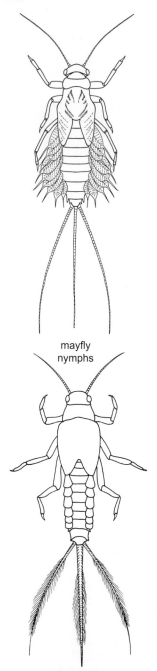

Figure 6.6b
Ephemeroptera

mayfly nymphs

insect is transformed into an adult. Three families are commonly collected: Ephemeridae, Heptageniidae, and Baetidae. Nymphs are found in a variety of aquatic habitats, from fast-flowing streams to the still waters of ponds, where some species burrow into the muck at the bottom. Adults are usually seen near water on vegetation and other objects, but at times they are attracted in large numbers to lights.

Mayfly nymphs can be collected with an aquatic net or can be handpicked from submerged rocks and vegetation. They should be killed and preserved in 80% alcohol or held and reared to the adult stage in aquariums or quart-sized jars filled halfway with water. Adult mayflies rarely live for more than a few days; nymphs often require an entire year to develop. Unless full-grown nymphs are collected, the collector should be prepared to keep the specimens in captivity for many months. Some species need well-oxygenated water, so a small electric pump to which air hoses can be attached may be necessary. Almost all mayfly nymphs are plant feeders, and an adequate supply of aquatic plants must be provided in the aquarium or rearing container. Some of the vegetation, a stick, or a rock should extend above the surface of the water to provide the subimago with a substrate to which it can cling after emergence.

Subimagos can be preserved in alcohol, as recommended for nymphs, but it is better to hold them in rearing cages until the final molt is completed. Both subimagos and adults can be captured with an aerial collecting net by sweeping or beating the foliage near the water. Mayflies are extremely fragile. Care must be taken when removing them from a net to protect their delicate appendages. Adults may be preserved in alcohol or pinned. However, pinned specimens usually shrivel badly, become brittle, and are difficult to handle.

6.7 Odonata [dragonflies, damselflies, Fig. 6.7]

The order name is derived from the Greek and means tooth, referring to the strongly toothed mandibles. The Odonata are ancient, with fossils in North America and Russia dating to the Permian. One fossil species living at that time had a wingspan of more than 70 cm. The Odonata are divided into several suborders, but only three survive today: Anisoptera (dragonflies), Zygoptera (damselflies), and Anisozygoptera. The Anisozygoptera consist of one small family found in the Himalayas and Japan. This group is considered a transitional element between the dragonflies and the damselflies. The adult body and immature characters resemble those of the Anisoptera, and wings are suggestive of the Zygoptera.

All Odonata are predaceous in the immature and adult stages. The dragonflies are generally large-bodied insects that usually keep their wings outstretched when at rest. The damselflies are generally smaller, more delicate insects that usually fold their wings rearward over the abdomen when at rest (Corbet et al. 1960). Dragonflies and damselflies are comparatively large and have two pairs of many-veined wings. The posterior wings are as large as or larger than the front pair. The antennae are short and bristlelike, the abdomen is long and slender, the mouthparts are adapted for chewing, and metamorphosis is simple. The aquatic nymphs respire with the aid of gills. Damselflies show their gills as leaflike structures at the apex of the abdomen. Dragonflies hide their gills in the form of ridges on the rectum. Oxygen is extracted from water drawn into the rectum. Nymphs display an elongate, scooplike, prehensile labium that can be extended forward rapidly to grasp prey.

The nymphs may be found in most freshwater habitats and occasionally in brackish water. Some nymphs can tolerate exposure to humid air, and members of at least one genus in Hawaii are terrestrial. Nymphs may be found clinging to aquatic vegetation or in the muck at the bottom of streams or ponds. Specimens are reared easily on a diet of young tadpoles, aquatic insects, or other small aquatic animals. A long stick placed in the rearing tank allows mature nymphs to crawl out of the water for the final molt to the adult stage. Metamorphosis usually occurs at night or early in the morning. Teneral adults avoid water. Nymphs should be preserved in alcohol; large specimens should be killed in boiling water and then transferred to alcohol.

adult damselfly

Dragonflies, especially the larger species, are fast fliers and are most easily caught when resting. Adults are typically diurnal, sometimes crepuscular, and occasionally nocturnal. Adults are strong fliers and use sight to locate prey. Prey are captured in flight. Individuals of some species have favorite resting places and fly regular routes. Collectors with patience to wait quietly may be rewarded by insects coming to them. Any strange and sudden movement, as with a collecting net, may cause the dragonfly to dart away. Try to swing the net from behind and a little below the insect. The specimen will be less apt to see the movement until it is too late. Captured specimens may be transferred directly to acetone for half a day, which preserves the colors. Specimens placed in a cyanide jar may lose some colors. Specimens may also be held alive for a day or two without food. The starvation eliminates the contents of the alimentary system before the insect is killed. Decomposing food in the body of a dry specimen may affect some colors. If held alive, adults should be confined to a small space and preferably kept in the dark so that they will not beat and damage their wings. Adults obtained by rearing should be held alive for a time to allow the colors to develop fully. Newly emerged, teneral adults should be held in cages large enough to permit the wings to expand fully and dry.

adult dragonflies

Figure 6.7
Odonata

Arthropod Collection and Indentification Techniques

cave cricket

camel cricket

mole cricket

field cricket

Jerusalem cricket

grasshopper

katydid

Figure 6.8
Orthoptera

For permanent collections, pin adults with the wings spread or place them in clear plastic envelopes with the wings folded above the dorsum. With pinned specimens, brace pins may be added on either side of the abdomen for support and to prevent "cartwheeling." This is especially important if the specimens are to be mailed.

6.8 Orthoptera [crickets, grasshoppers, katydids, Fig. 6.8]

The order name is derived from the Greek and refers to straight (orthos) wings (pteron). The Orthoptera are a cosmopolitan order of terrestrial insects with more than 20,000 described species. Members of the Orthoptera are anatomically diverse. The mandibles are well developed, and the mouthparts are adapted for chewing. The antennae are multisegmented and sometimes longer than the body. The pronotum is large. The forewings (called *tegmina*) are thickened and are characterized by numerous veins; the hind wings also contain many veins but are membranous, fanlike, and folded when in repose. Most Orthoptera display hind legs enlarged and adapted for jumping. The abdomen has 11 segments, and the cerci are well developed. Metamorphosis is simple, and the nymphs resemble the adults. Many Orthoptera communicate acoustically; in this respect they differ from the mantids and cockroaches. Older classifications of the Orthoptera included groups such as the cockroaches, mantids, and walkingsticks. Currently, these groups are separated, and the Orthoptera consist of about 15 families.

Some crickets are found in domestic situations, but most orthopterans inhabit vegetation such as grass and shrubs upon which they feed. Grasshoppers usually move by walking or jumping, but most adults can fly, often exposing brightly colored hindwings. To capture the adults in a net, watch where they land and bring the net down quickly over the spot. If the tip of the bag is held upright, the grasshopper usually can be coaxed up into the bag. Both adults and nymphs can be captured by sweeping vegetation. Flowerpot cages are suitable for rearing grasshoppers.

Most species of katydids are nocturnal. Many are attracted to light and usually can be caught easily by hand. Crickets are active at night and may be attracted to baits, such as molasses or oatmeal, that has been prepared as a paste and spread in a thin layer. The paste is made by mixing dry oatmeal with a little sugar, dry milk, and water. This same paste dried can be used to feed crickets in captivity, although the diet should be supplemented occasionally with bits of fruit and lettuce.

Adults and nymphs of Orthoptera can be killed and preserved in alcohol, but this is not recommended. Colors are lost in fluids such as alcohol during

Part 2: Classification of Insects and Mites

long-term storage. If specimens must be stored in alcohol, then a 5% solution of formalin should be added to the alcohol. Specimens should be preserved dry, such as layered in cellucotton. Drying should occur over a period of several hours. This can be accomplished with an oven, a heater, or exposure to the sun. Specimens exposed to the sun should be protected from ants and flesh flies. In tropical regions it may be necessary to embalm Orthoptera. Larger specimens sometimes decompose. For this reason, the internal organs should be removed. Dissect the internal structures (hindgut, fat body, internal reproductive system) from the abdomen by making a longitudinal incision along the ventral surface of the abdomen and removing the tissue with forceps. A wad of cotton, moistened with napthalene, may be placed in the abdominal cavity to retain the shape of the abdomen. For a permanent collection, pinned adults are preferable. For exhibits or demonstrations, winged forms may be mounted with one or both pairs of wings spread. For most collections the specimens should be mounted with legs, antennae, and wings folded close to the body to conserve space.

6.9 Blattodea (Blattaria) [cockroaches, Fig. 6.9]

The Blattodea (Latin, blatt = cockroach) are dorsoventrally compressed insects with the head concealed by the pronotum when viewed from the dorsal aspect. The mouthparts are adapted for chewing, and the antennae are filiform and typically display more than 30 segments. Blattodea are cursorial insects, with the hind legs similar in shape and size to the middle legs; many species are extremely rapid runners. Some species lack wings; when wings are present, the forewings are modified into moderately sclerotized tegmina that protect the membranous hind wings. The abdomen contains 10 apparent segments; the cerci usually display many segments. A few species of cockroaches are parthenogenetic. The female manufactures an ootheca or egg case. In species that deposit the ootheca on the substrate, the egg case is rather hard. Some species are ovoviviparous, with a membranous ootheca containing eggs retained within the female's brood pouch. A few species are viviparous, with the ootheca incompletely developed. The metamorphosis of cockroaches is simple, and nymphs and adults are found in the same habitats. Nymphs resemble the adult in shape but can differ considerably in coloration.

Cockroaches are omnivorous insects that feed upon plant and animal material. More than 50 species of cockroaches live in North America, of which only four are common indoors. These domestic cockroaches are nocturnal and usually avoid light. Some outdoor species may be attracted to light at night; some species are attracted to pitfall traps baited with molasses. Oatmeal trails may also be effective (Hubbell 1956). This attraction for sweets sometimes leads cockroaches to invade abandoned beehives to feed on the remaining honey. During the day, most cockroaches hide under the loose bark of trees, beneath

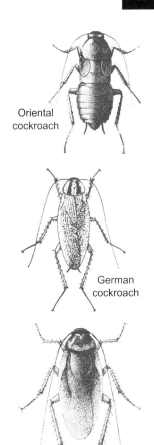

Oriental cockroach

German cockroach

American cockroach

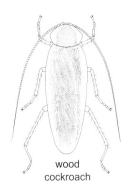

wood cockroach

Figure 6.9
Blattodea

rotting logs, or in similar fairly moist habitats. Some species of cockroaches can be found in leaf litter. They can be reared with little difficulty, and many species are available from biological supply houses. Specimens can be layered in cellucotton or pinned. Teneral or recently molted specimens should be allowed to harden before pinning. Large-bodied specimens should have pins placed on both sides of the abdomen to prevent the specimen from "cartwheeling" on the pin.

6.10 Mantodea [mantids, Fig. 6.10]

Mantids are sometimes called *praying mantids* or *soothsayers* (Greek, manti = soothsayer) because their forelegs are held in a supplicatory position resembling prayer. The Mantodea have been placed with the Orthoptera or with the Blattaria in other classifications. Nearly 2,000 species have been described. The group is cosmopolitan in distribution but predominantly tropical or subtropical. The mantid body size is typically moderate to large. The head rotates, and live specimens can give a turn to the head that seems quizzical. The pronotum is elongate, narrow, and maneuverable on the mesothorax. The forelegs are raptorial, with the forefemur and tibia extended forward to capture prey. The prey is held between the femur and tibia with opposable spines on the venter of the femur and tibia when the forelegs are retracted toward the head. The middle and hind legs are slender and are not modified for running or jumping. The abdomen is 11-segmented, and the cerci consist of many segments. Eggs are protected within an ootheca.

All mantids are predaceous as nymphs and adults. Mantids are usually found on vegetation, where they prey on other insects. During the autumn and winter, mantid oothecae can be found attached to shrubs or grass stems. If collected, the oothecae should be kept outdoors or in an unheated garage or porch until spring to prevent the eggs from hatching prematurely. However, if living insects (such as *Drosophila* flies or aphids) are available to feed the young nymphs, then eggs can be incubated at any time. Adult mantids usually die in the fall soon after the eggs are laid, but sometimes specimens can be kept alive in rearing cages for several weeks indoors by feeding. Mantids are voracious feeders, which suggests that they can be important biological control agents in the garden. Mantid oothecae are sold by garden suppliers, who advertise that the mantids will control many species of garden pests. Indeed, the nymphs and adults consume large numbers of insects, but only those readily visible on plant foliage. Soil-dwelling or stalk-boring insects are not attacked by mantids. Mantids appear formidable, but with a little experience they can be safely captured by hand. Care should be exercised when handling mantids because their raptorial front legs can cause a painful prick. Specimens may also be collected from vegetation with a sweep net, but some species are difficult to dislodge.

praying mantis

Figure 6.10
Mantodea

Adults and nymphs of mantids may be preserved in alcohol, but for a permanent collection, adults should be pinned. Soft-bodied specimens preserved on pins should have the body supported with card stock mounted on the pin beneath the specimen. Large-bodied specimens should have pins placed on both sides of the abdomen to prevent the specimen from "cartwheeling" on the pin.

6.11 **Phasmatodea** (Phasmida) [walkingsticks, leaf insects, Fig. 6.11]

Phasmids are called *walkingsticks* because they are slow-moving insects with elongate bodies that resemble twigs or sticks (Greek, phasm = phantom). Some tropical species are called *leaf insects* because their bodies are dorsoventrally flattened and expanded laterally so that the insect resembles a leaf. Phasmids are cosmopolitan in distribution, and more than 2,500 species have been described. Several species are more than 30 cm long. The phasmid's body is often spinose, and the head is typically prognathous. Ocelli are usually absent, and the mouthparts are of the chewing type (mandibulate). The prothorax of phasmids resembles that of mantids, in that it is movable on the mesothorax. However, wings are absent from most species of phasmids. Winged species are most common in tropical areas; when present, the forewings are modified into tegmina and the hind wings are broad with a uniform branching pattern. The legs are gressorial, with all legs similar in appearance and adapted for walking; the coxae are small and separated. The abdomen is 11-segmented; cerci are not segmented, but sometimes they are long and clasperlike in males. Metamorphosis is simple, and the nymphs resemble the adults. Eggs are not enveloped by an ootheca. Instead, they are scattered individually on the ground. Phasmids may remain in the egg stage for more than a year.

The biology and ecology of Phasmatodea have been reviewed (Bedford 1978; Carlberg 1986). All species of phasmids are phytophagous and may be located on trees or shrubs upon which they feed. Their slow movements and resemblance to twigs or leaves and cryptic coloration (usually green or brown) makes these insects difficult to detect. Sweeping or beating trees and shrubs may yield some individuals. In spring or early summer, newly hatched nymphs may be found, which in the most common species, *Diapheromera femorata* (Say), are only about 0.5 cm long. The nymphs mature in about 6 weeks, attaining a length of 8-10 cm. Their slow-moving, plant-feeding habits enable them to adapt readily to the environment of a rearing cage. Usually phasmids are leaf-feeding insects. However, in captivity the newly molted nymphs have been observed eating their cast exuviae.

walking sticks

Figure 6.11
Phasmatodea

Adults and immatures may be collected by hand and placed into alcohol or a killing jar. Specimens should be preserved dry. Because phasmids are extremely long-bodied, additional brace pins may be necessary to support the abdomen and prevent the specimen from shifting on the pin.

6.12 Grylloblattodea (Grylloblattaria) [rock crawlers, Fig. 6.12]

These rare insects, found along the edges of glaciers, have features resembling both crickets (Latin, gryll = cricket) and cockroaches (Latin, blatt = cockroaches). They derive their common name from their habit of moving about slowly on rocks in cool habitats. The order consists of one family and about 20 cryptozoic species found in cold, wet habitats of North America, Japan, and Russia. Rock crawlers are soft-bodied, elongate, slender, mandibulate insects from 15 to 30 mm long. They are characterized by a prognathous head with compound eyes that are reduced or absent; the ocelli are absent and the antennae are long with many segments. The prothorax is large, wings are absent, and the legs are cursorial and adapted for running with large coxae. The cerci are long and segmented; the ovipositor is strongly exserted.

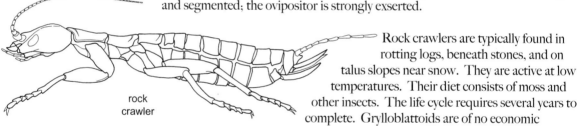

Figure 6.12 Grylloblattodea

Rock crawlers are typically found in rotting logs, beneath stones, and on talus slopes near snow. They are active at low temperatures. Their diet consists of moss and other insects. The life cycle requires several years to complete. Grylloblattoids are of no economic importance, but they are interesting because they are unusual and rarely found in collections. These insects are probably best preserved in 80% alcohol or glycerol.

6.13 Dermaptera [earwigs, Fig. 6.13]

The order name is derived from Greek (derma = skin; pteron = wing) and refers to the skinlike appearance of the forewings. The common name is derived from the erroneous superstition that earwigs crawl into the ears of sleeping people. The order Dermaptera is cosmopolitan and currently consists of about 1,200 named species. Dermaptera are most common in tropical and warm temperate regions. They were placed among the Orthoptera in nineteenth century classifications, but Dermaptera represent a distinct group that is probably related to the Grylloblattodea. Dermaptera are elongate insects that measure up to 50 mm long. The head is prognathous, mouthparts are of the chewing type, and the prothorax remains free from the mesothorax. The adults usually have two pairs of wings. The forewings are short, leathery, and elytriform or modified into tegmina. The hind wings are large, membranous, fanlike or circular, and folded under the forewings when the insect is at rest. The legs are cursorial, with three segments forming the tarsus. The abdomen is often telescopic, and the cerci are modified into forceps that can pinch if the earwig is handled. Metamorphosis is incomplete, and the nymphs resemble the adults.

Earwigs are typically nocturnal insects that are sometimes collected at lights. During the day, earwigs usually hide in cracks and crevices, under the bark of trees, or in rubbish on the ground. One species, *Anisolabis maritima* (Géné), is commonly collected under stones or driftwood along the Atlantic and Pacific coasts. The eggs are laid in a burrow in the ground, and the female cares for them until they hatch. The families of earwigs found in North America include the Forficulidae, Chelisochidae, Labiduridae, and Labiidae.

Earwigs feed as scavengers on decaying organic matter, but some species may also feed upon the living tissues of flowers, ripening fruits, and vegetables. Earwigs occasionally feed upon aphids and other small insects. Earwigs living in cooler habitats are predominantly herbivorous; those living in warm temperate and tropical regions are predominantly predaceous. Predatory species feed on a variety of insects but seem to prefer soft-bodied larvae. Despite their almost omnivorous feeding habits, or perhaps because of them, earwigs are not strongly attracted to bait traps. It is usually necessary to search by day in likely habitats and to collect the specimens by hand. Adults and nymphs may be killed and preserved in alcohol. However, for a permanent collection, the adults should be pinned or card-mounted. Pinned specimens may require a card placed under the body for support.

Figure 6.13
Dermaptera

6.14 Isoptera [termites, Fig. 6.14]

The order name is derived from Greek (isos = equal; pteron = wing) and refers to the size and shape of the wings. The common name *termite* is derived from Latin *termes*, which was the medieval name for wood worms. Termites are also called *white ants*, but they may be distinguished from true ants because termites have a broad waist and beaded antennae; winged forms have both pairs of wings the same size and shape. Termites are a cosmopolitan order including about 2,000 species. They are most abundant in tropical and subtropical regions. Termites are small, soft-bodied, polymorphic, cryptozoic insects. The antennae are moniliform or filiform; the mouthparts are mandibulate. Primary reproductives possess four membranous, similar-sized, net-veined wings that are shed; nymphs, soldiers, and workers lack wings. The abdomen shows 10 apparent segments. The cerci are one- to five-segmented, and sclerotized external genitalia are absent from most species. Metamorphosis is incomplete in termites, but each species displays several morphs.

The primary food of termites is cellulose, which is broken down by symbiotic bacteria or flagellate protozoans. Termites feed incidentally on

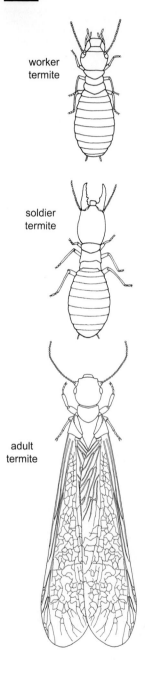

Figure 6.14
Isoptera

exuviae, dead nestmates, and excrement. The association with symbionts is vital. If these microorganisms are removed from the gut of the termite, the animal will eventually die of starvation. These microorganisms live in the hindgut of the termite. This relationship is precarious because the hindgut and the microorganisms are shed during the molting process. Fortunately, the newly molted individuals receive a new colony of microorganisms by feeding at the hindgut of other termites. This process is called *proctodaeal feeding* or *trophallaxis*.

Termites display a highly developed social system. Castes include primary reproductives (a king and a queen), secondary reproductives (neotenics), soldiers (male and female), and workers (male and female). Reproductives swarm periodically in great numbers. Bonding occurs in the swarm, with a royal pair (king and queen) then excavating a nuptial chamber. Copulation occurs within the chamber, and a new colony is established. After the dispersal flight from the parent colony, the wings of the adults are usually shed, leaving only short stubs that are visible under low magnification. The royal pair is long-lived, perhaps as long as 50 years.

Wingless forms are sterile adults of the worker and soldier castes. Workers provide food for the colony, construct new tunnels and chambers, and care for the egg-laying queen. In many groups of termites, the queen's abdomen is so swollen that she is immobile. This condition, called *physogastry*, is typical in some groups. Soldiers defend the colony against attack. Their head is greatly enlarged and usually displays powerful biting mandibles. In some species of termites the mandibles are supplemented with a beak through which a fluid may be ejected to repel enemies. Soldiers of some species display highly sclerotized heads that are modified to block passage through tunnels of the colony. Such heads are termed *phragmotic*.

Termites are responsible for the economically significant destruction of wood and wood products. Based on the habitat, termites construct several types of homes, including mounds (termitaria), arboreal nests, and subterranean nests. They are a common landscape scene in some tropical and subtropical areas of the world. Termitaria are constructed of earth mixed with salivary-gland secretions and excrement and are striking in their architectural diversity and range of size. Some species of termites construct termitaria nearly 7 m tall. Termitaria are exceedingly hard and must be penetrated with iron or steel implements. Arboreal termites include species that live in dead trees, wooden buildings, or furniture not in contact with soil. Subterranean termites live in the soil, sometimes creating mounds protruding a few meters above the surface. Colonies of subterranean termites may be located by digging into these mounds, by prying apart rotting stumps, or by turning over rotting logs and searching in the soil beneath.

Colonies of arboreal termites are more difficult to find, because often there is no external evidence of an infestation. Careful examination may show the entrance holes made by the reproductive adults as they entered the wood, but these holes usually are sealed with cementlike plugs secreted by the termites. If the termites are working close to the outer edge of the wood, surface blisters in the paint or a flaking of the surface of unpainted wood may be a clue to the presence of a colony.

For quantitative studies of termites, Pearce (1990) suggests using a cardboard coaster sandwiched between two glass plates held together with a clip. The cardboard is moistened and the entire trap buried in the soil. The termites feed on the moistened cardboard, and the entire trap can be lifted from the soil to assess damage and to collect termites without disturbing the trap.

A pick, shovel, or mattock is necessary for opening termite nests and splitting wood to expose specimens. An aspirator can be used to collect termites, but care should be exercised because many termites are fragile and easily damaged. Some termites possess volatile chemicals that can be irritating to the respiratory system of the collector. In such instances, a blow-type aspirator should be used. A moistened camel's-hair brush is effective in capturing small-bodied termites. Featherweight forceps are excellent for capturing termites. Representatives of each of the castes should be collected because identification often cannot be made from the workers alone. Many individuals of each caste should be collected to appreciate variation in size and coloration characters. Ideally, the king and queen should be captured, but this may be difficult. Do not overlook insects of other orders that often live in the colony and that may mimic termites in color and general appearance. All castes of termites should be killed and preserved in 70-80% alcohol. Photographs of the nests, casts of the nest architecture, or diagrams of tunnel patterns are also useful kinds of information for specialists.

6.15 Embiidina (Embioptera) [webspinners, footspinners, Fig. 6.15]

These insects have been called Embioptera, but the correct name is Embiidina (Greek, embi = lively), probably due to the fact that they can run backward very rapidly. The common names stem from their ability to spin silk with glands on the feet. Embiidina constitute a widespread group of about 200 named species and an indeterminate number of undescribed species. Embiidina are well represented in tropical and warm temperate regions. Webspinners are elongate, mostly small insects with a prognathous head and chewing mouthparts. The antennae are filiform, and ocelli are absent. Females are wingless; males are winged or wingless, and sometimes both conditions occur in males of one species. When wings are present, the wing venation is reduced. Legs display three tarsomeres. Males, females, and all instars of nymphs spin silk from glands

webspinner

Figure 6.15
Embiidina

located on the forebasitarsus. The silk issues from hollow, tubular structures on the ventral surface of the basitarsus and second tarsomere. The abdomen appears 10-segmented, but the 11th segment is reduced; the cerci are two-segmented and tactile. All species are gregarious, with nymphs and adults occupying silken galleries. Rapid rearward movement of webspinners is enabled through large depressor muscles in the hind tibiae. Species feed on bark, moss, lichen, dead leaves, and other plant material. Cannibalism may occur, and males are sometimes eaten by females after copulation. Eggs and early instar nymphs are guarded by the female.

Male webspinners are sometimes seen at lights during the night, but females are wingless and are found in the galleries. Galleries may be located among debris and in cracks and fissures or on the surface of bark. Specimens should be preserved in 70-80% alcohol.

6.16 Plecoptera [stoneflies, plaited-winged insects, Fig. 6.16]

The scientific name of the order is derived from Greek (plekein = to fold; pteron = wing) and refers to the habit of fanlike folding of the hindwing beneath the forewing in repose. The common name *stonefly* refers to the habitat in which specimens are often collected. The Plecoptera are a cosmopolitan order containing about 1,000 named species. Most genera of stoneflies are found in the Holarctic, where the order radiated or numerically expanded. Adult stoneflies are soft-bodied, moderate-sized, and dorsoventrally compressed, with a body seemingly loosely jointed. The head shows two to three ocelli, and the antennae are long, many-segmented, and slender. The mouthparts are of the chewing type but are sometimes reduced. The prothorax is large and mobile; the mesothorax and metathorax are smaller and subequal in size. Wings are membranous, with numerous veins and crossveins; a few species are brachypterous or apterous (usually males). Legs bear three tarsomeres. The abdomen has 10 segments; the cerci contain one to many segments. Metamorphosis is incomplete.

Most species of stoneflies are diurnal as adults and found near water. Immatures live in water, and the nymphs are called *naiads*. A few species found in the Southern Hemisphere apparently exist in damp terrestrial situations. Plecoptera naiads typically prefer clean, cold, moving freshwater. They demonstrate an intolerance of pollution and narrow tolerances of water temperature and related parameters. The life cycle of most species lasts one year, but a few species live longer. Some species apparently modify their life cycle in response to changing environmental conditions. Adults are short-lived. Males die soon after copulation; females die soon after egg laying. Evidence suggests that females must feed as adults for their eggs to develop. Females deposit eggs on the surface of water during flight when the apex of the abdomen touches the water surface. Some species broadcast or deposit eggs beneath the surface of the water or upon submerged objects. Stonefly eggs are small, with an adhesive coating activated by water. Females may lay more than one batch of eggs; total egg production may reach 1,000.

Stonefly naiads possess long antennae and cerci; their mouthparts are adapted for chewing. Some North American species display branched gills on the thorax and at the base of the legs. Naiads of species found elsewhere may display respiratory gills on the mouthparts, cervix, thorax, or abdomen or within the anus. In North America, stonefly naiads resemble mayfly naiads. However, stonefly naiads differ in the location of the gills and possess two tarsal claws instead of one. Stonefly naiads are found in running water, under stones, or clinging to submerged piles of drifted leaves and other debris. Specimens can be collected with an aquatic net, or a dipper or by hand and dropped into vials of alcohol or held for rearing. To rear Plecoptera, the food preferences of the naiads must be determined. The species most commonly collected are plant feeders, but some species are carnivorous. To reduce the time needed for rearing, some collectors schedule their collecting trips to coincide with the season of adult emergence for a particular species. The adults almost always emerge at night or early in the morning. By visiting streams at night, mature naiads can be collected as they crawl out of the water just before their final molt. These specimens should be held in temporary cages until the teneral adults expand their wings and dry.

Most insects hibernate or become inactive in cold weather, but adults of many species of Plecoptera emerge during the winter. Their dark forms against the snow make them conspicuous and easy to collect. Species that mature in spring or summer often fly to lights at night. During the day, adults are usually found resting on vegetation or other objects along stream banks and can be collected by sweeping or beating the foliage. Most stoneflies are poor fliers but agile runners. Because they generally are slow to take flight, specimens often can be picked up by hand or brushed with a flick of the finger from foliage directly into a vial of alcohol.

Traditionally, adults and naiads have been preserved in alcohol. This curatorial practice arose because specimens shrivel when they are pinned after air drying. Although distortion is an obvious problem, some collectors prefer this method of preservation. Specimens will remain lifelike and not distorted if they are critical-point dried. Wings, genitalia, and other structures critical for identification may be dissected and mounted on microscope slides or placed in microvials.

Figure 6.16
Plecoptera

6.17 Psocoptera [booklice, barklice, Fig. 6.17]

Figure 6.17
Psocoptera

The name Psocoptera is nonsensical because it is derived from Greek and means "chewing wings." The name stems from the attempts of Shipley (1904) to standardize the endings of names of insect orders. Unfortunately, Shipley's name stuck. The alternative order name for Psocoptera is Corrodentia. The common names of Psocoptera are more descriptive. The terms *booklice* and *barklice* refer to the habitats of some frequently encountered species. The Psocoptera form a cosmopolitan order of small to moderate-sized, soft-bodied insects currently containing about 1,800 species. The head is large and movable, the antennae are long and filiform, and the mouthparts are mandibulate; ocelli are present in winged forms but absent in wingless forms. The prothorax is reduced in winged forms; the meso- and metathorax are fused in wingless forms. The wings are membranous with venation reduced, and the pairs are held together when active; at rest the wings are held obliquely over the body. Legs display two or three tarsomeres. The abdomen consists of nine segments, and cerci are absent. Members of the Psocoptera resemble psyllids or aphids but differ from them in having chewing mouthparts. Psocoptera undergo incomplete metamorphosis.

Some usually wingless species found on old books or papers have received the common name *booklice*. In damp locations in houses and granaries, booklice may attain colossal numbers. Most species of Psocoptera occur outdoors on the trunks and leaves of trees and shrubs, on lichen-covered stones, and on fences. Psocids feed on mold, lichen, pollen, cereals, and starchy materials. Unfortunately, they also feed on dead insects and can cause considerable damage in an insect collection unless preventive measures are taken (see Section 3.5.6).

Specimens can be collected by sweeping or beating vegetation, by drying leaf debris in a Berlese funnel, or by picking specimens from tree trunks or rocks with an aspirator or fine brush moistened in alcohol. Adults and nymphs of most species should be preserved in 70-80% alcohol. Species with scales on their wings should be preserved dry on minuten pins or layered in cellucotton.

6.18 Zoraptera [zorapterans, Fig. 6.18]

Zoraptera constitute one of the smallest orders of Insecta, consisting of one family and about 30 named species. Geographically, zorapterans are widespread. The scientific name is derived from Greek (zoros = pure; a = without; pteron = wing) and is somewhat misleading. When the group name was proposed, entomologists believed that zorapterans were all wingless. Subsequently, it has been learned that some zorapterans are winged and that winged adults cast off their wings. Adults are minute to small-bodied, with a hypognathous head and chewing mouthparts. The adult antenna is moniliform with nine segments; the nymph displays eight antennal segments. The prothorax is well developed in all specimens, but the mesothorax and metathorax are not differentiated in primitively wingless forms. The wings of zorapterans are membranous with reduced venation and they are shed at the base. Wingless forms lack compound eyes or ocelli; winged forms possess compound eyes and ocelli. The legs display tarsi with two segments. The abdomen is composed of 11 segments, the ovipositor is vestigial or absent, and cerci are present but not segmented. Development is gregarious, usually under planks, in piles of old sawdust, in rotting logs, under bark, or in association with termites. Zorapterans are apparently fungivorous or necrophilous or both. Only two species are found in North America; they occur from Pennsylvania and Iowa to Texas and eastward.

Zorapterans have no economic importance. They may be collected with a Berlese funnel, aspirated, or captured with a camel's-hair brush while sifting through sawdust or debris. Zorapterans should be preserved in 70-80% alcohol because they are exceedingly small and not strongly sclerotized.

Figure 6.18
Zoraptera

6.19 Mallophaga [chewing lice, wool-eaters, Fig. 6.19]

The name Mallophaga is derived from Greek (mallos = wool; phagein = to eat) and refers to the habit of these insects to feed on mammals. Mallophaga are external parasites of birds and mammals, although mostly of birds. Species of Mallophaga are not known to attack humans. About 3,000 species of chewing lice have been described. Morphologically, chewing lice are small-bodied and dorsoventrally compressed or flattened. The mouthparts are adapted for chewing, and all species are wingless. Metamorphosis is simple, and the immatures resemble the adults. All stages occur on the host, and chewing lice are rarely found away from the host. Six families are commonly found in North America. Of these, the Philopteridae (on poultry) and the Trichodectidae (on cattle, horses, and dogs) are perhaps the most often collected.

Chewing lice are collected from living or dead hosts. When examining hosts, the collector should use gloves, preferably surgical or rubber dishwashing gloves. This precaution seems prudent when we consider the many kinds of disease that may be transmitted via contact with rodents or other mammals. Birds also transmit diseases. All parts of the host should be examined for chewing lice because different species of Mallophaga may be found in different places on the host.

Arthropod Collection and Indentification Techniques

Figure 6.19
Mallophaga

Chewing lice should be preserved in 70-80% alcohol. Formalin should not be used to preserve specimens. Use forceps or a fine brush moistened in alcohol to remove the lice from the feathers or fur of the host. Do not use a sucking-type aspirator; a blow-type aspirator may reduce the chance of contacting respiratory infection. Dusting with desiccant powders may prove effective for driving the parasites off the host.

Place collected lice in small vials of alcohol, using a separate vial for each species and for specimens from different hosts. In labeling the specimens, always include data reporting the part of the host's body from which the parasites were taken. Many species of lice can be found on one bird or mammal. Chewing lice species are frequently site-specific. For permanent collections, the lice should be cleared with a 10% solution of NaOH or KOH, stained if necessary, and mounted on microscope slides.

Collecting equipment and bags used to take hosts to the laboratory for inspection should be cleaned meticulously to avoid mistakes in associating hosts with their parasites. When in doubt, specimens should be discarded rather than collected, curated, and provided with erroneous information.

6.20 Anoplura [sucking lice, Fig. 6.20]

Figure 6.20
Anoplura

The name Anoplura is derived from Greek and is more tricky than some other Greek combinations (an = without; hoplon = weapon; oura = tail). The name refers to the lack of cerci or caudal appendages on the abdomen. Anatomically, sucking lice resemble Mallophaga, in that both groups are small-bodied and dorsoventrally flattened. The head of sucking lice is somewhat conical, and the compound eye is reduced to a facet or absent. Mouthparts consist of three sclerotized piercing stylets that are concealed when not in use; the maxillary and labial palpi are reduced or absent. All of the thoracic segments are fused, and wings are never present. The tarsi are one-segmented, each with a large claw opposable to a thumblike process on the apex of the tibia. This structure serves to grasp hairs. The abdomen is nine-segmented. The female lacks an ovipositor, and cerci are absent from both sexes. Metamorphosis in Anoplura is simple, and the nymphs resemble adults. All stages are found on the host, including eggs that are glued to hair.

Anoplura suck blood and are parasites of mammals only. Species that attack humans include the crab louse, *Pthirus pubis* (Linnaeus), the head louse, *Pediculus humanus capitis* De Geer; and the body louse, *P. humanus humanus* Linnaeus. The body louse is an important vector of typhus and other diseases. Three families in the order are found in the United States: the Pediculidae

Part 2: Classification of Insects and Mites

(to which the three species mentioned here belong), the Echinophthiridae (on marine mammals), and the Haematopinidae (on horses, cattle, sheep, hogs, and other animals).

Anoplura must be collected from the host. The methods of collection and preservation are the same as for Mallophaga. Techniques for mounting Anoplura were described by Kim et al. (1986).

6.21 Thysanoptera [thrips, fringe-winged insects, Fig. 6. 21]

The order name is derived from Greek (thyasnos = fringe; pteron = wing) and alludes to the long marginal fringe found in winged specimens. The common name *thrips* is singular and plural; we do not refer to a "thrip." Thrips are a cosmopolitan order consisting of about 4,000 named species. They are related to Hemiptera. Thrips are small and slender, but some species may attain a length of 12 mm. The compound eyes usually have large facets; ocelli are present in winged adults only. Another curious feature of thrips is that the right mandible is absent. The left mandible is adapted for rasping; the labrum and labium form a cone and the maxillae are adapted for piercing. Wings, when present, are narrow with reduced venation and display a long marginal fringe; wings are held over the abdomen in repose, but they are not folded. The tarsi are one- or two-segmented, and the pretarsus forms an eversible bladder that is used for adhesion. The abdomen of thrips is 10-segmented; an ovipositor is present or absent, but cerci are always absent. Metamorphosis is complex among thrips and shows a transition between simple and complete metamorphosis. Thus, nymphal instars 1-2 are active, feeding, and wingless or wing buds are internal; instar 3 (prepupa) and instar 4 (pupa) are quiescent and nonfeeding, with external wing buds.

Parthenogenesis and viviparity are recognized in some species. Thrips are predominantly phytophagous, a few species are fungivorous, and a few species are predaceous. Eggs are inserted into plant tissue, under bark, or into crevices. Thrips are typically multivoltine. Most thrips feed on flowers, pollen, leaves, buds, twigs, and other parts of plants. Specimens can be collected by sweeping vegetation or by shaking or beating shrubbery over a white or light-colored pan. Another method involves collecting infested plant parts (usually indicated by curled leaves or deformed buds) and placing the vegetation in a paper bag. These samples can be examined later and the thrips removed with an aspirator or a fine brush dipped in alcohol. Alternatively, the sample can be run through a Berlese funnel. Collect winged and wingless forms if both are present. Thrips

thrips

Figure 6.21
Thysanoptera

Arthropod Collection and Indentification Techniques

water strider

giant water bug

can be collected in 60-70% alcohol, but AGA (see Appendix I) is recommended because it leaves the appendages well extended and makes the specimens easier to mount on microscope slides.

For critical study and for permanent collections, thrips should be mounted on microscope slides. Specimens need not be stained before mounting, but some dark-colored and spore-feeding thrips must be treated with cold 10% sodium hydroxide to lighten the color and remove the body contents. Do not use potassium hydroxide because that compound is too caustic and destroys the wings before the harder parts of the body are sufficiently treated. In some species, the wings are held close to the body by paired setae on either side of the abdomen when the insect is at rest. To spread the wings, insert a needle between the wings and the body at the point where the thorax and abdomen join, and gently tease the wings out to a horizontal position. The antennae also should be extended because the number of antennal segments and the position of sensilla on the segments often are critical for identification to species or even to family.

6.22 Hemiptera [true bugs, Fig. 6.22]

The order name Hemiptera is derived from Greek (hemi = half; pteron = wing) and alludes to the condition of the forewing in which the basal part is thickened and the distal part is membranous. The etymology of the word *bug* is uncertain. The term *bug* is used commonly to refer to all insects. However, in the strict sense only species of the Hemiptera are the "true bugs."

The hemipteran head is usually prognathous, the antenna displays four or five segments, and the beak typically displays three or four segments. Forewings are modified into hemelytra, and the tarsi hold two or three segments with two apical claws. Members of the order Hemiptera are diverse in habits and habitats. Most species are terrestrial, but many Hemiptera live in or on the water. Most

water boatman

Figure 6.22
Hemiptera

plant bug   assassin bug

backswimmer

ambush bug

Hemiptera are plant feeders, but some species are predaceous on other insects. A few species imbibe the blood of humans and other vertebrate animals and are important vectors of diseases in the tropics.

Collecting methods for Hemiptera depend on the type of habitat. Most species are collected in nets; aquatic nets are used for species associated with water, and sweep nets are used for most plant-feeding and terrestrial predaceous species. Hemiptera living near water may be shot with a stream of alcohol from a laboratory squeeze bottle and then captured with forceps. Terrestrial specimens collected in sweep nets can be taken easily in an aspirator and later transferred to a killing jar. Some species are attracted to lights. Sifting or using a Berlese funnel on leaf litter or on soil around plant roots usually will yield some Hemiptera. Some species are found in association with ants and may mimic the ants in appearance. Species of Cimicidae (bed bugs) are ectoparasites of birds, bats, and other animals, including humans. Cimicids usually feed on their host at night and hide in cracks and crevices during the day. Cimicids are virtually wingless and can be collected easily with small forceps.

With the possible exception of some fragile plant bugs of the family Miridae, most immature and adult Hemiptera can be killed and preserved in 70-80% alcohol. Adults should not be stored in alcohol for long periods because specimens fade. Soft-bodied specimens may be critical-point dried, with excellent results. For permanent preservation, adults should be pinned. Mirids and most Hemiptera can be collected satisfactorily with killing jars, but never put large specimens in the same jar with mirids or other delicate bugs. Small killing tubes, such as those used for parasitic Hymenoptera, are ideal for relatively small Hemiptera. A rumpled piece of tissue paper placed in the jar will help prevent damage to the specimens. Specimens too small or slender for direct pinning should be glued to points on double mounts.

6.23 Homoptera [scale insects, mealybugs, whiteflies, leafhoppers, aphids, Fig. 6.23]

In the Homoptera (Greek, homo = same, uniform; pteron = wing) the forewings are uniform in texture: entirely membranous, as in the aphids, or almost leathery, as in the leafhoppers. The hind wings (absent in male scale insects) are entirely membranous. When at rest, members of the Homoptera hold their wings folded rooflike over the back. Both groups lack cerci. Metamorphosis is characterized as simple, and the immatures resemble the adults in most species. However, more complex patterns of development are seen in some Homoptera. For instance, members of the family Aleyrodidae pass through a pupal stage; males of scale insects pass through a prepupal or pupal stage (or both); the life stages of the Phylloxeridae are even more complex (Stoetzel 1985).

Arthropod Collection and Indentification Techniques

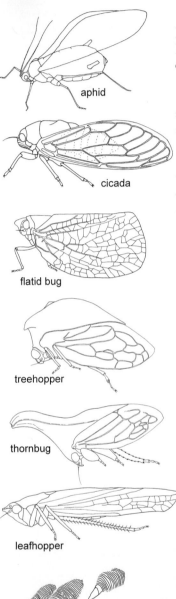

Figure 6.23
Homoptera

Anatomically, the Homoptera are exceedingly diverse. Wingless Homoptera, such as scale insects, are sometimes difficult to recognize as insects. In most Homoptera that are easily recognized as insects, the head is usually opisthognathous.

All Homoptera are terrestrial and feed on plants. Some Homoptera (such as whiteflies, psyllids, and leafhoppers) transmit plant diseases and are regarded as serious economic pests. Other Homoptera (such as soft scales, armored scales, mealybugs, and aphids) can develop into numerically large, highly dense populations. Homoptera developing under such conditions cause problems through the accumulation of honeydew, which provides a substrate for the development of fungi. Cicadas can reach epidemic proportions and cause damage to trees and create intolerable levels of sound.

Many Homoptera can be collected with sweep nets. Winged aphids are often attracted to yellow pan traps. However, because swept or trapped specimens lack associated host data, we recommend collecting specimens directly from plants whenever possible. Host plants should be identified by competent botanists and the information provided with the collecting data presented on labels. Collect winged and wingless forms if both are present, and note the color of the living specimens. Natural color may be lost in a preservative or when the specimens are cleared and mounted on microscope slides.

Most specimens of Homoptera can be collected and preserved, at least temporarily, in 70-80% alcohol. Exceptions include diaspidid scales and whitefly pupae, which should be collected dry on the host plant and placed between pieces of absorbent paper. Do not use plastic bags because they accumulate moisture, which ruins specimens. Aphids, whiteflies, and scale insects should be stained or bleached as needed and then mounted on microscope slides for critical study.

Adult cicadas, leafhoppers, and other relatively hard-bodied Homoptera can be collected in killing jars and then pinned. Specimens smaller than 10 mm should be glued carefully to card points. When collecting gall-forming psyllids and aphids, include a sample of the gall. If securely anchored with additional pins, the gall can be kept in the box with the pinned adults. Galls should be treated with a fumigant (naphthalene) to ensure that all of the occupants are dead.

Collectors may be interested in studying the life history of Homoptera or accumulating all of the life stages of some species. Aphids and many other plant-feeding Homoptera can be maintained easily in flowerpot cages. The cages allow all stages of the insect to be observed and collected. Also, the living specimens provide a source of food for mantids and other predaceous insects maintained in captivity.

Collecting Homoptera can often be rewarding in other ways. For instance, Homoptera are attacked by many species of parasitic Hymenoptera. Scale insects and mealybugs are particularly susceptible to attack by chalcidoid wasps (Tachikawa 1981; Viggiani 1984); leafhoppers are attacked by dryinid wasps (Olmi 1984a, b); the egg stage of many Auchenorrhyncha is attacked by species of mymarid wasps (Huber 1986). In some instances, hyperparasites can be reared from Homoptera (Viggiani 1990). Careful dissections of hosts under a microscope can often confirm suspicions of hyperparasitic habits. When parasitic wasps are reared from Homoptera, the remains of the host should be preserved with the parasite. Accurate host associations are very important in our understanding of parasitic wasps and their importance in biological control. All collectors can make valuable contributions to the taxonomy of parasitic Hymenoptera through the careful preservation of parasites and their homopterous hosts.

6.24 Coleoptera [beetles, Fig. 6.24]

The order name is derived from Greek (koleos = sheath; pteron = wing) and alludes to the thickened forewings that protect the hind wings. This order contains about 300,000 described species of insects and is the largest of the insect orders. More than 25,000 described species occur in the United States, with many undescribed species. Approximately 120 families are recognized, which are grouped into three suborders. This order is a favorite among collectors, and many amateurs specialize in collecting select families of beetles.

click beetle

Beetles have mouthparts adapted for chewing and antennae that are exceedingly variable in segmentation and shape. The prothorax is well developed; the mesothorax is generally reduced; the abdomen is broadly joined to the thorax. Beetles usually have two pairs of wings, although some beetles lack them and other species have highly modified wings. The forewings are called *elytra* (Greek, elytron = cover, sheath). Elytra are thickened, usually hard or leathery, lack veins, and often are sculptured or display pits and grooves. Elytra usually meet in a straight line dorsally along the middle of the back. The hindwings are membranous and are folded under the forewings when the insect is at rest. Several anatomical types of larvae occur within the order.

longhorned beetle

Figure 6.24a
Coleoptera

Beetles have invaded almost every conceivable aquatic and terrestrial habitat. They feed on plants and fungi. Plant-feeding beetles may be found in association with every part of a plant, from flowers to roots. Some beetles are external feeders; other species mine leaves or bore into the stalk. Many phytophagous species are serious economic pests. In contrast, some beetles are predators of other insects and are viewed as beneficial insects. Predaceous species of Coccinellidae are used in biological control. Some species of beetles

Arthropod Collection and Indentification Techniques

develop as parasites, and many species of beetles are scavengers. Some beetles are inquilines (welcome or unwelcome guests) in nests of social insects (termites and ants). Dermestid beetles feed on dead insects and can be a problem in insect collections unless protective measures are taken.

Collecting methods for beetles are almost as diverse as their habits and are limited only by the ingenuity of the collector. Pitfall and other bait traps using fermenting fruit, decaying meat, or excrement attract silphids, staphylinids, nitidulids, scarabs, and other scavengers. Small specimens may be taken with "window" traps (Peck & Davies 1980). Some scarabs, cerambycids, and many other beetles are attracted to light traps or light sheets. A sifter or Berlese funnel can be used effectively to collect soil-dwelling beetles or beetles found in moss and leaf litter (Bell 1990; Chandler 1990; Dybas 1990; Newton 1990; Peck 1990; Tashiro 1990). An aspirator may be helpful when collecting in grass clumps, bark, and crevices. Species living in ponds and streams can be captured by hand or with an aquatic net. Do not overlook beetles living along the shore in sand and under seaweed or other debris. Some beetles may be captured with an aerial net, but more often beetles can be taken with sweep nets. Sweeping or beating vegetation will gather leaf-feeding and some predaceous beetles. Many species of beetles reside near the base of vegetation. A net of strong construction whose hoop includes a straight surface may be helpful in dislodging specimens from such habitats (Belkin 1962). A tool for getting under bark or into wood is necessary for collecting species that bore into logs or plant stems.

Immature beetles (grubs or larvae) should be killed by placing them in boiling water for 1–5 minutes, depending on the size of the specimens. They can then be preserved in 70-80% alcohol. The larvae of some beetles are fatty and should be placed for 20-30 minutes in 30% alcohol with a few drops of acetic acid that has been heated to 70°C. Larvae can be critical-point dried and stored in vials or gelatin capsules or mounted on cards.

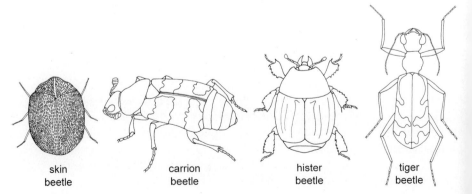

Figure 6.24b Coleoptera

Part 2: Classification of Insects and Mites

Adult beetles may be killed and preserved in alcohol, but we recommend that they be killed in alcohol or in a killing jar and then mounted on pins. Most beetles should be pinned with a no. 1 pin; large beetles should be pinned with a no. 2 or 3 pin. Some pinned specimens may become greasy with time. These specimens can be treated with carbon tetrachloride, xylene, or a similar solvent. The specimens should not be removed from the pin while undergoing treatment, but the labels should be removed from the pin. Specimens of most beetles should have the genitalia removed and prepared (see Section 3.2.8.7) after removal from alcohol before the specimen is pinned. Specimens smaller than 5 mm should be glued to points or cards, using care not to conceal characters on the ventral surface of the body. It is generally inadvisable to use minutens or to spread the wings of beetles. Very thin or very flat cleared beetles may be mounted on microscope slides.

diving beetle

Figure 6.24c
Coleoptera

Many Coleoptera are easily reared; this is often desirable because adults may then be associated with immature stages. Mealworms and other larvae that infest stored grain and flour are particular easy to rear and can be used to feed predaceous insects held in captivity.

6.25 Strepsiptera [twisted-winged parasites, Fig. 6.25]

The order name is derived from Greek (strepsis = turning; pteron = wing) and refers to the twisted-wing condition found in males. The Strepsiptera form a widespread order of about 300 species. Strepsipterans are sometimes grouped with the Coleoptera near the Rhipiphoridae. Adults exhibit strong sexual dimorphism. Females of most Strepsiptera lack eyes, antennae, and legs; the head and thorax are fused, and the general body shape is larviform. Females of some free-living species (some Mengeidae) display compound eyes, a developed head, antennae, and chewing mouthparts. All males resemble a generalized insect, the forewings are reduced (elytraform), and the hind wings are enlarged and membranous with radial venation. Strepsiptera are morphologically unique among insects in lacking a trochanter in the adult leg.

adult

triungulan

protruding from host abdominal segments

Figure 6.25
Strepsiptera

The immature stages of Strepsiptera are parasitic on Thysanura, Blattodea, Mantodea, Orthoptera, Hemiptera, Diptera, and Hymenoptera. Postembryonic development of parasitic species is hypermetamorphic: the first-instar larva is called a *triungulinid*. The free-living first instar resembles the triungulin of Coleoptera but differs in lacking a trochanter in all legs, and the antennae and mandibles are not well formed. The triungulinid is very active and searches for a host. After a host is found, the triungulinid enters the host's body and then molts into a legless, wormlike form that feeds and pupates within the integument of the host. The adult male is winged and leaves the host, but the female remains in the host with its body protruding from between the host's abdominal segments. After producing large numbers of the tiny triungulinids, the female dies.

Strepsiptera males may sometimes be found on flowers, apparently searching for females in their hosts. A few species are found under rocks. However, the best way to collect strepsiterans is to capture and rear parasitized hosts (bees, wasps, flies, bugs). Hosts may be recognized by the small saclike females protruding from the often-distorted abdomens of their hosts. Males of some species are attracted to light. Males of some Stylopidae are attracted to hosts parasitized by females. These males can be collected by placing a parasitized (stylopized) host in a screened cage (MacSwain 1949). Specimens should be preserved in 70-80% alcohol or mounted on microscope slides; males may be mounted on minutens.

6.26 Mecoptera [scorpionflies, hangingflies, Fig. 6.26]

The order name is derived from Greek (meco = long; ptera = wing). Some Mecoptera are called *scorpionflies* because the upturned genitalia of some males resemble the tail of a scorpion. The name *hangingflies* pertains to one family (the Bittacidae), which resemble craneflies and hang suspended from vegetation by their front and middle legs. Mecoptera are moderate-sized (to 25 mm long), slender-bodied insects with relatively long wings. The head is prolonged into a rostrum beneath compound eyes, and the mouthparts are mandibulate and positioned at the end of the rostrum. Most Mecoptera are winged; some are brachypterous and a few are apterous. Winged species have four wings similar in size, shape, and venation; the wings are not folded when in repose. The metathorax is fused with the first abdominal tergum, but the first sternum is free. The abdomen is 11-segmented; cerci are two-segmented in the female and one-segmented in the male (or rarely absent). Metamorphosis is complete in Mecoptera. The larvae of some species possess compound eyes; the pupa is decticous and exarate.

Scorpionfly adults and larvae feed primarily on living and dead insects, but some adults are attracted to nectar and fermenting fruit. Some species apparently feed on moss. Males are predaceous, but females do not take living prey. Larvae of scorpionflies resemble the larvae of sawflies and some Lepidoptera but differ from the latter by the absence of crochets (tiny hooks) on the prolegs and from the former by having seven or more ocelli.

Figure 6.26 Mecoptera

The larvae of most scorpionflies are found in the soil or in leaf litter and can be collected in a sifter. Most adults are found in heavily wooded areas and can be collected by sweeping or beating the vegetation. Adults are not strong fliers and are relatively easy to net. Adults of the so-called snow scorpionflies of the family Boreidae emerge during winter and often can be picked from the surface of snow with forceps. Members of the Panorpidae (common scorpionflies) should be held until the gut is voided before placing them in a killing jar (Beirne 1955). Some Mecoptera are collected in light traps, but most species are active only during the day, and flight traps such as the Malaise trap are usually more effective.

Larvae and adults can be killed and preserved in 70–80% alcohol, but critical-point dried specimens are superior. Adults should be mounted on pins or card points. The wings of the adults may be spread if desired, but it is customary to leave them in a natural position.

6.27 Neuroptera [alderflies, antlions, dobsonflies, fishflies, lacewings, owlflies, snakeflies, Fig. 6.27]

The order name is derived from Greek (neuro = nerve; pteron = wing) and refers to the reticulate nature of the wing venation. The common names are generally applied to specific families or groups of Neuroptera based on their appearance or habits. The Neuroptera are a relatively small, cosmopolitan order of endopterygote, neopterous insects that are best represented in tropical regions. Specimens vary in size from small to very large, with a wingspan of more than 100 mm. Neuroptera are soft-bodied; the compound eyes are well developed, but ocelli are usually absent. The antennae are long, multisegmented, and sometimes display a club-shaped enlargement at the apex. Mouthparts are of the mandibulate biting type. The prothorax is movable, and the entire thorax is loosely organized. Neuroptera usually display four large, membranous wings that are subequal in size; venation is abundant and netlike. The wing coupling mechanism is simple; when the insect is at rest, the wings are held rooflike over the back. The abdomen is 10-segmented (except in Chrysopidae); cerci are absent. The immatures do not resemble the adults, and pupation usually occurs inside a silken cocoon. The pupa is decticous and exarate; pupal mandibles are well developed.

Classification of Neuroptera varies among entomologists. Here we recognize three suborders and 14 families. Neuroptera here are considered in the broadest sense to include the dobsonflies (Megaloptera) and snakeflies (Rhaphidioidea). Neuropterans such as dobsonflies and fishflies (Corydalidae), alderflies (Sialidae), brown lacewings (Hemoerobiidae), green lacewings (Chrysopidae), and antlions (Myrmeleontidae) are frequently collected.

Figure 6.27
Neuroptera

Figure 6.27b
Neuroptera

Neuroptera are predaceous as larvae and adults. As a consequence they are highly beneficial to human agricultural efforts. In fact, green lacewings are sold commercially for the control of various pests. Many Neuroptera will bite if handled incautiously when collected. The larvae of several members of the order are aquatic and commonly concealed under stones in streams. These larvae can be collected by hand, in an aquatic net, or with a dipper. Adults of aquatic immatures usually remain on vegetation near water. They are relatively poor fliers and usually can be collected directly into killing jars. Most immatures and adults of terrestrial forms are found on vegetation and can be collected in a sweep net. Many adults of terrestrial and aquatic immatures are attracted to lights. Immature myrmeleontids (called antlions) are found partially buried at the bottom of small pits that they dig in sand or dust to trap ants or other insects. These larvae can be collected by blowing strongly on the pit or captured by scooping out the bottom of the pit with a spoon. The rarely collected mantispid larvae are parasitic on spider eggs.

Adult Neuroptera may be killed with cyanide or ethyl acetate. Specimens may be preserved in alcohol, but we recommend mounting the specimens on pins or points. The wings may be spread or left folded over the body in repose. Green specimens lose their color when exposed to light for prolonged periods of time. Immature specimens of medium to small size should be killed and preserved in 70–80% alcohol; large-bodied immature specimens should be killed in boiling water and then transferred to alcohol. Larvae can be critical-point dried and mounted on cards or preserved in vials.

6.28 Trichoptera [caddisflies, Fig. 6.28]

The order name is derived from Greek (thrix = hair; pteron = wing) and refers to the hairy appearance of the wing. Caddisflies are soft-bodied insects with long, slender antennae and two pairs of membranous wings. The wings are clothed with setae and held rooflike over the back when the insect is at rest. The larval mouthparts are adapted for chewing; the adult mouthparts are adapted for feeding on liquids. About 5,000 species in 20 families are known for the order.

Figure 6.28a
Trichoptera

Trichoptera larvae are aquatic and display a pair of hooklike appendages at the apex of the abdomen. Many larvae live in characteristic cases constructed of pebbles, sand grains, twigs, or other materials found in ponds and streams. The abdominal hooks are used to drag the case about as the larva feeds. When the larva is ready to pupate, the animal attaches the case to a rock or other fixed object in the water. Case-making larvae are mostly detritus or plant feeders. Some caddisfly larvae do not make cases but instead spin silken webs, which are used to capture food drifting in the stream. A few species do not build cases or webs but are free-living and predaceous. When collecting immature Trichoptera, preserve the case in alcohol along with the larva, pupa, or both.

Adult caddisflies are weak, usually crepuscular (dawn or dusk) fliers. They are generally found during the day resting on vegetation, bridges, or other objects near ponds and streams. Specimens can be collected with a sweep net or captured directly with a killing jar. Large numbers of adults are often attracted to lights, particularly if the lights are near water. Some collectors report that blue light is more attractive than other colors (Borror et al. 1989). Larvae and pupae collected from the habitat are valuable because when reared to adult, the immature and adult stages can be associated.

free-living caddisfly larva

Adults killed and preserved in 70–80% alcohol are satisfactory, but specimens mounted on pins are preferred. Soft-bodied species tend to shrivel when dried. Critical-point drying leaves soft-bodied specimens in lifelike condition. The wings of pinned specimens can be spread or left in repose. The abdomen should be bent forward on specimens whose wings are not spread. This permits examination of genitalic characters. Some species can be identified by genitalic characters visible in the late pupal stage. Adults too small for direct pinning should be mounted on minutens.

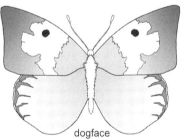

Figure 6.28b
Trichoptera

6.29 Lepidoptera [butterflies, skippers, moths, Fig. 6.29]

The order name is derived from Greek (lepis = scale; ptera = wing) and refers to the wings usually covered with scales (flattened setae). The common name *butterfly* applies to most of the diurnal Lepidoptera with clubbed antennae; the common name *skipper* refers to diurnal Lepidoptera of the family Hesperiidae, with fast, erratic flight and whose antennae are hooked at the apex; the common name *moth* is applied to nocturnal Lepidoptera whose antennae are not clubbed. The Lepidoptera are a cosmopolitan order of Holometabola consisting of about 150,000 described species placed in about 80 families. More than 11,000 species have been described in North America, but many species await description.

Lepidoptera are among the most popular insects collected by amateur entomologists. The mouthparts of adult Lepidoptera are adapted for imbibing nectar and water; a few primitive Lepidoptera have vestigial mandibles. The mouthparts of most Lepidoptera are formed into a long, spirally coiled haustellum or proboscis. The coil actually is derived from elongate maxillary galeae; the maxillary palpi are small or absent, but the labial palpi are well developed. Adult Lepidoptera all possess large, multifaceted compound eyes; some Lepidoptera have two ocelli, but many lack ocelli. (Sometimes scales must be removed from the head to find the ocelli.) The antennae of adult Lepidoptera

Figure 6.29a
Lepidoptera

Arthropod Collection and Identification Techniques

are elongate and multisegmented but otherwise diverse in appearance. The antennae may be covered with scales or display featherlike branches. The adult head is freely movable; the prothorax is small and membranous and often displays a pair of articulated sclerites called *patagia*. Most Lepidoptera have four membranous, scale-covered wings; the wing-coupling mechanisms are diverse. Legs are usually long and tapered, but the posterior or anterior pair may be reduced in some groups; the tarsi display five segments. The adult abdomen contains 10 segments, and cerci are not present.

The larvae of Lepidoptera are commonly called *caterpillars*. The name *caterpillar* is derived from Latin and literally means "hairy cat," probably in reference to the many caterpillars that are covered with setae. Caterpillars usually have a well-differentiated, strongly sclerotized head and mandibles adapted for chewing. The thorax consists of three thoracic segments, each bearing a pair of legs with five segments; each leg terminates in a single claw. The larval abdomen also consists of 10 segments, and there are "prolegs," which usually are displayed on segments 3–6 and 10. The prolegs are apically truncate, with the truncate surface called the *planta*. Minute, sclerotized hooks called *crochets* are found along the margin of the planta. These hooks are used as holdfast structures and are diagnostically important. Lepidoptera pupae are decticous and exarate or adecticous and obtect.

Most lepidopterous larvae feed on live plants. Some species feed on or in the leaves, and other species bore into the stems, seeds, or other parts of the plant. The larvae of many species are host-specific and will starve rather than feed on a plant other than the preferred species. Larvae of some Lepidoptera are predaceous (Montgomery 1982, 1983). A few larvae are scavengers on dead plant or animal matter, including woolen clothing. The larvae of many Lycaenidae are associated with ants and can be found in ant nests (Hinton 1951).

Most adult Lepidoptera fly actively, but a few (such as female bagworms) are wingless or possess wings too short for use in flight. Most species of butterflies are diurnal, and they are often collected with an aerial net or captured as they come to rest on flowers, leaves, bait stations, or the ground. Many brightly colored butterflies are found congregating around puddles. During the dry season in tropical regions, colorful butterflies are often found on moisture-laden domestic livestock manure. Collecting during the early morning and late afternoon is not highly productive. Most species of diurnal butterflies are taken from midmorning through midafternoon. Warm, sunny days are best for collecting butterflies, particularly when the wind is not strong.

Figure 6.29b
Lepidoptera

Adult moths are predominantly nocturnal and many species may be taken during the first hours after sunset. However, a few moth species are active during the day, and some nocturnal moths may be flushed from their resting places and netted during the day. The most productive and efficient method for collecting moths involves catching or trap-ping them when they are attracted to light. Bait collecting also is a standard technique for some groups (see Section 1.10).

Lepidoptera are easily reared by amateur entomologists, and the commercial breeding and hybridization of some butterfly and moth species has become widespread. Special equipment or skills are not necessary to rear most foliage-feeding species from the egg stage to adult. Many adult female moths will lay eggs when confined in a small plastic box, a glass jar, or other container with a piece of the food plant or some crumpled paper. For large moths, a paper bag or shoebox can be used as a rearing container. Keep a piece of damp paper towel or sponge in the container to maintain high humidity, especially if the specimens are reared in an air-conditioned or heated building. A water-soaked raisin will provide nourishment for the female and moisture for species that feed as adults.

The rearing of butterflies may sometimes be difficult. Most butterflies require large, brightly illuminated cages containing an ample supply of the natural food to induce them to oviposit. Females caught outdoors will nearly always be fertile, and their eggs will hatch in a week or 10 days unless the species is one that overwinters in the egg stage. If the food plant is unknown, a trial-and-error procedure must be used by offering the newly hatched larvae bits of leaves from a variety of plants that may be possible hosts.

Rearing cages made of screening are appropriate for large moths and most butterflies. Caterpillars require little air, and most species may be reared successfully in closed plastic or metal containers. Excessive moisture is not a problem if the containers are kept at a reasonably constant temperature and the bottom is lined with a clean paper towel or blotting paper. Only dry foliage should be used as food, and tight containers keep the leaves fresh for several days. Caterpillars need only the water they get by eating fresh leaves. Cleanliness is essential. Uneaten leaves and the accumulation of frass at the bottom of the container must be removed at regular (often daily) intervals. When mature, the larvae should be provided with suitable conditions for pupation. For example, species that pupate in the ground need several inches of slightly dampened peat, sand, or sawdust. Species that pupate in leaf litter often will do well if allowed to crawl into the folds of a crumpled paper towel. The larvae of some moths will pupate in the corrugations of cardboard.

Figure 6.29c
Lepidoptera

Preserving the larval instars is often desirable so that they may be identified by positive association with adults reared from the same brood. Larvae should be killed by immersing them for about a minute in boiling water. Larval specimens may be stored in glycerine. Traditionally, they were preserved in 70–80% alcohol. Large specimens in alcohol should have the alcohol changed after the first 24 hours because the body fluids will have diluted it. As an alternative to alcoholic preservation, larvae may be critical-point dried, freeze-dried, or inflated and then mounted on pins.

Avoiding the loss of wing and body scales is a difficult problem in collecting adult Lepidoptera. Killing the specimen quickly is essential for preserving scales on the body and wings. Rapidly acting killing agents such as cyanide are recommended but dangerous. Ethyl acetate is not as desirable as a killing agent for Lepidoptera because it is volatile and killing jars must be constantly recharged to work effectively. Some collectors pinch the thorax of the specimen while it is in the net. This is accomplished by holding the thorax of the specimen between the thumb and index finger and applying just enough pressure to stun the specimen. In this way the specimen does not flutter its wings or struggle when placed inside the killing jar. The jar should be clean, and absorbent paper should be placed in the bottom. Excessive moisture affects the scales. For large specimens, a widemouthed killing jar is recommended. Do not place large beetles or other large insects in the same jar with butterflies and moths, and do not crowd too many Lepidoptera into one jar. Some collectors prefer to remove the specimens soon after they have been killed and place each specimen temporarily in a separate, labeled envelope; this procedure is especially useful for butterflies. Such papered specimens later can be relaxed and their wings spread. However, it is always advisable to pin specimens as soon as possible after killing them to minimize damage.

For permanent collections, adult Lepidoptera should be pinned with the wings spread in the standard manner (see Section 3.2.7.2) so that all important features of wing pattern and structure may be seen. Never glue adults to points or place them in alcohol. Pin small microlepidoptera with minutens, spread the wings, and then double mount. Papered specimens or specimens left for long periods in killing jars must be placed in a relaxing chamber before the wings can be spread. Even fresh specimens are easier to handle if placed overnight in a relaxing chamber (see Section 3.2.1).

6.30 Diptera ["true" flies, mosquitoes, Fig. 6.30]

The order name is derived from Greek (dis = twice; pteron = wing) and alludes to the presence of two membranous wings used in flight. Perhaps more than any other order of insects, families of Diptera are called by common names: mosquitoes, gnats, midges, punkies, no-see-ums, horse flies, bat flies, snipe flies, robber flies, house flies, bottle flies, and so forth. This no doubt is a consequence of the long-term intimate relationship between flies and humans. Incidentally, by convention we separate the word "fly" from its descriptor when referring to dipterous insects. We combine the word "fly" with its descriptor when referring to nondipterous insects whose names contain that word (e.g., scorpionflies, sawflies). Names notwithstanding, the Diptera constitute a cosmopolitan order of holometabolous insects that consists of about 150,000 described species. The order contains about 120 families, and in North America over 18,000 species have been described. Flies are found in virtually all habitats of all zoogeographical realms; they are omnipresent!

Adult Diptera are relatively soft-bodied and recognized by a typically large, manipulable head with large and multifaceted compound eyes. Most flies have three ocelli, some species have two ocelli, and a few species are anocellate. The antennae of flies are highly variable and range from moniliform and multisegmented to three-segmented and aristate. The mouthparts of Diptera are adapted for sucking and form a proboscis or rostrum. A piercing-type proboscis is found in predatory and bloodsucking species; a sponging-type proboscis is found in house flies; a few primitive flies have nonfunctional mandibles. The prothorax and metathorax of Diptera are reduced; the mesothoracic wings are membranous and used in flight; the metathoracic wings have become modified into clublike halteres that are used as balancing organs. Other orders may have species with only one pair of wings, but no other insects have knobbed halteres. (The halteres or hamulohalteres of winged male scale insects are not knobbed and usually are tipped with one or more hooked setae. Most winged male scale insects also differ from flies in having a single, long, stylelike process at the apex of the abdomen.) A few species of flies lack wings. The legs of flies are variable in shape and structure. The abdomen has 11 segments in the ancestral or primitive condition and 10 segments in the derived condition (10 and 11 fuse to form the proctiger). The distal segments of "higher" Diptera are often modified into a telescopic "postabdomen." Cerci are present at the apex of segment 10; the female cerci are primitively two-segmented (most Nematocera, Brachycera) or reduced to one segment (throughout Diptera); male cerci consist of one segment on the proctiger. Male genitalia are complex.

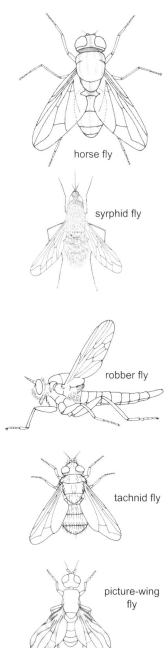

horse fly

syrphid fly

robber fly

tachnid fly

picture-wing fly

Figure 6.30a
Diptera

Arthropod Collection and Indentification Techniques

house fly

fly larva (maggot)

flesh fly

black fly

The larvae of higher Diptera are frequently called *maggots*. All larvae are apodous, but some species have one or more pairs of prolegs. In some primitive families, the larvae possess a distinct head capsule and the pupae are free-living. In more advanced flies, as in the muscoid families, a head is not apparent on the larvae. Larvae often pupate inside the last larval exuviae, which is called a *puparium*. The pupae of Diptera are adecticous and exarate or obtect.

The biological importance of Diptera cannot be overemphasized: bloodsucking species transfer many diseases, including malaria; house flies transfer enteric diseases; the nuisance value of biting flies cannot be measured. The larvae of some species cause myiasis. In terms of agriculture, the impact of some Diptera is notable. Tephritid fruit flies are serious pests of fruit. Quarantine and eradication programs in some areas cost tens of millions of dollars yearly in response to the threat posed to agriculture by fruit flies. Leaf-mining and gall-forming Diptera cause conspicuous damage to foliage and other plant parts, but their economic impact is questionable. In contrast to these negative impacts on humans and their activities, many species of Diptera are beneficial as pollinators; other species are predators or parasites of insect pests. Most species of flies probably are neither beneficial nor harmful to humans, but often are collected and studied because of the unusual ecological habitats they occupy.

Habitats of Diptera are numerous and varied, but most dipterous larvae live in aquatic or semiaquatic habitats. Perhaps the most unusual habitats involve the seeps of crude petroleum in which larvae of some ephydrid larvae (shore flies) develop or the hot springs in which some stratiomyid larvae are found. A rain-filled can by the side of the road may harbor mosquito larvae (Culicidae), and stones in a fast-flowing stream may hold black fly larvae (Simuliidae). A dipper or small aquatic net is useful in collecting such larvae. A sweep net or aerial net can be used effectively to collect the adults, which generally remain near the water. Marsh fly adults (Sciomyzidae) are slow-flying and rest on emergent vegetation along the margins of freshwater. They often perch with their head directed downward. Marsh fly larvae are obligate feeders on freshwater snails, fingernail clams, land snails, and slugs.

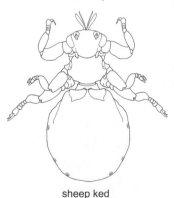
sheep ked

Figure 6.30b
Diptera

adult mosquito

mosquito larva (wiggler)

mosquito pupa (tumbler)

Some Diptera larvae live in plant tissues, where they mine leaves, form galls, or feed in the stems or roots. By placing the host plant, or infested parts of the host plant, in a rearing container, the larvae can be kept until they pupate and the adult flies emerge. Similarly, adult tachinids and other parasitic flies can be obtained by keeping the parasitized host insect in a rearing container. Some species pupate in the soil. To rear these species, place a few centimeters of moist sand or moss in the bottom of the cage into which the mature larvae can burrow for pupation.

Adult flies can be swept from vegetation and collected into a killing jar. The net should remain dry, and the fabric of the net should be made of a lightweight tight-weave material. Large specimens can be collected into a killing jar with a minimum of handling. Small specimens can be collected from the net with an aspirator and then transferred to a killing jar. Alternatively, a small wad of cotton laden with ethyl acetate can be blown down the collecting tube of the aspirator into the bottle. Care should be exercised because the setal and scale patterns may be important in identification. Flies that are scavengers on decaying animal and vegetable matter may be collected in bait traps. Chemical baits that attract some specific groups of flies are available commercially. Light traps or flight traps such as the Malaise trap (see Section 1.6.3), used with or without a bait, usually are extremely successful in capturing adult flies. Some flies are ectoparasites of bats, birds, and other animals. Many of these parasitic flies are wingless and can be collected readily from the host with forceps or an aspirator.

Most flies should be killed dry and mounted. Large specimens should be mounted on pins; small specimens should be mounted on minutens or card points. Mounting should be accomplished within a few hours after the specimens have been killed. If not, the specimens can be held in the freezer compartment of a refrigerator until they can be mounted. Remember that newly emerged adults may be teneral and should be held alive until the wings and colors fully develop. Some species may require more than a day to develop. Small flies, if not mounted on minutens or points very soon after collecting, can be placed in alcohol and run through the Cellosolve-xylene series (see Section 3.2.3) or critical-point dried (see Section 3.2.7.6). Flies should not be kept dry-layered in boxes or envelopes because their heads, antennae, and legs are easily detached. Dipterous larvae should be killed in boiling water and preserved in alcohol. For permanent collections, mosquito larvae preferably are mounted on microscope slides. Some adult flies also can be preserved satisfactorily in alcohol, but most should never be placed in alcohol. If adults of many families are immersed in liquid, the scales, setae, or bristles critical for identification may detach and the specimens will become useless. The larval exuviae and puparium of reared specimens can be preserved in alcohol or glycerine or mounted with the adult.

6.32 Siphonaptera [fleas, Fig. 6.31]

cat flea

Figure 6.31
Siphonaptera

The order name is derived from Greek (siphon = tube; aptera = wingless) and refers to fleas as wingless siphons. This name seems appropriate for reasons explained later. Fleas represent an order of about 1,400 species of holometabolous, endopterygote Neoptera, divided into about 20 families. Adults may be recognized by the body, which is wingless and typically less than 6 mm long. Fleas are laterally compressed and display bristles and cuticular projections that are directed rearward. These structures are sometimes modified into combs called *ctenidia* that may be found on the gena, prothorax, or metathorax. The compound eyes are absent or replaced by one large lateral ocellus on either side of the head. Mandibles are not present, and the mouthparts are of the piercing-sucking type. The piercing elements are derived from the sclerotized epipharnyx and the laciniae of the maxilla; the maxillary and labial palpi are well developed in adult fleas. The antennae are short, typically with three segments. The antennae are held in depressions called *fossae* on the side of the head. All fleas are apterous. The legs are spinose, with large coxae, and the hind legs are adapted for jumping. The abdomen contains 10 segments, with tergum I reduced and sternum I absent. A complex sensory apparatus called the *sensillum* is positioned at the apex of the abdomen. Larval fleas display 13 body segments, lack appendages, and show a wormlike appearance. The head capsule is well sclerotized but lacks eyes. The mandibles are brushlike and the antennae are one-segmented. The terminal body segment displays anal struts. The larva undergoes three instars before constructing a pupal cocoon. The pupa is adecticous and exarate; pupal wing buds are visible in some species, which suggests that fleas had winged ancestors.

Adults of all flea species are external parasites of mammals and birds. Fleas such as the cat flea, *Ctenocephalides felis* (Bouché), and the dog flea, *C. canis* (Curtis), are well known as household pests. However, fleas are most important to humans because some species can transmit bubonic or sylvatic plague, endemic typhus, and other serious diseases. Some species serve as intermediate hosts of tapeworms.

Adult fleas reflect many features that are adaptations for a parasitic existence on the bodies of their hosts. Adult fleas have lost the need for wings because they live in the fur and feathers of their hosts. Moving around on the body of a host with fur and feathers is difficult, so the flea's body has become greatly compressed laterally to facilitate movement between hairs and feathers. To protect the flea from damage, its body has become heavily sclerotized. The flea is armed with combs or bristles that engage fur and resist dislodgment from the host during grooming. The antennae are short and fitted into grooves for protection from damage. The piercing-sucking type mouthparts are an adaptation for feeding on warm-blooded vertebrates.

Flea larvae live in the nests of their hosts. The larvae do not suck blood because their mouthparts are not adapted for that kind of feeding. Instead, flea larvae feed on organic debris (including dried blood) within the nest or habitation of the host. Fleas commonly infest animals that live in nests or burrows and seldom are found on cattle, deer, or other hoofed animals. Rodents and other small mammals are frequently heavily infested with fleas, which can be collected by shooting or trapping the host and immediately picking or brushing the fleas from the fur. Because fleas are active, place the host in a jar or bag with a few drops of liquid killing agent to stun or kill the fleas before examining the fur. Placing the host animal in a refrigerator for an hour also will slow the activity of the fleas. If immediate examination of the host is not possible, the animal should be placed in a bag (each host animal in a separate bag) to confine the fleas, which tend to leave a dead host. We must emphasize that bags and other containers must be scrupulously cleaned between collections. Otherwise, erroneous collection data can be included with collections. In many ways, erroneous collection data are worse than no collection data.

Birds also have fleas, but the fleas that attack birds usually are collected most readily by examining a nest soon after it has been abandoned. In every instance, the species of bird that made the nest should be noted on the data label. The nest can be teased apart or placed in a Berlese funnel or sifter to collect the adult fleas and immature stages. Because the immatures cannot be identified with certainty without associated adults, some of the larvae or pupae should be placed in jars with a quantity of the nest material and reared. For successful rearing, maintain the humidity as high as would normally be found in a nest. The burrows or nests of mammals also should yield many adult and immature fleas.

Siphonaptera adults and larvae may be killed and preserved satisfactorily in 70–80% alcohol. They should not be preserved in a formalin-based liquid fixative because these fixatives make the process of clearing specimens difficult. Permanent collections display adult fleas mounted on microscope slides. Be sure to include the name of the host on the slide label.

6.32 Hymenoptera [sawflies, ants, wasps, bees, Fig. 6.32]

The order name is derived from Greek (hymen = membrane; pteron = wing) and refers to the four membranous wings. The Hymenoptera are a cosmopolitan order with about 125,000 described species. The Hymenoptera are subdivided into two suborders, the Symphyta (sawflies) and Apocrita (bees, wasps, ants, parasitic Hymenoptera). The adult head is mobile, usually hypognathous but

Arthropod Collection and Indentification Techniques

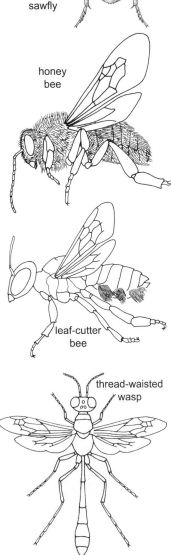

Figure 6.32a
Hymenoptera

sometimes prognathous; the head is not fused with the pronotum. The compound eyes typically are large and multifaceted; the eyes are reduced to a few ommatidia in some ants and parasitic species. Three ocelli usually are present, but these are absent from some species. Antennal segmentation and shape are variable, but the antenna typically is *geniculate* (elbowed). The antenna frequently shows sexual dimorphism, with the funicle sometimes *ramose* (branched) in the male and apical segments sometimes differentiated into a club in the female. Mandibles always are present in adult Hymenoptera, and the mouthparts usually are adapted for biting-chewing; some bees show adaptations for chewing-sucking. Most Hymenoptera have four membranous wings, with the forewing usually substantially larger than the hind wing; the forewing and hind wing are connected during flight by hooks called *hamuli*. The forewing venation is complex and extensive in large-bodied species; it is drastically reduced in small to minute parasitic species. The hind wing venation forms a few cells at most; otherwise the venation is reduced. Apterous species are distributed throughout the Hymenoptera. Reproductive ants develop wings but shed their wings after nuptial flight; the worker and soldier castes of ants are wingless. Among the Apocrita the first abdominal segment is incorporated into the thoracic region and called a *propodeum*. The propodeum and thorax collectively are called the *mesosoma* or alitrunk (in ants). The propodeum usually is separated from the remainder of the abdomen by a constriction of the second abdominal segment (called the *petiole*) or segments two and three (called the *postpetiole* in ants). The abdomen behind the petiole is called the *metasoma* or *gaster* (in ants). The abdomen of Hymenoptera has 10 segments in the ancestral condition; the first segment (propodeum) and second segment (petiole) typically lack the sternal elements. Female Hymenoptera display a sclerotized, elongate, tubular ovipositor (sawflies, parasitic Hymenoptera) or sting (*aculeata*); the cerci are one-segmented or modified into two setose patches (*pygostyli*). The male genitalia are complex and sometimes diagnostically important.

Hymenoptera larvae are extremely disparate in terms of their anatomy. The larvae of Symphyta are caterpillarlike, with thoracic legs and prolegs. The prolegs lack crochets, a character that may be used to separate these larvae from the larvae of Lepidoptera. The larvae of Apocrita are apodous or vermiform. Some species display hypermetamorphic development. That is, the first-instar larva does not resemble subsequent instars. The number of instars among Hymenoptera is variable. Some egg parasites may have only two instars; some sawflies may have seven instars. The pupae of Hymenoptera are adecticous and usually exarate, but some species display an obtect condition.

The biology of Hymenoptera is exceedingly diverse. Species are phytophagous, parasitic, predaceous, and gall-forming. Among the apocritous larvae, the gut is not connected to the anus until feeding has been completed. Polyembryonic development has been shown in some species, and eusociality has evolved several times among the bees and ants. Parthenogenesis is universal within the order; females possess the diploid ($2n$) complement of chromosomes and males possess the haploid (n) complement of chromosomes.

The Hymenoptera, as is the case with the Diptera, have beneficial and noxious species. Hymenoptera are highly beneficial in terms of crop pollination. Many bees are important pollinators of plants, and some are valued for the honey they produce. The parasitic species are of ever-increasing importance as biological control agents of agricultural pests. The plant-feeding forms include some of the most destructive defoliators of forest trees.

Methods of collecting Hymenoptera vary because their habits and habitats are so diverse. Sawfly adults, so-called because of their sawlike ovipositor, may be captured by interception traps such as the Malaise trap or swept from low vegetation. Most sawfly larvae are external feeders on vegetation, some species mine the leaves or leaf petioles, and a few species cause galls. Outbreaks of sawflies may occur sporadically in an area and offer collectors an excellent opportunity to obtain large samples. From such collections, some specimens can be preserved as immatures and others can be held in rearing cages until adults emerge. Procedures used to rear Lepidoptera (see Section 1.15) can be used to rear sawflies. Sawfly larvae may be preserved in alcohol, inflated, or freeze-dried. Many sawflies attach their cocoons to leaves or twigs on the host plant; larvae of some species drop to the ground and pupate in the soil or leaf litter. For this reason, 5–8 cm of moist soil or peat moss should be kept on the bottom of the rearing cages. The adults that emerge may be killed in alcohol but later should be removed and pinned, preferably with the wings extended.

A simple but novel method of collecting paper wasps is to tie a widemouth jar containing alcohol to a long pole. The pole is used to position the jar immediately beneath the nest opening. As wasps approach the opening, they become intoxicated and fall into the alcohol (Bentley 1992). Care must be taken not to touch or bump the nest.

The parasitic Hymenoptera represent a significant portion of the Hymenoptera, yet they are poorly understood. More than 50,000 species have been described, but several times that number remain undescribed. These insects are multitudinous and attack all stages of other insects. Many species of parasitic wasps can be found in Malaise traps. Smaller species associated with soil or duff can be recovered from Berlese samples. Some species are myrmecophilous and are found in ant nests. Some species have only been taken with a sweep net. Collectors can make a valuable contribution to entomology by

Figure 6.32b
Hymenoptera

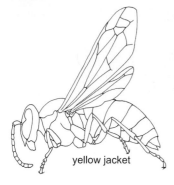

yellow jacket

ant

Figure 6.32c
Hymenoptera

correctly associating parasites with their hosts. This necessitates rearing adult parasitic Hymenoptera from their hosts. The value of accurate host associations cannot be overemphasized. Policy of the Biosystematics and Beneficial Insects Institute precludes accepting parasitic Hymenoptera for identification unless the host insect has been identified at least to genus. Rearing parasites from their hosts is also a method of associating the sexes with certainty, because males and females of many species differ greatly in appearance. The collection labels should include the scientific name of the host insect and the life stage of the host from which the parasite emerged. If the specimens are determined to be polyembryonic (more than one individual produced from one egg), this also should be noted on the label. When possible, the remains of the host should be retained and mounted below the parasite on the pin. Parasitic wasps too small for direct pinning should be glued to points or rectangular cards. Microhymenoptera should not be mounted on minutens.

Mounting parasitic Hymenoptera involves arranging the antennae, legs, and wings so that characters on the body needed for identification are not obscured. Care must be exercised when mounting large-bodied wasps such as ichneumonids. They possess long wings and antennae that, if pointed upward, may be broken when the pin is handled. Tiny specimens (under 2 mm), such as trichogrammatids, may be cleared and mounted on microscope slides, preserved in alcohol, or critical-point dried and mounted on cards. Taxonomists in the Systematic Entomology Laboratory prefer that specimens of the Ichneumonoidea be killed and preserved in 95% alcohol; other parasitic Hymenoptera may be preserved satisfactorily in 70–80% alcohol.

Gall wasps and the galls from which they emerge should be preserved together, because identification of many species is based on the gall. If galls are placed in a pinned collection, make sure that they are pinned securely, with additional brace pins on each side. Use care when labeling specimens associated with a particular gall. Not all insects that emerge are gall producers; some may be parasites of the gall wasps or inquilines.

Mound-building ants can be collected by pushing an alcohol-filled vial into the excavated center of the ant mound. The vial serves as a pitfall trap for the ants. If ants are not active outside the nest, a small trowel can be used to disturb the ants. Once the ants are active, avoid getting stung or bitten. Ant nests also may yield other kinds of insects besides ants; many of them mimic ants. In collecting ants from a nest, use an aspirator or Berlese funnel and try to obtain specimens of each caste. It may be necessary to return to a nest periodically through the year to find the males or winged females. Ants may be collected and preserved in alcohol; but for a permanent collection, larger specimens should be pinned, and specimens under 5 mm should be glued to points.

Part 2: Classification of Insects and Mites

Aculeate or stinging wasps and bees may be caught in Malaise traps or by sweeping. Mesh size and other features of Malaise trap design influence the kinds and numbers of Hymenoptera collected (Darling & Packer 1988). See Section 1.3.5 on how to remove stinging insects from a net. If specimens are taken from a nest, try to collect the nest. If this is not practical, make a note or sketch of the approximate measurements of the nest. Do not overlook the insects that the wasps may have stored in cells in the nest.

SUGGESTED READING
Insect Orders
General taxonomic:
- Essig 1942, 1958
- Jacques 1947
- Peterson 1948
- Chu 1949
- Matheson 1951
- Brues et al. 1954
- Usinger 1956
- Borror & White 1970
- Richards & Davies 1977
- Hollis 1980
- Parker 1982
- Ross et al. 1982
- Arnett 1985
- Scott 1986
- Borror et al. 1989

Arachnids:
- Rohlf (1957)
- Evans et al. (1964)
- Cooke (1969)

Ticks and mites:
- Lipovsky 1951, 1953
- Strandtmann & Wharton 1958
- Furumizo 1975
- Jeppson & Kiefer 1975
- Krantz 1978
- McDaniel 1979
- Evans 1992

Arthropod Collection and Indentification Techniques

Summary

Insects have been on earth for an estimated 280 million years, mites even longer. Considering both numbers of individuals and numbers of species, insects and mites represent a dominant form of life. Clearly, there are more insects and mites alive than any other group of metazoan animals. In terms of biological diversity, insects occupy more habitats and adopt more lifestyles than any other group of organisms; mites rank second. Insects represent humankind's principal competitors for food and fiber; mites often cause considerable damage to plants and food. Insects and mites both directly and significantly impact the health and welfare of humans and other animals. They are at once both beneficial as scavengers, biological control agents, and environmental indicators and detrimental as humans' most damaging pests. Considering their geological age, numbers, biological diversity, and economic importance, the study of insects and mites is compelling. This study can be personally rewarding through understanding the diversity of nature and professionally rewarding by solving the problems these organisms cause. In a philosophical sense, anyone who collects insects or mites as a pastime, studies their life histories, or uses keys to identify them is a student of entomology or acarology. Every student can make a contribution to the understanding of these fascinating and ubiquitous animals. We hope that this text will provide some of the tools necessary to facilitate such a study.

Glossary

GLOSSARY

– A –

ABDOMEN. Noun. (Latin, *abdomen* = belly). The third or posteriormost division of the insect body. Typically composed of 11 segments in the primitive condition, with a tendency toward reduction to 9–10 visible segments in higher orders. The abdomen of the adult stage lacks legs but contains the organs of reproduction, respiration, digestion, and excretion.

ACEPHALOUS. Adj. (Greek, *a* = without; *kephale* = head). Relating to a headless condition. Typically applied to the larval stage of some insects, including Diptera and higher Hymenoptera.

ACULEATE. Adj. (Latin, *aculeus* = prickle). 1. A member of the order Hymenoptera, the female members of which bear a sting. 2. A structure invested with spines or minute bristles under the scales of the wing (microlepidoptera).

ACUMINATE. Adj. (Latin, *acuminatus* = point). Tapering to a long point.

ADECTICOUS. Adj. (Greek, *a* = without; *dekticos* = biting). Insects in the pupal stage that lack functional mandibles capable of cutting through the pupal cocoon. Compare Decticous.

ADECTICOUS PUPA. A form of holometabolous development in which mandibles in the pupal stage are immobile and cannot be used for biting. Compare Decticous Pupa.

ADFRONTAL AREA. Lepidoptera larvae: An oblique sclerite laterad of the frons that extends from the base of the antenna to the epicranial suture. A median epicranial suture is formed by the union of oblique sclerites from each side of the head; an epicranial sulcus (notch) is formed when the sclerites do not fuse.

ADFRONTAL SETAE. Lepidoptera larvae: A pair of setae found on adfrontal areas.

ADFRONTAL SUTURE. Lepidoptera larvae: The suture formed by the medial fusion of adfrontal sclerites.

ADVENTITIOUS VEIN. A wing vein that is not part of the regular plan but appears in one species or a single individual; or the result of two or more true veins in line with one another.

Glossary

AEDEAGUS. Noun. (Greek, *aidoia* = genitals; late Latin. Pl., aedeagi). The penis or intromittent organ of copulation in male insects.

AESTIVATION. Noun. (Latin, *aestivus* = of summer). Dormancy in summer. Typically occurs during periods of continued high temperatures in temperate areas or during a dry season in tropical areas. Also compare Diapause.

ALLOMONE. Noun. (Greek, *allos* = other; *hormaein* = to excite). A chemical substance (produced or acquired by an organism) that, when contacting an individual of another species in the natural context, evokes in the receiver a behavioral or physiological reaction adaptively favorable to the emitter. Compare Kairomone, Pheromone.

ALULA. Noun. (Latin, *ala* = wing. Pl., alulae). In some Diptera and Coleoptera: the expanded axillary membrane of the wing. In Diptera a pair of membranous scales above the halteres. The outermost lobe at the base of the wing.

AMETABOLOUS. Adj. Having no metamorphosis, i.e., primitively wingless, emerging from the egg in a form essentially resembling the adult, apart from the small size and lack of developed genitalia.

AMPLEXIFORM COUPLING. Term applied to a type of wing-locking mechanism in Lepidoptera in which there is no frenulum and the large humeral lobe of the hind wing projects under the forewing.

ANAL. Adj. (Latin, *anus* = anus). In the direction of the anus. Pertaining to the last abdominal segment, which bears the anus. Pertaining to the wing area behind the cubitus and to the veins belonging to it. Compare Oral.

ANNULATE. Adj. (Latin, *annulus* = ring). 1. Ringed, but not demonstrating primary or secondary segmentation. 2. Bearing rings, or arranged in the form of an annulus or ring. 3. A narrow circle or ring of a color different from the adjacent region. Alternate: Annulated.

ANNULUS. Noun. (Latin, *annulus* = ring. Pl., annuli). 1. A ring of membrane, sclerite, or pigment surrounding a joint, segment, spot, or mark. 2. A ringlike marking or a ring of hard cuticle. 3. The sclerotized ring of the head into which the basal segment of the antenna is inserted. 4. An antennal sclerite. Alternate: Antennalis.

ANTENNA. Noun. (Latin, *antenna* = sailyard. Pl., antennae). The paired, mobile, segmented sensory appendages located on the head of most arthropods. Missing from Chelicerata; two pairs found in Crustacea; one pair in Insecta. Among hexapods, missing from Protura; present in Diplura and Collembola. In the Insecta, the antenna is the anteriormost appendicular structure of the postembryonic organism, above or before the mouth opening.

ANTERIOR. Adj. (Latin, *anterior* = former). Pertaining to structure at or near the head of the insect, or facing in that direction. Ant. Posterior.

ANUS. Noun. (Latin, *anus* = anus). 1. The opening at the posterior end of the digestive tract, through which the excrement is passed. 2. Coccidae: A more or less circular opening on the dorsal surface of the pygidium, in various locations, depending on the location of the circumgenital gland orifices.

APHIDOPHAGOUS. Adj. (Middle Latin, *aphidis* = a plant louse; Greek, *phagein* = to devour). Pertaining to or descriptive of insects that feed on aphids, typically as predators. Example: some Coccinellidae.

APHIS. (Acronym: Animal and Plant Health Inspection Service). The branch of the U.S. Department of Agriculture responsible for port inspection and the detection and eradication of unwanted organisms in the United States.

APICAD. Adv. (Latin, *apex* = summit; *ad* = toward). Toward the apex or away from the body. Ant. Basad.

APICAL. Adj. (Latin, *apex* = summit). At, near, or pertaining to the apex of any structure.

APOCRIA. Suborder of Hymenoptera in which segment I of the abdomen is fused to the thorax to form the propodeum in the adult and in which the larvae are apodous.

APODAL. Adj. (Greek, *a* = without; *pous* = feet). Pertaining to organisms without feet. Typical Diptera larvae. Alternate: Apodous.

APODEME. Noun. (Greek, *apo* = away; *demas* = body). Any rigid, integumental process that projects into the body to form the insect endoskeleton. Apodemes provide surface for the attachment of muscles. See also Phragma.

APODOUS. Adj. (Greek, *a* = without; *pous* = feet). Without feet; legless. See also Apodal.

APOLYSIS. Noun. In molting, the separation of the old cuticle from the underlying epidermal cells.

APPENDAGE. Noun. (Latin, *ad* = to; *pendere* = to hang). Any structure of the body that is attached to the head, thorax, or abdomen by a joint. Example: antennae, legs, wings.

APPENDIX. Noun. (Latin, *ad* = to; *pendere* = to hang). Any structural attachment of the body, particularly a short, stumplike vein.

APTEROUS. Adj. (Greek, *a* = without; *pteron* = wing). Pertaining to insects without wings; a wingless condition.

APTERYGOTA. Old name for primitively wingless and ametabolous insects.

APTERYGOTE. Adj. (Greek, *a* = without; *pteron* = wing). Belonging to the Apterygota, insects whose ancestors were wingless.

AQUATIC. Adj. (Latin, *aquaticus* = water). Organisms living in fresh water or salt water; saltwater forms are usually described as marine. Compare Terrestrial.

ARBOREAL. Adj. (Latin, *arbor* = tree). Descriptive of organisms living in, on, or among trees.

ARCUATE. Adj. (Latin, *arcuatus* = curved). Pertaining to a structure that is arched or bowlike. Alternate: Arcuatus.

AREOLE. Noun (Latin, *areola* = small space). Lepidoptera: The closed radial cell of the forewing between veins R_3 and R_4.

ARISTA. Noun. (Latin, *arista* = awn). A specialized bristle or process on the antenna; usually dorsal, rarely apical.

AROLIUM. Noun. (Greek, *arole* = protection. Pl., arolia). A cushionlike medial pad on the pretarsus of many insects. The typical median terminal lobe of the pretarsus, between the bases of the claws. Orthoptera: The terminal pad between the claws.

ARTICULATE. Adj. (Latin, *articulus* = joint). Jointed or segmented.

ASPIRATOR. Noun. (Latin, *aspirare* = to breathe toward). A handheld device for collecting small, terrestrial arthropods. Designs are variable, but typically constructed of a test tube or glass vial that is closed at one end and sealed with a cork or rubber stopper at the other end. Two small glass or metal tubes, of equal diameter, project through the stopper and penetrate the atrium of the larger vial. One small tube is covered with muslin or screen on the end inside the vial; the opposite end of the tube is placed in the mount and lung power is used to generate an air current that draws air and the insect into the tube via the aperture at the end of the other small tube. The device is popular for collecting small insects, particularly microhymenopera, some Hemiptera, and small Diptera. Syn. Pooter.

ATROPHY. Noun. (Greek, *a* = without; *trophe* = nourishment). A term applied usually to tissue, muscles, appendages, parts, or organs that seem diminished in size or lack growth, implicitly from insufficient nourishment or exercise. A structure that is reduced in size or withered.

ATTENUATE. Adj. (Latin, *attenuare* = to thin). Gradually tapering toward the apex or away from the body; appearing drawn out or slender. Alternate: Attenuated.

AUDITORY ORGANS. Organs by which an insect can hear sounds or perceive vibrations.

– B –

BASAD. Adv. (Latin, *basis* = base). Toward the base of a structure and away from the tip or apex. Ant. Apicad.

BASAL. Adj. (Latin, *basis* = base). Pertaining to the base or point of attachment to or nearest the body. Ant. Distal proximal.

BASITARSUS. Noun. (Greek, *basis* = base; *tarsos* = sole of foot. Pl., basitarsi). The proximal or basal segment of the tarsus.

BIFID. Adj. (Latin, *bis* = twice; *findere* = to split). Cleft or divided into two parts or lobes; forked.

BIFURCATE. Adj. (Latin, *bis* = twice; *furca* = fork). A forked structure or one that is partly divided into two parts.

BIRAMOSE. Adj. (Latin, *bis* = twice; *ramus* = branch). Pertaining to a structure that has two branches or doubled appendages.

Glossary

BLADE. Noun. (Anglo-Saxon, *blaed* = leaf). 1. Any thin, flat structure like a leaf, sword, or knife. 2. The lacinia of many insects, such as bees. 3. The large, flattened surface of the insect wing.

BRACHYPTEROUS. Adj. With short or abbreviated wings.

BUCCAL CAVITY. The mouth opening.

BULLA. Noun. (Latin, *bulla* = bubble. Pl., bullae). 1. A blister or blisterlike structure. 2. Ephemeridae: A stigmalike enlarged part of the costal area of the wing near the apex with many crossveins. The weak spots on some of the wing veins where they are crossed by furrows. 3. A shieldlike sclerite closing the tracheal aperture in lamellicorn beetles.

– C –

CALLOW. Noun. (Anglo-Saxon, *calu* = bald). 1. The condition immediately following molting when the integument is colored as when hardened. 2. A term sometimes applied to newly emerged (neonate) worker ants. Compare Teneral.

CALLUS. Noun. (Latin, *callum* = hard skin. Pl., calli). 1. A hard lump or moundlike, rounded swelling of the integument, such as a swelling at the base of the wing articulating with the thorax. 2. Heteroptera: The thickened or raised spots on the thorax, especially of Pentatomidae.

CAMERA LUCIDA. An image-deflecting device attached to the eyepiece of a microscope that projects the microscopic image to a mirror and then onto a piece of paper for drawing.

CAMPODEIFORM. Adj. (Greek, *kampe* = caterpillar; *eidos* = form; Latin, *forma* = shape). A term used to describe elongate larval forms that are characterized by well-developed thoracic legs and a prognathous head. The term calls to mind stages that resemble the genus *Campodea* (Thysanura).

CANNIBALISTIC. Adj. (Spanish, *canibales*, from *caribes* = brave and daring men). Pertaining to insects that feed on members of the same species.

CAPITATE. Adj. (Latin, *caput* = head). Structure with a terminal knoblike enlargement or with a head. Reference to a type of antenna in which the club is abruptly enlarged at the apex and forms a spherical mass.

CARINA. Noun. (Latin, *carina* = keel. Pl., carinae). An elevated cuticular ridge that is not necessarily high or acute. Syn. Keel.

Arthropod Collection and Indentification Techniques

CARNIVOROUS. Adj. (Latin, *caro* = flesh; *vorare* = to devour). Feeding upon the flesh of other animals. Compare Phytophagous. See also Cannibalistic.

CATERPILLAR. Noun. (Late Latin, *cattus* = cat; *pilosus* = hairy). 1. Lepidoptera: The polypod or eruciform larva. 2. Mecoptera, Symphytous Hymenoptera: The larval stage with abdominal prolegs.

CAUDAD. Adv. (Latin, *cauda* = tail). Toward the tail end, posterior end, or anus of the insect body. Compare Cephalad.

CAUDATE. Adj. (Latin, *cauda* = tail). 1. With tail-like extensions or processes. 2. Hymenoptera (Apocrita): Specialized body form of some endoparasitic ichenumonid larvae, characteristically segmented, with long, flexible, caudal appendages. Function of caudal appendages not established, but sometimes progressively reduced in later instars and lost in the last instar.

CELL. Noun. (Latin, *cella* = compartment). 1. Any closed area in an insect wing bounded by veins. In the conventional system, cells derive their names from the vein forming the upper margin (for example, all cells just below the radius are called radial cells); cells are numbered from the base outward. 2. In a nest or honeycomb: The small chambers in which eggs are placed and the larvae develop.

CEPHALAD. Adv. (Greek, *kephale* = head). Toward the anterior end or head of the body. Compare Caudad.

CEPHALIC. Adj. (Greek, *kephale* = head). Belonging to or attached to the head; directed toward the head.

CEPHALOTHORAX. Noun. (Greek, *kephale* = head; *thorax* = chest). 1. The anterior body tagma of Arachnida and Crustacea; analagous, perhaps homologous, with insect head and thorax.

CERCUS. Noun. (Greek, *kerkos* = tail. Pl., cerci). An appendage (generally paired) of the tenth abdominal segment. Usually regarded as sensory and characteristically slender, filamentous, and segmented. Not segmented in Orthoptera, Phasmatodea, and Dermaptera. Incorrectly applied to appendicular structures on the ninth segment of Coleoptera larvae or eighth and ninth segments of Hymenoptera. See also Urogomphus, Pygostylus.

CHAETA. Noun. (Greek, *chaite* = long hair. Pl., chaetae). A cuticular projection from the integument of insects, typically hairlike or a bristle. The form, number, and arrangement of chaetae are sometimes considered taxonomically useful. Functionally diverse and serving as sensory receptors and components of stridulatory devices.

Glossary

CHAIN OF CUSTODY. In forensic entomology, a process used to maintain and document the chronological history of the evidence. (Documents should include name or initials of the individual collecting the evidence, each person on entity subsequently having custody of it, date the item was collected or transferred, agency and case number, victim's or suspect's name, and a brief description of the item.

CHARACTER. Noun. (Greek, *charassein* = to engrave). A general term used in taxonomic entomology to describe features of insects or insect parts.

CHELICERA. Noun. (Greek, *chele* = claw; *keras* = horn. Pl., chelicerae). 1. The pincerlike first pair of appendages of adult Chelicerata. Structures regarded as homologous with the second antennae of Crustacea. 2. Pinching or grasping claws of Phymatidae in Heteroptera.

CHEMORECEPTOR. Noun. (Greek, *chemeia* = transmutation; Latin, *recipere* = to receive). A sense organ having a group of cells sensitive to chemical properties of matter.

CHEWING MOUTHPARTS. Appendages of the head that are opposable in operation and adapted for the mastication of particulate matter or matrixlike material. Compare Piercing-sucking mouthparts.

CHITIN. Noun. (Greek, *chiton* = tunic). 1. A colorless, nitrogenous polysaccharide, linearly arranged as beta-linked *N*-acetyl-glucosamine units. Widespread in arthropods and plants. First isolated from fungi by Braconot in 1811 and called fungine; first isolated from insects (beetle elytra) in 1823 by Odier. A chemical compound intermediate between proteins and carbohydrates; the chemical formula is $(C_{32}H_{54}N_4O_{21})_x$.

CHLOROFORM. Noun. (CHC_{13}). A clear, colorless, highly flammable, and volatile liquid with an etherlike odor and a burning, sweet taste. Miscible with alcohol, ether, benzene, and oils. Used as a killing agent but stiffens specimens; forms phosgene (carbonyl chloride, COC_{12}, a severe respiratory irritant when stored in dark-colored glass jars.

CHORDA. Noun. (Greek, *chorde* = string). In the lepidopterous wing, the stem of veins R_{4+5}; the part of veins R_{4+5} separating the areole from the basal cell in Lepidoptera.

CHORDOTONAL ORGAN. A scolopophorous sense organ found between the tibia and basitarsus of some Coleoptera, Diptera, and Hemiptera. Used for the detection of substrate vibrations that enable an insect to hear sounds. The cellular elements form an elongate structure attached at both ends to the body wall but not necessarily containing sense rods (scolops). See also Johnston's Organ, Typmanal Organ.

CHORION. Noun. (Greek, *chorion* = skin). The outer shell or covering of the insect egg.

CLAVATE. Adj. (Latin, *clava* = club). Clublike or becoming thicker toward the apex.

CLAVUS. Noun. (Latin, *clavus* = nail). 1. The club of an antenna. 2. The knob or enlargement at the apex of the stigmal or radial veins of the Hymenoptera. 3. The sharply pointed anal area of the hemelytra in Heteroptera, next to the scutellum when folded. 4. A rounded or fingerlike process in the genitalia of male Lepidoptera.

CLAW. Noun. (Anglo-Saxon, *clawu* = claw). A hollow, sharp multicellular organ; generally paired and at the apex of the insect leg.

CLUB. Noun. (Old Norse, *klubbe* = club). Insect antenna: Apical segments of the flagellum that are enlarged or thickened such that they are morphologically differentiated.

CLYPEUS. Noun. (Latin, *clypeus* = shield). The anterior sclerite of the insect head below the frons and above the labrum. Highly variable in size and shape. Separated from the frons by the frontoclypeal suture and separated from the labrum by the clypeolabral suture or membrane. Provides area for attachment of cibarial dilator muscles. In some Diptera, divided into an anteclypeus and postclypeus.

COARCTATE. Adj. (Latin, *coarctare* = to press together). 1. Contracted or compacted. A term applied to a pupal form in which the future adult is concealed by a thickened, usually cylindrical case or covering (often the hardened skin of the larva). 2. Compressed such that the abdomen and thorax are forced together. 3. Meloid beetles: The third phase of hypermetamorphic development in which the equivalent of the sixth larval instar is heavily sclerotized, develops rudimentary appendages, but is immobile. Comopare Obtect.

COARCTATE PUPA. Diptera: A pupa that remains enclosed in the old larval integument.

COCOON. Noun. (French, *cocon* = cocoon). A covering, composed of silk or other viscid fibers, spun or constructed by the larval stage. In some Lepidoptera: A casing of earth, wood fragments, or other debris. Principally serves as protective device for the pupa.

COLLAR. Noun. (Middle English, *coler* = collar). 1. General: Any structure between the head and thorax. 2. Hymenoptera: The neck. 3. Diptera: May mean the neck, the sclerites attached to the prothorax, the prothorax, or its processes (antefurca). 4. Coleoptera: The narrow, constricted anterior part of the pronotum, generally set off by a groove.

Glossary

COMB. Noun. (Anglo-Saxon, *comb* = comb). 1. Hymenoptera: Brood cells constructed of wax by bees and used to house individual larvae and to store honey. 2. Termite nests: A spongy, dark reddish-brown material made by the workers from excreta from which they construct fungal beds. 3. Lepidoptera (Lycaenidae): Serrate distal margin of rostellum of valva. 4. A row or rows of close-set, short bristles on the distal end of some leg segments.

COMMENSAL. Noun. A species that benefits from commensalisms.

COMMENSALISM. Noun. Symbiosis in which members of one species are benefited while those of other species are neither benefited nor harmed.

COMPLETE METAMORPHOSIS. Metamorphosis in the Holometabola that has four stages: egg, larva, pupa, and adult. Each stage is entirely different from the others.

COMPOUND EYE. Paired, lateral aggregation of separate visual elements (ommatidia) located on the head. See also Ocellus.

CONTIGUOUS. Adj. (Latin, *contiguus*, from *contingere* = to touch on all sides). Pertaining to adjacent structures; structures with points, margins, or surfaces in contact but not united or fused.

CORIUM. Noun. (Latin, *corium* = leather). 1. Heteroptera: The elongate middle section of the hemelytra that extends from the base to the membrane below the embolium. 2. The membrane of the joints in segmented appendages.

CORNEOUS. Adj. (Latin, *corneus* = horny). Of a horny or chitinous substance; resembling horn in texture.

CORNICLE. Noun. (Latin, *cornu* = a horn). Paired, secretory structures on the abdomen of aphids. Typically located on the posterior margin of the fifth segment; less commonly located on the sixth segment. Variable in shape. Absent or ringlike in Pemphigidae; elongate and cylinderlike in most families. Surface ranging from smooth to sculptured; apex with membranous lid and sclerotized plate that serves as a valve manipulated by abdominal musculature. Cornicles produce defensive wax secretions, alarm pheromones, but not honeydew.

COSTA. Noun. (Latin, *costa* = rib). 1. An elevated ridge that is rounded at its crest. 2. The thickened anterior margin of any wing, but usually of the forewings. 3. The vein extending along the anterior margin of the wing from the base to the junction with the subcostal vein.

COXA. Noun. (Latin, *coxa* = hip. Pl., coxae). The basal segment of the insect leg. The coxa articulates with the pleural wall of the thorax and is attached to the trochanter. Coxae are paired, ventrolateral in position, and found on each of the thoracic body segments.

COXITE. Noun. (Latin, *coxa* = hip. Pl., coxites). 1. The basal segment of any leglike appendage such as found on the abdominal sternites. 2. A rudimentary abdominal limb in Thysanura displayed in the form of paired lateral plates.

CREMASTER. Noun. (Greek, *kremastos* = hung). 1. The apex of the last segment of the abdomen. 2. The terminal spine or hooked process of the abdomen of subterranean pupae. Used to facilitate emergence from the earth. 3. An anal hook by which some pupae are suspended.

CREPUSCULAR. Adj. (Latin, *crepusculum* = dusk). Pertaining to animals active or flying at dusk or (more rarely) at dawn. Compare Diurnal, Nocturnal.

CROCHET. Noun. (French, *crochet* = small hook). Curved spines or hooks on the prolegs of caterpillars and on the cremaster of pupae.

CROP. Noun. (Middle English, *croppe* = craw). The dilated portion of the alimentary canal behind the gullet. The crop receives and holds food before its passage through the digestive tract.

CROSSVEIN. Noun. Typically short veins between the lengthwise veins and their branches; numerous crossveins exist in net-veined wings. Veins that normally extend more or less crossways in the insect wing.

CRYPTIC. Adj. (Greek, *kryptos* = hidden). Pertaining to a hidden or concealed condition, particularly in relation to protective camouflage.

CTENIDIUM. Noun. (Greek, *ktenos* = comb. Pl., ctenidia). A comblike row of short spines (bristles) most evident on parasitic insects, including Siphonaptera. Found all over the body of Polyctendidae; found only on the prothorax and sometimes the metathorax of Siphonaptera. Aid in retention and movement through host fur.

CUNEUS. Noun. (Latin, *cuneus* = wedge). Heteroptera: The small triangular area at the end of the embolium of the hemelytra. Odonata: The small triangle of the vertex between the compound eyes.

CURSORIAL. Adj. (Latin, *cursor* = a runner). 1. Adapted to running habits. 2. Pertaining to legs that are long and tapered.

Glossary

– D –

DEBRIS. Noun. (Old French, *debruisier* = to break). The rubbish or remains of anything that has decomposed or physically broken down with time. Compare Detritus, Duff.

DECTICOUS. Adj. (Greek, *dektikos* = biting). Insects in the pupal stage that have functional mandibles capable of being used to free the insect from the pupal cocoon. Compare Adecticous.

DECTICOUS PUPA. Exarate pupa in which the mandibles are modified to help the insect to escape its cell or cocoon. Compare Adeticous Pupa.

DESICCATION. Noun. (Latin, *desiccatus* from *desiccare* = to dry up). Excessive drying by natural loss of moisture or artificial means.

DETRITIVORE. Noun. (Latin, *detritus* = rubbed or worn away; *vorare* = to devour). An organism that feeds upon detritus.

DETRITIVOROUS. Adj. Pertaining to an organism that feeds on detritus. Feeding upon fur and feather detritus.

DETRITUS. Noun. (Latin, *detritus* = rubbed or worn away). Material that remains after disintegration; rubbing away or the destruction of structure; fragmented material. Any disintegrated or broken matter. Compare Debris, Duff.

DIAPAUSE. Noun. (Greek, *dia* = through; *pausis* = suspending; *diapauein* = to cause to cease). A condition of restrained development and reduced metabolic activity that cannot be directly attributed to unfavorable environmental conditions. Regarded by entomologists to involve a resting period of an insect, especially of larvae in winter. Coompare Hibernation, Quiescence.

DIGIT. Noun. (Latin, *digitus* = finger). 1. Insects: Any fingerlike structure. 2. Chelicerates: Distal portion of a chela or chelicera. 3. Immature Diptera: Prothoracic spiracles. Alternate Digitus.

DIMORPHISM. Noun. (Greek, *dis* = twice; *morphe* = shape). A genetically controlled, non-pathological condition in which individuals of a species are characterized by distinctive or discrete patterns of coloration, size, or shape. Dimorphism can be a seasonal, sexual, or geographic manifestation. See also Polymorphism.

DISCAL. Adj. (Latin, *discus* = disk). On the disk or main surface of any part of the body. Ant. Marginal.

DISCAL CELL. Any large cell in the central part (disk) of the wing; usually the first M_2 in Diptera and cell R in Lepidoptera.

DISK. Noun. (Latin, *discus* = disk). 1. The upper, central surface of any anatomical structure or part. 2. The central area of a wing. 3. Orthoptera: The obliquely ridged outer surface of the hind femur in Saltatoria. Alternate Disc.

DISTAL. Adj. (Latin, *distare* = to stand apart). 1. Near or toward the free end of any appendage. 2. The part of a structure farthest from the body. Ant. Basal, Proximal.

DIURNAL. Adj. (Latin, *diurnae* = pertaining to day). Pertaining to activity patterns during daylight only. Compare Crepuscular, Nocturnal,.

DORSAD. Adv. (Latin, *dorsum* = back; *ad* = toward). In the direction of the dorsum or back of an insect. Ant. Ventrad.

DORSAL. Adj. (Latin, *dorsum* = back). Of or belonging to the upper surface. Ant. Ventral.

DORSUM. Noun. (Latin, *dorsum* = back). The dorsal surface or plate. Ant. Ventrum.

DRONE. Noun. (Anglo-Saxon, *dran* = drone). Hymenoptera: A male bee or ant.

DUFF. Noun. (Middle English, *dogh* = dough). The partly decayed vegetable matter on the forest floor. Compare Debris, Detritus.

– E –

ECDYSIS. Noun. (Greek, *ek* = out; *dyein* = to enter; *ekdysai* = to strip). The process of shedding the integument during molting.

ECTOPARASITE. Noun. (Greek, *ektos* = outside; *parasitos* = parasite). A parasitic animal that lives on the external parts of its host. Examples: fleas and lice. Syn. Ectoparasitoid. Compare Endoparasite.

ELATERIFORM LARVA. 1. Any larva that resembles an elaterid (wireworm) larva in form. 2. A slender larva, moderately heavily sclerotized, lacking elaborate ornamentation or sculpture and bearing three pairs of thoracic legs.

Glossary

ELYTRON. Noun. (Greek, *elytron* = sheath, from *elyein* = to roll around. Pl., elytra). 1. Coleoptera: Anterior leathery or chitinous wings that cover the hind wings. In repose, elytra typically meet and form a straight line along the middle of dorsum. 2. Orthoptera: The tegmina. 3. Heteroptera: The hemelytra.

EMBOLIUM. Noun. (Greek, *embolos* = wedge). The differentiated costal part of the corium in the forewing or hemelytron of some Heteroptera.

EMPODIUM. Noun. (Greek, *en* = in; *pous* = feet. Pl., empodia). 1. A bristle between the pulvilli of the foot in some Diptera. 2. The single, padlike median structure sometimes present between the insect claws.

ENDOPARASITE. Noun. (Greek, *endon* = within; *parasitos* = parasite). Any organism that develops as a parasite within the body of another organism at the expense and to the detriment of the "host." Examples: many Hymenoptera and some Diptera. Syn. Endoparasitoid. Compare Ectoparasite.

ENDOPTERYGOTA. Noun. (Greek, *endon* = within; *pterygion* = little wing). Insects with complete, complex metamorphosis in which the wings develop internally. Syn. Holometabola; Oligoneoptera; Oligoneuroptera. Compare Exopterygota.

EPISTOMAL SUTURE. (Greek, *epi* = upon; *stoma* = mouth). 1. A suture across the face and separating the frons from the clypeus. The suture forms a strong internal ridge (sulcus). Typically transverse and straight, but arched in some groups of insects and absent in other groups. Syn. Frontoclypeal Suture.

ERUCIFORM LARVA. (Latin, *eruca* = caterpillar; *forma* = shape). Shaped like a caterpillar. Characterized by a well-developed head capsule, thoracic legs, and abdominal prolegs. Seen in Lepidoptera and Symphyta.

ESOPHAGUS. Noun. (Greek, *Oisophagos* = gullet). The part of the foregut between the mouth and the crop. Alternate: Oesophagus.

EXARATE PUPA. A form of pupal development in which body appendages (legs and wings) are free from the body. Characteristic of the lower Endopterygota but not restricted to that group. Syn. Free Pupa. Compare Obtect Pupa.

Arthropod Collection and Indentification Techniques

EXOPTERYGOTA. Noun. (Greek, *exo* = outside; *pterygion* = little wing). The Heterometabola (Hemimetabola). A division of the Pterygota in which postembryonic immature insects pass through a simple (sometimes slight) metamorphosis and wing pads develop externally on the body. A pupal stage found in only a few forms (Thysanoptera). Compare Endopterygota. See also Paurometabola.

EXOPTERYGOTE. Adj. (Greek, *exo* = outside; *pterygion* = little wing). Pertaining to insects whose wings develop externally in all stages; insect with incomplete metamorphosis. Compare Endopterygote.

EXOSKELETON. Noun. (Greek, *exo* = outside; *skeletos* = hard). The entire body wall, to the inner side of which muscles are attached; the outside skeleton in insects.

EXUVIAE. Plural noun. (Latin, *exuere* = to strip off). 1. General: The integument of a larva or nymph that has been cast from the body during molting. 2. Diaspinae: The larval integument when cast from the body and incorporated into the scale cover. Incorrectly given as "exuvium" when a singular form of the noun is intended. Alternate Exuvia.

– F –

FACE. Noun. (Latin, *facies* = form, shape, face). 1. Upper or outer surface of any part or appendage. 2. Anterior aspect of the head between compound eyes from mouth to vertex. Usually applied to insects in which the head is vertical. 3. Ephemeroptera: A fusion of front and vertex. 4. Hymenoptera: Generally, area between antennae and clypeus; in bees face extends between the eyes to base of antennae. 5. Diptera: Area between base of antennae, oral margin, eyes, and cheeks.

FACET. Noun. (French, *facette* = small face). Any small face or surface, such as the parts, areas, or lenslike divisions of the compound eye. See also Ommatidium.

FAMILY. Noun. (Latin, *familia* = household). In zoological classification, a level in the taxonomic hierarchy below the order and above the genus. All zoological family names end in "idae;" all zoological subfamily names end in "inae." The family-level taxon must include a type genus that contains a type species.

FEMUR. Noun. (Latin, *femur* = thigh. Pl., femora). The third and usually the stoutest segment of the insect leg. Articulated with the body via the trochanter and bearing the tibia at its distal margin.

Glossary

FIXATION. Noun. (Latin, *fixus* = fixed). A histological procedure intended to: (1) terminate life process quickly with minimum of distortion to cytological detail; (2) prevent autolysis; (3) prevent microbial action; (4) increase the refractive index of tissue. See also Histology.

FLAGELLUM. Noun. (Latin, *flagellum* = whip. Pl., flagella). 1. Part of the antenna beyond the pedicel (second segment). 2. Any whip or whiplike process. 3. The tail-like process of a spermatozoon.

FORAMEN MAGNUM. The opening on the posterior surface of the head that gives passage to structures that extend from the head into the thorax. Syn. Occipital Foramen.

FORCEPS. Noun. (Latin, *forceps* = tongs). 1. Hook or pincerlike processes at the apex of the abdomen of many insects. Depending upon the insect species, forceps are used in copulation, defense, predation, or sensory reception. 2. An instrument for grasping or holding delicate objects, similar to tweezers.

FOREGUT. Noun. One of the three principal divisions of the alimentary canal. Originates at the mouth (buccal cavity) and terminates at the so-called gizzard. The foregut epithelium is formed from ectodermal invagination and is lined with chitin. Compare Hindgut, Midgut.

FOSSORIAL. Adj. (Latin, *fossor* = digger). Structures modified for digging or burrowing.

FRASS. Noun. (German, *fressen* = to devour). Solid larval excrement or macerated plant material fashioned by wood-boring insects that is often combined with excrement. Compare Meconium.

FREE PUPA. See Exarate Pupa.

FRENULUM. Noun. (Latin, *frenulum*, diminutive of *frenum* = bridle. Pl., frenula). The spine (simple in males, compound in females) at the base of the hind wings in many Lepidoptera. The frenulum projects beneath the forewing to unite the wings in flight.

FRONS. Noun. (Latin, *frons* = forehead). The sclerite of the head between the arms of the epicranial suture. Typically, the frons bears the median ocellus.

FRONTOCLYPEAL SUTURE. See Epistomal Suture.

FUNICLE. Noun. (Latin, *funiculus* = small cord). 1. Antennal segments between the ring segments (anelli) or pedicel and the club (clavus). 2. A small cord; a slender stalk.

FURCA. Noun. (Latin, *furca* = fork. Pl., furcae). 1. A forklike process. 2. The anal appendage used by Collembola for leaping. 3. The forked processes of the sternum of higher insects; an endosternite. 4. Genitalia of male Lepidoptera: A structure often consisting of paired halves.

– G –

GALEA. Noun. (Latin, *galea* = helmet. Pl., galeae). In the generalized maxilla: The lateral sclerite attached to the distal margin of the stipes. Sometimes appearing two-segmented and often hoodlike. Subject to great modifications in Hymenoptera and Diptera; forms the proboscis (coiled tongue) in Lepidoptera. See also Maxilla.

GALL. Noun. (Anglo-Saxon, *gaella* = gall). An abnormal growth of plant tissue. Caused by stimuli external to the plant, generally by insects, sometimes by parasitic fungi or other diseases of the plant.

GALL WASP. 1. A member of the hymenopterous genus Cynips, which causes gall formation in plants. 2. In general, any hymenopterous insect that induces tumors on a plant.

GASTER. Noun. (Greek, *gaster* = stomach; Late Latin, belly). Apocritous Hymenoptera: The posterior seven to eight segments of the abdomen, behind the constricted second segment. In Symphyta the abdomen is broadly attached to the abdomen; in Parasitic Hymenoptera and Aculeata the first abdominal segment has become separated from the remainder of the abdomen by a sclerotized annular constriction. The first (anteriormost) segment is called the *propodeum*; the second segment, which forms the constriction, is called the *petiole*; the remaining segments of the abdomen are collectively called the *gaster*.

GENA. Noun. (Latin, *gena* = cheek. Pl., genae). The "cheek" or sclerotized area of the head below the compound eye extending to the gular suture. Odonata: The area between the compound eyes, clypeus, and mouthparts. Diptera: The space between the lower border of the compound eye and the oral margin, merging into the face at the front and limited by the occipital margin behind.

GENERIC NAME. The name of a genus. Always a single word (simple or compound), written with a capital initial letter and employed as a noun in the nominative singular.

GENICULATE. Noun. (Latin, *geniculum* = little knee). Elbowed or sharply bent. A term used to characterize the antennae of ants and some other insects.

GENITALIA. Plural Noun. (Latin, *gegnere* = to beget). The internal and external sexual organs of insects.

GILL. Noun. (Middle English, *gile*, of Scandanavian origin). A respiratory organ in the aquatic immature stages of many insects. Usually hollow, thin-walled, lamellar, or filamentous projections from various regions of the body. Gills permit oxygen and other gasses in solution to pass from the water into the body. Compare Trachea.

GLOSSA. Noun. (Greek, *glossa* = tongue. Pl., glossae). 1. The median lobe of the maxillae formed by the fusion of the two paraglossae and the two glossae in bees. 2. The paired inner lobes of gnathobases of the labium or fused second maxilla. 3. Loosely used as a synonym for the tongue, especially applied to the coiled structure of the Lepidoptera. See also Ligula.

GREGARIOUS. Adj. (Latin, *grex* = flock). 1. Nonsocial insects that live in societies or communities. 2. Several parasitic insects of one species that develop simultaneously on or in one host. Compare Solitary.

GRESSORIAL. Adj. Having legs fitted for walking.

GRUB. Noun. Scarabaeiform larva; an apodous larva having a tiny head, few sense organs, and a fleshy, rounded body, e.g., larve of bees (Hymenoptera: Apoidea) and some Coleoptera.

GYNADROMORPH. Noun. (Greek, *gyne* = female; *aner* = male; *morphe* = form). An insect that has some male features and some female features. A sexual mosaic or intersex.

GYNADROMORPHISM. Noun. (Greek, *gyne* = female; *aner* = male; *morphe* = form). The expression or development of secondary sexual characters or features of both sexes in one individual.

– H –

HABITAT. Noun. (Latin, *habitare* = to inhabit). 1. Ecology: The area within which an organism is found, but not a particular location. The natural region that an organism inhabits or where it was found or taken.

HAIR. Noun. (Anglo-Saxon, *haer* = hair). 1. An epidermal outgrowth unique to mammals. 2. A term applied in insects to any slender, flexible filament largest at the base and often tapering toward the apex. See also Seta.

HALLER'S ORGAN. A sensory organ on tarsus 1 of ticks (acarines).

HALTERES. Plural noun. (Greek, *halter* = weight). Modified hind wings of Diptera that are club-like and consist of a base (scabellum), stem (pedicel), and knob (capitulum). Halteres oscillate during flight and serve as balancing organs to maintain stability during flight. Alternate: Balancing organs.

HAMULUS. Noun. (Latin, *hamulus* = little hook. Pl., hamuli). 1. Any small hook or hooklike process. 2. Orthoptera (treecrickets): Hooklike processes of the genitalia. 3. Siphonaptera: Movable sclerites from the lateral walls of the aedeagal palliolum. 4. Hymenoptera: a. The small hooklike structures along the anterior margin of the hind wing that link with the posterior margin of the forewing during flight. b. Penis valves (gonocoxites + gonostyli + volsellae).

HAUSTELLATE. Adj. (Latin, *haurire* = to drain). Pertaining to mouthparts modified for sucking. A term applied frequently to mouthparts of Homoptera-Hemiptera and Lepidoptera. See also Suctorial.

HEAD. Noun. (Anglo-Saxon, *heafod* = head). The first or anteriormost tagma of the insect body. The head bears the mouth structures and antennae and articulates with the thorax. Syn. Cranium. Compare Abdomen, Thorax.

HEMIMETABOLOUS. Adj. (Greek, *hemi* = half; *metabole* = change). Insects with an incomplete metamorphosis. The individual develops with gradual changes in size and shape from a first-instar nymph to an adult. Seen in exopterygote insects. Commpare Holometabolous, Paurometabolous.

HEMOLYMPH. Noun. (Greek, *haima* = blood; *lympha* = water). The watery bloodlike or lymphlike nutritive fluid of the lower invertebrates. Insect blood.

HERBIVOROUS. Adj. (Latin, *herba* = green crop; *vorare* = to devour). 1. Pertaining to organisms that feed upon plant tissue. 2. Leaf feeding. 3. Feeding on growing plants.

HIBERNATION. Noun. (Latin, *hibernus* = wintry). A period of suspended development in animals occurring during seasonal low temperatures and short day length. 2. Inactivity during the winter. Compare Diapause.

HINDGUT. Noun. The portion of the intestine extending from the midgut valve to the anus, including the Malpighian tubules and anal glands. Divided into several anatomical regions: ileum, colon, rectum, anus. Syn. Proctodaeum. Compare Foregut, Midgut.

HISTOLOGY. Noun. (Greek, *histos* = tissue; *logos* = discourse). The study of the tissues of organisms at the cellular and subcellular levels.

Glossary

HOLOMETABOLOUS. Adj. (Greek, *holos* = entire; *metabole* = change). Insects with a complete transformation during metamorphosis. Egg, larval, pupal, and adult stages are distinctly separated by profoundly different morphs at each stage. Compare Hemimetabolous, Paurometabolous.

HOLOTYPE. Noun. (Greek, *holos* = entire; *typos* = pattern). The single specimen selected by the author of a species as its type or the only specimen known at the time of description. The name bearer of a species. The type.

HOMONYM. Noun. (Greek, *homos* = same; *onyma* = name). 1. The same name for two or more different zoological entities. 2. One of two or more names applied to different taxa of animals. Rules for the treatment of scientific-name homonyms are given in the International Code of Zoological Nomenclature. See also Synonym.

HONEYDEW. Noun. 1. A sweetish excretion produced by insects such as aphids, whiteflies, and scale insects. 2. An exudate from the surface of some galls.

HOST. Noun. (Latin, *hospes* = stranger, guest, host). An organism that supplies nutrition or protection essential for the development of another organism (termed a parasite).

HYALINE. Adj. (Greek, *hyalos* = glass). Clear, transparent, or partly so; glasslike, waterlike in color. Alternate: Hyalinus.

HYPANDRIUM. Noun. (Greek, *hypo* = beneath; *aner* = male). 1. The apical abdominal sternal sclerite of many insects (sternum X of mayflies, sternum IX of Psocoptera). 2. The hypoproct sensu Needham; the subgenital plate sensu Snodgrass; the ninth abdominal sternum of male insects sensu Crampton.

HYPERMETAMORPHOSIS. Noun. (Greek, *hyper* = above; *meta* = after; *morphosis* = shaping). 1. The process in which endopterygote insects as larvae change shape or substance during successive instars as a normal consequence of development. Examples are found in (but not restricted to) Coleoptera (Meloidae), Strepsiptera, Diptera (Acroceridae, Bombyliidae), Lepidoptera (Epipyropidae), and Hymenoptera (Eucharitidae, Perilampidae). 2. The process in which insects pass through a larger-than-expected number of stages, with supernumerary stages interpolated between the mature larva and the adult (archaic).

HYPERPARASITE. Noun. (Greek, *hyper* = above; *para* = beside; *sitos* = food). An organism that develops as a parasite upon another parasite. Syn. Secondary Parasite. Compare Tertiary Hyperparasitism.

HYPOGNATHOUS. Adj. (Greek, *hypo* = under; *gnathos* = jaw). In entomological usage, pertaining to insects with the head vertically oriented and the mouth directed downward. Compare Opisthognathous, Orthognathous, Prognathous.

HYPOPHARYNX. Noun. (Greek, *hypo* = under; *pharyngx* = pharynx). 1. A tonguelike sensory structure projecting from the oral cavity. 2. The upper surface of the labium, which serves as an organ of taste.

HYPOPYGIUM. Noun. (Greek, *hypo* = under; *pyge* = rump. Pl., hypopygia). 1. Adults: The posterior portion of the abdomen; the anus; the ventral sclerite of the anal opening. 2. Diptera: The male sexual organs and terminal segments of the abdomen; propygium. 3. Coleoptera: The last segment behind the elytra.

HYPOSTOMA. Noun. (Greek, *hypo* = under; *stoma* = mouth). 1. Diptera: Part of the head between the antennae, compound eyes, and mouth. 2. Hemiptera: The lower part of the face. 3. Ticks: A dartlike structure arising from the median ventral surface of the basis capituli. 4. The upper lip or labrum of the Crustacea. Alternate: Hypostome.

– I –

IMAGO. Noun. (Latin, *imago* = image. Pl., imagines, imagos). The adult-stage or sexually mature insect.

INQUILINE. Noun. (Latin, *inquilinius* = tenant). 1. An insect that lives as a "guest" of other insects. Typically, an insect that habitually lives within the nests of other species, chiefly bees, ants, wasps, or termites. 2. An insect that lives within the gall developed by another species of insect. Compare Commensal, Parasite.

INSTAR. Noun. (Latin, *instar* = form). A postembryonic immature insect between molts. In the larva, instars are numbered to designate the various periods (for example, the first-instar larva is the stage between the egg and the first molt). See also Apolysis, Ecdysis, Stadium.

INTEGUMENT. Noun. (Latin, *integumentum* = covering). The outer covering of the insect body. Consists of several layers, including a living epidermal layer and secreted layers that have different functions.

INTERCALARY VEIN. (Latin, *intercalaris* = inserted). Any added or supplementary wing vein.

INTERSEX. Noun. (Latin, *inter* = between; *sexus* = sex). An insect whose sexual characteristics include male and female features. See also Gynadromorph.

INTERTIDAL. Adj. (Latin, *inter* = between; Anglo-Saxon, *tid* = tide). Occurring on the beach between high-water and low-water levels.

INVAGINATION. Noun. (Latin, *in* = into; *vagina* = sheath). A pouch or sac formed by an infolding or indrawing of the outer surface. An infolding or projection toward the inside of the body.

– J –

JOHNSTON'S ORGAN. An auditory organ of the scolopophorous type located in the second segment (pedicel) of the antenna in most larger orders of insects. Compare Tympanal Organ.

JOINT. Noun. (Old French, *joindre* from Latin, *jungere* = to join). 1. A point or area of articulation between two sclerites. 2. The area of fusion between limb segments. 3. Nonsclerotized cuticula between adjacent sclerotized regions of the integument.

JUGAL LOBE. An area of the forewing projecting posteriad and contacting the hind wing of the insect. Syn. Jugal Region, Fibula.

JUGATAE. Adj. (Latin, *jugum* = yoke). Lepidoptera having a jugum wing-coupling apparatus; any Lepidoptera with a jugum.

JUGUM. Noun. (Latin, *jugum* = yoke. Pl., juga). 1. Heteroptera: Two lateral lobes of the head, one on each side of the tylus. 2. Some Lepidoptera and Trichoptera: A lobe or process at the base of the forewings, overlapping the hind wings and holding the wings together in flight.

– K –

KAIROMONE. Noun. (Greek, *kairos* = opportunistic). A chemical produced or acquired by one organism that mediates behavioral or physiological response in another organism that is favorable to the receiver but not the emitter. See also Allomone, Pheromone.

KEEL. Noun. An elevated ridge or caarina.

KEY. Noun. (Middle English, *key* = key). A taxonomic device by which objects are identified based on decisions for suites of characters or character states. Types: Dichotomous Key, Pictorial Key, Tabular Key.

– L –

LABELLUM. Noun. (Latin, *labellum* = small lip. Pl., labella). 1. The sensitive, ridged, apical mouth structures of some Diptera. 2. A prolongation of the labrum covering the base of the rostrum in Coleoptera and Hemiptera. 3. A small spoon-shaped lobe at the apex of the glossa in the honeybee.

LABIAL GLANDS. The "salivary glands" of insects, opening by a median duct between the base of the hypopharynx and the labium or on the hypopharynx.

LABIAL PALP. (Pl., palpi). The segmented sensory appendages of the insect labium; attached to the palpiger and shorter than the maxillary palp. Compare Maxillary Palp.

LABIUM. Noun. (Latin, *labium* = lip. Pl., labia). The second maxilla, or "lower lip," of the insect mouth. A compound, bilaterally symmetrical sclerite that forms the floor of the mouth in mandibulate insects. Located behind the first maxilla and opposed to the labrum. Sometimes referred to as the "tongue."

LABRUM. Noun. (Latin, *labrum* = lip. Pl., labra). The upper lip, which covers the base of the mandible and forms the roof of the mouth.

LACINIA. Noun. (Latin, *lacinea* = flap. Pl., laciniae). 1. A bladelike sclerite. The mesal endite sclerite attached to the maxillary stipes, or the inner lobe of the first maxilla that is articulated to the stipes and bears brushes of setae or spines. 2. Diptera: A flat lancetlike piercing structure that is never jointed. 3. Psocoptera: The styliform appendage of Ribaga; a hard, elongated rod, slightly bifurcated at its free end and ensheathed by the galea. See also Maxilla.

LAMELLA. Noun. (Latin, *lamella* = small plate. Pl., lamellae). A thin plate or leaflike process; a parademe.

LARVA. Noun. (Latin, *larva* = ghost. Pl., larvae). An immature stage of holometabolous insect; the developmental stage following the egg stage and preceding the pupal stage. Differing fundamentally in form from the adult. In a strict zoological sense, the immature form of animals that undergo metamorphosis. Compare Naiad, Nymph. See also Caterpillar, Grub, Maggot, Slug.

LATERAD. Adv. (Latin, *latus* = side; *ad* = toward). Toward the side and away from the median line.

LATERAL. Adj. (Latin, *latus* = side.) Relating, pertaining, or attached to the side.

Glossary

LEAFMINER. An insect (usually a larva) that makes a tunnel between the upper and lower surfaces of a leaf.

LEG. Noun. (Middle English, *legge* = leg). Cuticular appendage projecting from the ventrolateral portion of the thoracic wall of the insect body. Six legs are present in the adult and nymphal Paurometabola; the number varies in larvae, depending upon the taxonomic group. Each leg is divided into component parts, including the coxa, trochanter, femur, tibia, and tarsus. Primarily an appendage responsible for locomotion, but other adaptations have evolved, including prey capture, courtship, and excavation. See also Cursorial, Fossorial, Gressorial, Raptorial, Saltatorial.

LIGULA. Noun. (Latin, *ligula* = little tongue. Pl., ligulae). The median sclerite of the labium. The ligula inserts onto the distal margin of the prementum. Usually unpaired (single); sometimes paired. Often regarded as synonymous with "glossa" and "tongue." Corresponds to the united laciniae of the right and left maxillae.

LITTORAL. Adj. (Latin, *litus* = seashore). Dwelling on the seashore or the edge of freshwater.

– M –

MACROPTEROUS. Adj. (Greek, *makros* = large; *pteron* = wing). Long- or large-winged. Wings that are not reduced in size and presumably fully functional in active flight. Compare Brachypterous, Micropterous.

MAGGOT. Noun. (Middle English, *magot* = grub). The larva of Diptera: Without a distinct head, legless, with body usually pointed anteriorly and blunt posteriorly.

MANDIBLE. Noun. (Latin, *mandibulum* = jaw). The anteriormost pair of oral appendages in the insect head. Typically the hardest part of the insect integument; usually stout and highly modified in shape. Toothlike in chewing insects; needle- or sword-shaped in piercing-sucking insects.

MASK. Noun. (French, *masque* = mask). Dragonflies: The modified extensible labium of the nymph, which at rest conceals the mouthparts.

MAXILLA. Noun. (Latin, *maxilla* = jaw. Pl., maxillae. Without qualifying adjective). The second pair of jaws in mandibulate insects. The most persistent mouthparts when the mouth is modified, and represented by some functional part in all insects in which the mouth structures are useful. Second maxillae: The labium or third pair of jaws in a mandibulate insect; composed of five parts. See also Stipes.

MAXILLARY PALP. (Pl., palpi). The palp carried by the stipes on its outer apical surface; consisting of one to seven segments. Presumably sensory in function. Compare Labial Palp.

MECONIUM. Noun. (Latin, from Greek, *mekonion* = poppy juice, from *mekon* = a poppy. Pl., meconia). 1. The liquid substance excreted from the anus by certain holometabolous insects after emergence from the chrysalis or pupa. 2. Parasitic Hymenoptera: The pelletlike excrement of larvae or prepupae. Compare Frass.

MEDIAL. Adj. (Latin, *medius* = middle). Referring to or at the middle of a structure.

MELANIN. Noun. (Greek, *melas* = black). Any one of a group of organic pigments that produces black, amber, and dark brown colors by deposition in the cuticle.

MEMBRANE. Noun. (Latin, *membrana* = membrane). 1. Any thin, transparent, flexible body tissue. 2. The wing tissue between the veins. 3. Heteroptera: The thin, transparent, or translucent apex of the hemelytra, as distinguished from the thickened basal part, the corium.

MEMBRANOUS. Adj. (Latin, *membrana* = membrane.) Tissue that is thin, pliable, and semitransparent; like a membrane.

MESAD. Adv. (Latin, *medius* = middle; *ad* = toward). Toward or in the direction of the median plane of the insect body.

MESAL. Adj. (Latin, *medius* = middle). Pertaining to, situated on, or in the median plane of the body.

MESONOTUM. Noun. (Greek, *mesos* = middle; *noton* = back. Pl., mesonota). The upper surface of the second (middle) thoracic segment (mesothorax) of the insect body.

MESOPLEURON. Noun. (Greek, *mesos* = middle; *pleura* = side. Pl., mesopleura). 1. The lateral surface of the mesothorax. 2. Diptera: The upper portion of the episternum of the mesothorax. 3. Hymenoptera: The sclerite below the insertion of the wings.

MESOTHORAX. Noun. (Greek, *mesos* = middle; *thorax* = chest). The second (middle) thoracic segment, which bears the middle legs and the anterior wings. See also Metathorax, Prothorax.

METAMORPHOSIS. Noun. (Greek, *meta* = change of; *morphe* = form). The transformation in shape or substance during successive stages of development.

Glossary

METANOTUM. Noun. (Greek, *meta* = after; *noton* = back). 1. The upper surface of the third (posterior) thoracic segment (metathorax). 2. Diptera: The oval, arched sclerite behind the scutellum; best developed in flies with a long, slender abdomen.

METAPLEURON. Noun. (Greek, *meta* = behind; *pleura* = side. Pl., metapleura). 1. In general, the lateral area of the metathorax. 2. Diptera: The pleuron of the metathorax. 3. Hymenoptera: The sclerite behind and below the insertion of the hind wings.

METATHORAX. Noun. (Greek, *meta* = behind; *thorax* = chest). The third thoracic segment, which bears the hind legs and hind wings. The segment is variable: sometimes distinct, sometimes closely united with the mesothorax, and sometimes appearing as a part of the abdomen. See also Mesothorax, Prothorax.

MICROPTEROUS. Adj. (Greek, *mikros* = small; *pteron* = wing). Small-winged. A condition in which the wings of an insect are disproportionately small and presumably less effective as instruments of flight. Compare Brachypterous Macropterous.

MICROTRICHIUM. Noun. (Greek, *mikros* = small; *thrix* = hair. Pl., microtrichia). 1. Minute hairlike structures found on the wings of some insects (for example, Mecoptera and Diptera). Microtrichia resemble small covering setae, but the absence of basal articulation distinguishes them. 2. Plecoptera: Cuticular projections of unknown function.

MIDGUT. Noun. The middle portion of the digestive system. Compare Foregut, Hindgut.

MOLT. Verb. (Latin, *moutare* = to change). 1. The process by which insects shed elements of the integument during larval, nymphal, or naiadal growth. Noun. 2. The period of transformation when the larva, nymph, or naiad changes from one instar to another. 3. The cast portion of the integument resulting from the processing of molting.

MULTIVOLTINE. Adj. (Latin, *multus* = many; Italian, *volta* = turn). Pertaining to organisms with many generations in a year or season. A term often applied to Lepidoptera. Compare Univoltine.

MYIASIS. Noun. (Greek, *myia* = fly). Disease or injury of animals, particularly humans, caused by the attack of dipterous larvae.

MYRMECOPHILE. Noun. (Greek, *myrmex* = ant; *philos* = loving). A commensal or parasite of ants that inhabits ant nests. Some myrmecophiles are tended by the ants; others prey upon the ants or their brood. Alternate: Myrmecophil.

Arthropod Collection and Indentification Techniques

MYRMECOPHILOUS. Adj. Ant-loving, being applied to insects that live in ant (Hymenopera: Formicidae) nests.

– N –

NACREOUS. Adj. (French, *nacre* = with a pearly luster). Resembling mother-of-pearl.

NAIAD. Noun. (Greek, *naias* = water nymph). The aquatic nymph of the Hemimetabola. A name given by Comstock to distinguish between forms living in water and the terrestrial nymphs of the Paurometabola. See also Nymph.

NAIL. Noun. (Anglo-Saxon, *naegel* = nail). A tarsal claw. Specifically the stout, apically pointed claws in predatory Heteroptera. See also Unguis.

NAKED. Adj. (Anglo-Saxon, *nacod* = naked). Not clothed; lacking a covering of setae. Applied to the pupa when it is not enclosed in a cocoon or other covering.

NECK. Noun. (Middle English, *necke* = neck). The cervix or slender connecting structure between head and thorax of insects that have the head free. Any contraction of the head at its juncture with the thorax.

NECROPHAGOUS. Adj. (Greek, *nekros* = dead; *phagein* = to devour). Pertaining to organisms that feed on dead or decaying matter.

NECROPHILOUS. Adj. (Greek, *nekros* = dead; *philein* = to love). Pertaining to organisms that are associated with dead and decaying plants or animals.

NECROSIS. Noun. (Greek, *nekrosis* = deadness). 1. The condition of decay. 2. The death of cells or tissues. 3. A diseased condition of plant tissues that causes them to turn black and decay.

NECTAR. Noun. (Greek, *nektar* = death overcoming; drink of gods). Sweet, frequently scented, substances secreted by flowers, nectarines, or other plant structures. Nectar provides nutrition for many insects, particularly Hymenoptera, Lepidoptera, and some Diptera.

NEOTENY. Noun. Retention of juvenile characters in the adult.

NEW SPECIES. A species that is purportedly unknown to science and does not have a scientific name. A term appended to the description of formally described and nomenclaturally validated taxa of the species level.

NOCTURNAL. Adj. (Latin, *nox* = night). Active during the night. Applied to insects that fly or are active at night. Ant. Diurnal.

NODULE. Noun. (Latin, *nodulus* = diminutive of *nodus* = knob). A small knot or swelling.

NOMENCLATURE. Noun. (Latin, *nomen* = name; *calare* = to call). The application of proper scientific names to any taxon based on the prescribed rules.

NOTUM. Noun. (Greek, *noton* = back. Pl., nota). The dorsal or upper surface of a body segment. Particularly applied to the thorax. Syn. Tergum. See also Dorsum. Ant. Sternum.

NYMPH. Noun. (Greek, *nymphe* = chrysalis). An immature insect that emerges from the egg in a relatively advanced stage of morphological development and that undergoes gradual metamorphosis. Nymphs differ from adults in displaying incomplete wings and genitalia. See also Larva, Naiad. Pupa.

– O –

OBTECT. Adj. (Latin, *obtectus* = covered over). Covered; within a hard covering.

OBTECT PUPA. A pupa with the appendages and body fused by a hardening of the exoskeleton. Termed the "theca" in older works. Lepidoptera: A more highly developed specialized pupa, smooth and rounded, with only the fourth, fifth, and sixth segments free. See Coarctate Pupa, Exarate Pupa.

OCCIPITAL. Adj. (Latin, *occiput* = back of head). Pertaining to the occiput, or the back part of the head.

OCCIPUT. Noun. (Latin, *occiput* = back of head). 1. General: The posterior portion of the epicranium between the vertex and the neck; rarely present as a distinct sclerite or clearly demarcated by benchmark sutures. 2. Diptera: The entire posterior surface of the head.

OCELLUS. Noun. (Latin, *ocellus* = little eye. Pl., ocelli). The simple eye of many adult insects, consisting of one biconvex lens. Typically three in number but reduced to one or two in some insects or absent. See also Compound Eye.

OLFACTION. Noun. (Latin, *olfacere* = to smell). Behavior and physiology: The perception of odors, or the sense of smell.

OMMATIDIUM. Noun. (Greek, *ommation* = little eye; *idion* = diminutive. Pl., ommatidia). One of the visual elements that compose the compound eye.

OMNIVOROUS. Adj. (Latin, *omnis* = all; *vorare* = to devour). Feeding generally on animal or vegetable food or on both.

OPERCULATE. Adj. (Latin, *operculum* = lid). Pertaining to a lidlike covering.

OPERCULUM. Noun. (Latin, *operculum* = lid. Pl., opercula). 1. A lid or cover. 2. The lidlike portion of the insect eggshell. Typically over the anterior pole that partly separates and elevates from the body of the eggshell to permit the escape of the nymph, naiad, or larva during eclosion. 3. Aleurodidae: The lidlike structure covering the vasiform orifice. 4. Cicadidae: One of the paired plates covering the timbals. 5. Lecaniine coccids: The anal plates. 6. Diptera: The chitinous lower part of the muscid mouth; the labrum-ephipharynx of Dimmock; the scutes covering the metathoracic stigmata.

OPISTHOGNATHOUS. Adj. (Greek, *opisthe* = behind; *gnathos* = jaw). 1. General zoological usage: Pertaining to animals with retreating jaws. 2. Entomological usage: Pertaining to one of the three principal orientations of the insect head in relation to the body. Characterized by a posteroventral position of the mouthparts resulting from a deflection of the facial region. Seen in most Homoptera. Compare Hypognathous, Prognathous.

ORAD. Adv. (Latin, *os* = mouth; *ad* = toward). Toward the mouth.

ORAL. Adj. (Latin, *os* = mouth.) Pertaining to the mouth. Any structure having to do with the mouth.

ORTHOGNATHOUS. Adj. (Greek, *orthos* = straight; *gnathos* = jaw). Literally, straight jaws. Pertaining to a head whose primary axis is at a right angle to the primary axis of the body. Compare Hypognathous, Opisthognathous, Prognathous.

OVARIOLE. Noun. (Latin, *ovarium* = ovary). An egg tube of the insect ovary. Functionally, three types of ovarioles have been identified in insects: panoistic, polytrophic, telotrophic.

OVARY. Noun. (Latin, *ovarium* = ovary). The enlarged basal portion of the female reproductive system, which typically consists of paired lateral ovaries, each of which is composed of several ovarioles.

OVATE. Adj. (Latin, *ovum* = egg). Egg-shaped. Traditionally viewed as shaped like the egg of a bird because egg shape in insects is highly variable.

OVIDUCT. Noun. (Latin, *ovum* = egg; *ducere* = to lead). The distal tubular portion of the female reproductive system that transmits the egg outside the body.

OVIPARITY. Noun. (Latin, *ovum* = egg; *parere* = to bring forth). Conventional form of development in most insects. The egg is fertilized within the body of the female, with embryonic development resulting in the development of one individual from one egg that has been fertilized by one sperm. Embryogenesis usually proceeds after the egg has been oviposited, but in some instances considerable embryogenesis may occur within the body of the mother.

OVIPAROUS. Adj. Reproducing by eggs laid by the female. See Ovoviviparous, Viviparous.

OVIPOSITOR. Noun. (Latin, *ovum* = egg; *ponere* = to place). The external tubular part of the female reproductive system through which eggs are passed. The ovipositor may be rigid and fixed in length or flexible and telescopic.

OVOVIVIPARITY. Noun. (Latin, *ovum* = egg; *vivus* = living; *parere* = to bring forth). A method of reproduction in which eggs are maintained in the common oviduct (vagina) until eclosion or eclosion occurs soon after oviposition. A common form of development in Diptera (Tachinidae) and some Coccoidea, and represented in most orders of insects. Typically, the wall of the vagina is extensively tracheated to provide oxygen for the developing embryo. There is a trend toward a reduction in the number of eggs laid per female in some groups that utilize ovoviviparity. Compare Parthenogenesis, Polyembryony, Viviparity.

OVOVIVIPAROOUS. Adj. Producing living young by the hatching of the ovum while still within the mother.

OVUM. Noun. (Latin, *ovum* = egg. Pl., ova). A female gamete.

– P –

PAEDOGENESIS. Noun. (Greek, *pais* = child; *genesis* = descent). Reproduction by an immature form. A method of reproduction in which ovaries become functional during the larval stage and eggs develop parthenogenetically. Some of the developing eggs result in larvae that themselves become parthenogenetic; other developing eggs result in normal adults. Seen in some Diptera and Coleoptera. Alternate: Pedogenesis. Compare Neoteny.

PALP. Noun. (Latin, *palpare* = to stroke. Pl., palpi). A paired, digitiform mouthpart appendage of the maxilla and labium that is tactile or chemosensory in function. See also Labial Palp, Maxillary Palp. Alternate: Palpus.

PARASITE. Noun. (Greek, *parasitos*; Latin, *para* = beside; *sitos* = food). An organism that lives in, on, or at the expense of another organism during at least part of its lifetime.

PARASITOID. Noun. (Greek, *parasitos* = parasite; *eidos* = form). 1. A transitional condition between predation and parasitism in the classical sense. The parasitoid larva is parasitic during the early stages and episitic during later development (Tachinidae). 2. A hymenopterous parasite that eventually kills its host.

PARATYPE. Noun (Greek, *para* = beside; *typos* = pattern). Nomenclature: Any specimen in a series from which a description has been prepared, other than the one specified as the type specimen or holotype of the species. Incorrectly, a specimen that has been compared with the type.

PARTHENOGENESIS. Noun. (Greek, *parthenos* = virgin; *genesis* = descent). Reproduction without fertilization. Development of individuals from egg cells without fertilization by a male gamete. A reproductive phenomenon common in some groups of arthropods, including the Acari and Insecta.

PATAGIUM. Noun. (Latin, *patagium* = boarder. Pl., patagia). Lepidoptera: Lobelike structures covering the base of the forewings; often used synonymously with tegula or squamula. Mosquitoes: A sausage-shaped body on each side of the prothorax in front of the first pair of spiracles.

PATHOGEN. Noun. (Greek, *pathos* = suffering; *genes* = producing). Any disease-producing microorganism.

PATHOGENIC. Adj. (Greek, *pathos* = suffering; *genes* = producing). Disease-causing or disease-producing. A term applied to organisms that cause or carry disease.

PATHOLOGICAL. Adj. (Greek, *pathos* = suffering; *logos* = discourse). A diseased or abnormal condition; unhealthy or arising from unhealthy conditions.

PATHOLOGY. Noun. (Greek, *pathos* = suffering; *logos* = discourse). The study of diseases.

PAUROMETABOLA. Noun. (Greek, *pauros* = little; *metabole* = change). A division of the Heterometabola characterized by a gradual development in which the young resemble the adults in general form and mode of life. Examples include Dermaptera, Orthoptera, Embioptera, Isoptera, Zoraptera, Corrodentia, Mallophaga, Anoplura, Heteroptera, most Homoptera.

PAUROMETABOLOUS. Adj. (Greek, *pauros* = little; *metabole* = change). Pertaining to organisms with metamorphosis in which the changes of form are gradual and inconspicuous.

Glossary

PECTEN. Noun. (Latin, *pecten* = comb). 1. Any comblike structure. 2. Hymenoptera: Curved, rigid setae on the base of the maxilla and labium; compact rows of tibial spines on fossorial wasps; rows of spines on the tarsomeres of pollen-gathering bees; any series of bristles arranged like a comb. 3. Mosquito larvae: Comblike teeth on the breathing tube. 4. Lepidoptera: Stiff, scalelike setae on the antennal scape.

PECTINATE. Adj. (Latin, *pecten* = comb). Comblike. Applied to structures, especially antennae, with even processes like the teeth of a comb.

PEDICEL. Noun. (Latin, *pediculus* = small foot). 1. Generally, a stalk or stem supporting an organ or other structure. 2. The second segment of the insect antenna, with intrinsic musculature and forming the pivot between the scape and funicle.

PEDIPALP. Noun. (Latin, *pes* = foot; *palpare* = to feel, Pl., pedipalpi, pedipalps). Chelicerata: The second pair of appendages on the cephalothorax; used in crushing prey, corresponding to the mandibles in Mandibulata.

PELAGIC. Adj. (Greek, *pelagos* = sea). Pertaining to organisms inhabiting the open sea; oceanic.

PENIS. Noun. (Latin *penis* = penis. Pl., penes). The flexible, partially membranous, intromittent, copulatory organ of the male insect. See also Aedeagus.

PENULTIMATE. Adj. (Latin, *paene* = almost; *ultimus* = last). Adjacent to the terminal segment; next to the last.

PERIPHERAL. Adj. (Greek, *peripherein* = to move around). Relating to the outer margin or distant from the center and near the circumference.

PERIPNEUSTIC. Adj. (Greek, *peri* = around; *pneustikos* = breathing). Pertaining to the respiratory system of insect larvae with spiracles in a row on each side of the body, with those of the wing-bearing segments closed.

PERITROPHIC MEMBRANE. The delicate membrane surrounding the food in the midgut. Present in most insects.

PETIOLE. Noun. (Latin, *petiolus* = small foot. Pl., petioles, petioli). 1. A stem or stalklike projection. 2. The slender segment between the thorax and abdomen in certain Diptera and apocritous Hymenoptera. In the latter, a pedicel formed of only one segment, or the first segment of a two-segmented pedicel in ants.

PHARATE. Adj. (Greek, *pharos* = loose mantle). A term applied to the condition of an instar within a previous cuticle before ecdysis.

PHARNYX. Noun. (Greek, *pharyngx* = gullet). The posterior portion of the mouth and upper part of the throat; a slight enlargement at the beginning of the esophagus. Diptera: Sometimes restricted to the space between the hypopharynx and subclypeal pump, and then equivalent to the subclypeal tube.

PHEROMONE. Noun. (Greek, *phereum* = to carry; *horman* = excite, stimulate). A chemical compound secreted by an animal that mediates the behavior of an animal belonging to the same species. See also Allomone, Kairomone.

PHORESY. Noun. (Greek, *pherein* = to bear, to carry). A form of symbiosis in which one organism is carried on the body of another, larger-bodied organism but the former does not feed on the latter. Commonly manifested as mites on insects and insects on insects.

PHRAGMA. Noun. (Greek, *phragma* = fence. Pl., phragmata). A partition, dividing membrane, or structure. A transverse partition of the endoskeleton; an internal ridge from the endocuticle to which a muscle is attached. Phragmata typically occur at the junction of tergites, pleurites, and sternites, especially in the thorax See also Apodeme, Suture.

PHYLOGENETIC CLASSIFICATION. A system of classification that purports to be natural, in that all subordinate taxa are members of a higher category (genus, family, order) by genetical relatedness or descent through evolutionary process.

PHYSOGASTRIC. Adj. (Greek, *physan* = to blow up; *gaster* = belly). Pertaining to females with a swollen or abnormally distended abdomen and mainly or entirely membranous with sclerites small and widely separated (for example, a queen termite).

PHYSOGASTRY. Noun. (Greek, *physan* = to blow up; *gaster* = belly). A phenomenon in which a female displays a pathological enlargement of the abdomen associated with the development of immatures within the body. Females give birth to larvae or adults via "birthing openings" or rupture of the integument. The condition is characteristic of pygmephoroid mites.

PHYTOPHAGOUS. Adj. (Latin). Feeding in or on plants.

PIERCING-SUCKING MOUTHPARTS. Mouthparts with mandibles or maxillae or both (modified into stylets) fitted for piercing plant or animal tissue.

PILE. Noun. (Latin, *pilleus* = felt cap). A setose or furlike covering of thick, fine, short, erect setae that gives a velvetlike appearance.

PILIFEROUS. Adj. (Latin, *pilus* = hair; *ferre* = to carry). Bearing a vestiture of setae or forming a pile.

PILOSE. Adj. (Latin, *pilosus* = hairy). Covered with soft down or short setae; covered with long, flexible setae; covered with long, sparse setae.

PLANIDIUM. Noun. (Greek, *planos* = wandering; *idion* = diminutive. Pl., planidia). The hypermetamorphic, migratory, first-instar larva of some parasitic insects. Morphologically characterized by a legless condition and somewhat flattened body that often displays strongly sclerotized, imbricated integumental sclerites and spinelike locomotory processes. A term most appropriately restricted to Hymenoptera (Euchartiidae, Perilampidae, some Ichneumonidae) and Diptera (Tachinidae). Incorrectly used interchangeably with Triungulin.

PLANTA. Noun. (Latin, *planta* = sole of foot. Pl., plantae). 1. The basal segment of the posterior tarsus in pollen-gathering Hymenoptera. 2. The ventral surface of the posterior tarsal segments or anal clasping legs at the end of the caterpillar's abdominal proleg.

PLASTRON. Noun. (French, *plastron* = breast plate). A film of gas held in place by setae or cuticular modifications. Frequently found on the body of aquatic insects.

PLATE. Noun. (Greek, *platys* = flat). 1. Any broad, flat surface, such as a sclerite. 2. Coccids: A thin projection of the pygidium.

PLEOMORPHIC. Adj. (Greek, *pleion* = more; *morphe* = form, shape). Pertaining to pleomorphism.

PLEOMORPHISM. Noun. (Greek, *pleion* = more; *morphe* = form, shape). Structure that is characterized by having many forms. Syn. Polymorphism.

PLEURON. Noun. (Greek, *pleura* = side. Pl., pleura). Lateral sclerites between the dorsal and sternal parts of the thorax. In general, the sides of the body between the dorsum and sternum.

PLEXUS. Noun. (Latin, *plexus* = interwoven). 1. Any complicated network of vessels, nerves, fibers, or tracheae. 2. A knot.

PLICA. Noun. (Latin, *plicare* = to fold). 1. General: A fold, convolution, or wrinkle. 2. A longitudinal plait of a wing. 3. The lamellate infolded thickening of the anterior and posterior margin of the abdominal segments.

PLUMOSE. Adj. (Latin, *pluma* = feather). Feathered, like a plume. Antennae that have long, ciliated processes on each side of each joint.

POLLEN BRUSH. A vernacular term for appendicular structures (typically setae) adapted for collecting pollen from the anthers of flowers. See also Scopa.

POLLINOSE. Adj. (Latin, *pollen* = fine flour). Covered with a loose, mealy, often yellow dust, like the pollen of flowers.

POLYEMBRYONY. Noun. (Greek, *polys* = many; *embron* = fetus). The form of reproduction in which several embryos develop from one fertilized egg. Seen in Strepsiptera and parasitic Hymenoptera, including Encyrtidae, Platygasteridae, and Dryinidae.

POLYGAMOUS. Adj. (Greek, *polys* = many; *gamos* = marriage). Pertaining to male sequestering or copulating with many females. Ant. Monogamous.

POLYGAMY. Noun. (Greek, *polys* = many; *gamos* = marriage). The condition in which one male copulates with or inseminates many conspecific females.

POLYMORPHIC. Adj. (Greek, *polys* = many; *morphe* = form). Pertaining to a population of individuals that occur in several forms. Differences may be apparent in sex, in season, in locality, or without apparent reason. Alternate: Polymorphous.

POLYMORPHISM. Noun. (Greek, *polys* = many; *morphe* = form). 1. The condition of having several forms in one species. 2. The occurrence of different forms of structure or different forms of organs in one individual during different periods of life. See also Dimorphism.

POOTER. Noun. A vernacular term coined by an English entomologist for an aspirating device used to collect small, highly mobile insects. Named in honor of F. W. Poos, an American entomologist who employed the device to collect Cicadellidae.

PORE. Noun. (Greek, *poros* = channel). 1. A large, isolated puncture. 2. A minute impression that perforates the surface. 3. Any small, round opening on the surface of a structure.

POSTABDOMEN. Noun. (Latin, *post* = after; *abdomen* = belly). The posterior segments (including the genital segments) of the female insect abdomen retracted within a genital sinus.

Glossary

POSTCLYPEUS. Noun. (Latin, *post* = after; *clypeus* = shield). 1. General: The posterior or upper part of the clypeus when any line of demarcation exists. 2. Odonata: The upper portion of a transversely divided clypeus. 3. Psocidae: A peculiar inflated structure behind the clypeus. Syn. Eupraclypeus, Nasus, Afternose, Paraclypeus, First Clypeus, Clypeus Posterior.

POSTERIOR. Adj. (Latin, *posterior* = latter). A term of position pertaining to a structure situated behind the axis. Toward the rear, caudal, or anal end of the insect. Ant. Anterior

PREDATOR. Noun. (Latin, *predaetor* = hunter). Any animal that overpowers, kills, and consumes other organisms (called prey) for food. Compare Parasite.

PREPUPA. Noun. (Latin, *prae* = before; *pupa* = puppet. Pl., prepupae). 1. A quiescent instar between the end of the larval period and the pupal period. 2. An active but nonfeeding stage in the larva of the Holometabola. 3. A full-fed larva.

PRETARSUS. Noun. (Latin, *prae* = before; Greek, *tarsos* = sole of foot. Pl., pretarsi). The terminal limb segment. In most adult insects and nymphs, the pretarsus usually comprises a pair of lateral claws (ungues) and reduced median parts (arolium, unguitractor plate, median claw). In most larvae, a simple clawlike segment, as a rule withdrawn into the terminal joint or absent.

PRIMORDIAL. Adj. (Latin, *primordium* = beginning). Primitive in development. The first or earliest point in time of a developing structure or lineage.

PROBOSCIS. Noun. (Greek, *proboskis* = trunk. Pl., proboscises, proboscides). The extended mouth structure of an insect. Typically applied to the extensile mouthparts of adult Diptera, the beak of Hemiptera (labium enclosing maxillary and mandibular stylets), the tongue of Lepidoptera (fused galeae and maxillae), and the mouth of long-tongued bees (labium).

PROGNATHOUS. Adj. (Greek, *pro* = before; *gnathous* = jaw). Entomological usage: Orientation of the insect head with a foramen magnum near the vertex and with mandibles directed anterior and positioned at the anterior margin of the head. When viewed in lateral aspect, the primary axis of the head is horizontal. The prognathous condition is displayed by some predaceous forms and other insects living in concealed situations such as between bark and wood or similar confined habitats. Compare Hypognathous, Opisthognathous.

PROLEG. Noun. (Latin, *pro* = for; Middle English, *legge* = leg). Any process or appendage that serves the purpose of a leg. Typically, the pliant, unjointed abdominal legs of caterpillars and some sawfly larvae. Syn. Abdominal Foot, False Leg, Pseudopod.

PRONOTUM. Noun. (Greek, *pro* = before; *noton* = back). The upper (dorsal) surface of the prothorax.

PROPODEUM. Noun. (Greek, *pro* = before; *podeon* = neck. Pl., propodea). Hymenoptera (Apocrita): The first abdominal segment that disassociates from the abdomen and becomes associated with the thorax. In Parasitic Hymenoptera and Aculeata, characterized by anterolateral spiracles and broad attachment to the metanotum anteriad and separated from the remainder of the abdomen by a narrow constriction (the petiole) posteriad.

PROPOLIS. Noun. (Greek, *pro* = for; *polis* = city). A gluelike or resinlike substance elaborated by bees to serve as a cement in situations where wax is not effective in binding a cell.

PROSOMA. Noun. (Greek, *pro* = before; *soma* = body). 1. The cephalothorax of arachnids. 2. The head in apocritous Hymenoptera.

PROTHORAX. Noun. (Greek, *pro* = before; *thorax* = chest). The first thoracic segment bearing the anterior legs but not wings. See also Mesothorax, Metathorax.

PROVENTRICULUS. Noun. (Latin, *pro* = before; *ventriculus* = small stomach). A typically musculated portion of the stomodaeum situated between the crop and circular muscles that provides an anterior constriction for the midgut. Frequently poorly developed or absent from fluid-feeding insects and characterized by well-developed longitudinal folds and acanthae in adult insects that feed on particulate matter. Variably developed in immature insects.

PROXIMAD. Adv. (Latin, *proximus* = next, *ad* = toward). Toward the proximal end of a structure or an appendage.

PROXIMAL. Adj. (Latin, *proximus* = next). The part of an appendage or structure nearest the body or an imaginary midline through the primary axis of the body. Near the base of a structure. Ant. Basal, Distal.

PRUINOSE. Adj. (Latin, *pruina* = hoarfrost). Covered with fine dust, as if frosted; with the brightness of a surface somewhat obscured by the appearance of a plumlike bloom but that cannot be rubbed off.

PSEUDOPOD. Noun. (Greek, *pseudes* = false; *pous* = foot. Pl., pseudopodia). A soft, flexible, footlike appendage characteristic of some dipterous larvae. A proleg.

Glossary

PTEROSTIGMA. Noun. (Greek, *pteron* = wing; *stigma* = mark. Pl., pterostigmata). An enlarged, pigmented area on the costal margin of the wing near its middle or at the apex of the radius. Found on both wings of Odonata; found on forewing of Psocoptera, Metagloptera, Mecoptera, and Hymenoptera.

PTEROTHORAX. Noun. (Greek, *pteron* = wing; *thorax* = chest. The closely fused meso- and metathorax in most winged insects, such as the wing-bearing thoracic segments in Hymenoptera, Lepidoptera, Thysanoptera.

PTERYGOTA. Noun. (Greek, *pteryx* = wing). A subclass of the Insecta; most members are winged, but some species are secondarily apterous. Compare Apterygota.

PTILINUM. Noun. (Greek, *ptilon* = feather). Schizophorous Diptera: A modification of the frons that forms a saclike, inflatable, pulsatile, cuticular organ on the head of teneral flies. The ptilinum is thrust through an arcuate ptilinal suture just above the base of the antenna. Used to aid emergence from the puparium and burrowing through the soil to reach the surface. After emergence, the ptilinum is retracted. A crescent-shaped sclerite between the ptilinial suture and the antennal socket is called a *lunule*.

PUBESCENT. Adj. (Latin, *pubescere* = to become mature). Covered with soft, short, fine, closely set setae.

PULVILLUS. Noun. (Latin, *pulvillus* = small cushion. Pl., pulvilli). Membranous, padlike structures between the tarsal claws. Cushions of short, stiff setae or other clothing on underside of tarsal joints, rarely fleshy lobes.

PUNCTATE. Adj. (Latin, *punctum* = point). Pertaining to a surface with impressed points, microscopic pits, or punctures.

PUPA. Noun. (Latin, *pupa* = puppet, young girl. Pl., pupae). A phase of complete metamorphosis during which larval anatomical features are destroyed and adult features are constructed. In holometabolous insects, the stage between larva and adult. See also Metamorphosis.

PUPARIUM. Noun. (Latin, *pupa* = puppet, young girl. Pl., puparia). 1. Cyclorrhaphous Diptera: The thickened, hardened, barrel-like third larval instar integument within which the pupa is formed. 2. The covering of certain coccids. 3. Stylopids: The integument of the seventh-instar larva in which the adult female is enclosed.

PYGIDIUM. Noun. (Greek, *pygidion* = narrow buttock, rump. Pl., pygidia). 1. The tergum of the last segment of the abdomen, whatever its numerical designation. 2. Coleoptera: The segment left exposed by the elytra. 3. Diaspididae: A strongly chitinized, unsegmented region terminating the abdomen of the adult female, following after the first four abdominal segments and not to be confused with the true pygidium of other insects.

PYGOSTYLUS. Noun. (Greek, *pyge* = rump; *stylos* = column. Pl., pygostyli). Hymenoptera: Appendicular sensory structures on gastral tergum VIII or IX. Sometimes regarded as homologous with cercus.

– Q –

QUADRANGLE. Noun. (Latin, *quadrangulus* = four-angled). Odonata: A cell in the wing of the Zygoptera bounded by M, Cu, arculus, and a cross-vein between M and Cu (similar in position to the triangle of the Anisoptera).

QUADRATE. Adj. (Latin, *quadrans* = fourth part). Four-sided.

QUIESENCE. Noun. A temporary slowing down of metabolism (for example, aestivation or hibernation) brought about by unfavorable conditions of temperature, humidity, or other climatic factors, by lack of some essential vitamins or other food substances, by failure to secrete particular hormones, or by a disturbance of the balance between opposing hormones. Compare Diapause.

– R –

RACE. Noun. (French, *race* = family). An anthropological term used to indicate a population of a species with constant characters that are not quite specific, usually occurring in a different faunal region from the type and thus geographical; sometimes incorrectly considered synonymous with subspecies.

RADIAL. Adj. (Latin, *radius* = ray). 1. Arranged like rays starting from a common center. 2. Of or pertaining to the radius or radial wing vein.

RAPTORIAL. Adj. (Latin, *raptor* = robber). Adapted for seizing prey; predacious. Modified for seizing prey, as of forelegs.

RECTUM. Noun. (Latin, *rectus* = straight). The posterior part of the terminal section of the proctodaeum, commonly applied to the posterior intestine as a whole.

Glossary

RETICULATE. Adj. (Latin, *reticulatus* = latticed). Covered with a network of lines, veins, or ridges; meshed, netted.

RETINACULUM. Noun. (Latin, *retinaculum* = tether. Pl., retinacula). 1. Lepidoptera: The loop into which the frenulum of the male is fitted. 2. Hymenoptera: Horny scales that move the sting or prevent its hyperextension from the body. 3. Coleoptera: The middle, toothlike process of the larval mandible. 4. Collembola: See Tenaculum.

RETRACTILE. Adj. (Latin, *retractus* = withdrawn). Pertaining to a structure that is capable of being produced and drawn back or retracted.

RETRACTOR. Noun. (Latin, *retrahere* = to draw back). Any structure used to withdraw; specifically, a muscle.

REVISION. Noun. (Latin, *revisio* = a seeing again). A comprehensive taxonomic treatment of a higher taxonomic category such as a genus, tribe, or family. Revisions include, but are not restricted to, descriptions of new taxa, redescriptions of poorly characterized taxa, clarification of nomenclatural problems, proposal of new synonymy, review of existing synonymy, assessment of previous taxonomic work, statements of geographical distribution, phenology, and summaries of biological information. Compare Synopsis.

RING JOINT. Insect antenna: The much shorter proximal segment(s) distad of the pedicel; ringlike in form and smaller than following segments. See also Annulus.

RIPARIAN. Adj. (Latin, *ripa* = riverbank). Descriptive of organisms frequenting rivers or their shores.

RIPICOLOUS. Adj. (Latin, *ripa* = riverbank). Pertaining to organisms inhabiting river banks.

ROSTRUM. Noun. (Latin, *rostrum* = beak). 1. General: A snoutlike prolongation of the head. 2. Hemiptera: The beak; a joined sheath formed by the labium to enclose the stylets or trophi. 3. Coleoptera: The rigid extension of the head in the Rhynchophora.

RUDIMENT. Noun. (Latin, *rudimentum* = the beginning). 1. The beginning of any structure or part before it has developed. 2. Structure not developed beyond an elementary or incomplete stage. 3. Vestige.

RUGOSE. Adj. (Latin, *ruga* = wrinkle). Wrinkled.

SALIVARY GLANDS. Glands that open into the mouth or at the beginning of the alimentary canal and secrete a digestive, irritant, or viscid material called *saliva*. See also Labial Glands.

SALTATORIAL. Adapted for jumping usually describing the hind legs.

SANGUINVOROUS. Adj. (Latin, *sanguis* = blood; *vorare* = to devour). Pertaining to insects that are blood-eating; blood-feeding.

SAPROPHAGOUS. Adj. (Greek, *sapros* = rotten; *phagein* = to eat). Pertaining to insects that feed on dead or decaying animal or vegetable matter.

SARCOPHAGOUS. Adj. (Greek, *sarx* = flesh; *phagein* = to devour). Flesh-eating; feeding on flesh.

SCALE. Noun. (Anglo-Saxon, *sceala* = shell). 1. General: Any small, flat, cuticular projection from the integument. Typically, a unicellular outgrowth of the body wall, wing, or appendage of various shapes; often a modified seta. 2. Isoptera: The stump of the shed wing. 3. Coccoids: The cover of a scale-insect, consisting of exuviae and glandular secretions; the waxy covering of a male lecaniid. 4. Lepidoptera: The flattened, highly modified, and ridged setae or macrotrichia on the wing; scales overlap and form the wing covering. 5. Diptera: The alula.

SCANSORIAL. Adj. (Latin, *scandere* = to climb). Formed for climbing. Applied to insects (such as lice) whose legs are adapted for climbing on mammalian hair.

SCAPE. Noun. (Greek, *skapos* = stalk). The basal segment of the antenna.

SCATOPHAGOUS. Adj. (Greek, *skat* = dung; *phagein* = eating). Feeding upon dung or excrement, merdivorous. Syn. Coprophagous.

SCAVENGER. Noun. (Middle English, *scavenger* = an officer with various duties). An animal that feeds on dead and decaying animal or vegetable matter. Compare Predator.

SCLERITE. Noun. (Greek, *skleros* = hard). Any hard portion of the insect integument separated from similar areas by suture, sulcus, or apodeme.

SCLEROTIZATION. Noun. (Greek, *skleros* = hard). The process of hardening and darkening (tanning) of the exoskeleton after ecdysis. Sclerotizing.

Glossary

SCOPA. Noun. (Latin, *scopa* = broom. Pl., scopae). 1. Any brush, tuft, mat, or pile of setae found on the body or appendages that is used to collect pollen. 2. Hymenoptera: Bees: The pollen brush located on the posteriomedial surface of the hind tibia and basitarsus. Megachilidae: The mat of setae on the metasomal venter. Ichneumonidae: A tuft of setae found on the hind coxa of females. Symphyta: An enlarged, sometimes setose, apicoventral flange on the gonostylus. Lepidoptera: Aa fringe of long scales along the posterior margin of abdominal segment VIII.

SCOPULA. Noun. (Latin, *scopula* = little broom, small brush. Pl., scopulae). 1. A brush. 2. Any small, dense tuft of setae, frequently bristlelike, stiff, of similar length and shape.

SCROBE. Noun. (Latin, *scrobis* = ditch). 1. A cuticular impression or groove formed for the reception, protection, and concealment of an appendage. 2. Orthoptera: The pits in which the antennae are set. 3. Rhynchophora: Grooves at the sides of the rostrum to receive the antennal scapes; also applied to grooves on the sides of mandibles. 4. Hymenoptera: The usually circular impressions on the frons, in which the scapes revolve. Syn. Scrobal Impressions.

SCUTELLUM. Noun. (Latin, *scutellum* = a little shield. Pl., scutella). 1. Any small, shieldlike sclerite. 2. Heteroptera: The triangular part of the mesothorax, generally placed between the bases of the hemelytra, but in some groups overlapping them. 3. Coleoptera: The triangular sclerite between the elytra. 4. Hymenoptera: Sclerite posterior of the transcutal suture. 5. Diptera: A subhemispherical part cut off by an impressed line from the mesonotum. Compare Scutum.

SCUTUM. Noun. (Latin, *scutum* = shield. Pl., scuta). 1. The second dorsal sclerite of the meso- and metathorax. 2. The middle division of the notum in ticks.

SEGMENT. Noun. (Latin, *segmentum*, from *secare* = to cut). 1. Any natural or apparent subdivision of the body. 2. Any subdivision of an arthropod appendage separated from similar structural elements by areas of flexibility, or associated with muscle attachments. Syn. Arthromere, Embryonic, Metamere, Podite, Somite.

SEJUGAL FURROW. A line that separates the propodosoma and hysterosoma in acarines.

SENSE ORGAN. 1. Any specialized, innervated structure of the body wall receptive to external stimuli; most insect sense organs are innervated setae. 2. Any structure by which an insect can see, smell, feel, or hear.

SENSILLUM. Noun. (Latin, *sensus* = sense. Pl., sensilla). 1. A simple sense organ or sensory receptor found on various appendages and tagmata of the insect body. 2. A complex, bilaterally symmetrical, sexually dimorphic structure found on the 10th tergum of Siphonaptera. Covered with short, tapering spines (termed *microtrichia*), circular pits that give rise to longer setae (termed *trichobothria*), and a domelike cupola. Functions as a compound sense organ.

Arthropod Collection and Indentification Techniques

SESSILE. Adj. (Latin, *sedere* = to sit). 1. Pertaining to any body structure or appendage broadly attached to or resting on a base without a constriction, stalk, or pedicel. 2. Pertaining to the abdomen broadly attached to the thorax.

SETA. Noun. (Latin, *seta* = bristle. Pl., setae). Hollow, cuticular structures developed as extensions of the epidermal layer of the integument. Often slender and hairlike. Syn. Bristle, Hair, Microtrichium. See also Chaeta.

SILK GLANDS. 1. Lepidopterous caterpillars: The greatly elongated, tubular labial or salivary glands. 2. Head glands or rectal glands in other silk-spinning insects. 3. Glands that secrete liquids that harden into silk on exposure to the air.

SOLITARY. Adj. (Latin). Occurring singly or in pairs, i.e., not in colonies.

SPECIES. Plural noun. (Latin, *species* = particular kind). 1. Species or kind. The primary biological unit, debatably an actual thing or a purely subjective concept. 2. A static moment in the continuum of life; an aggregation of individuals similar in appearance and structure, mating freely, and producing young that themselves mate freely and bear fertile offspring resembling each other and their parents, including all varieties and races. Abbreviated "sp." for one species, "spp." for two or more species.

SPECIMEN. Noun. (Latin, *specere* = to look, to behold). A single example of anything (for example, the individual unit of a collection, a single insect of any species, or a part of an insect properly preserved).

SPERMATHECA. Noun. (Greek, *sperma* = seed, *theke* = case. Pl., spermathecae). A sac, duct, or reservoir within the female that receives spermatozoa during copulation. Variable in number and morphology and capable of storing spermatozoa for periods up to several years.

SPERMATOPHORE. Noun. (Greek, *sperma* = seed; *pherein* = to bear, carry). A packet or capsule composed of sperm mixed with secretions from accessory glands. Typically manufactured by elements of the male reproductive system and transmitted to the female during copulation. Provides nutrients to the female and/or serves as a physical or chemical barrier to subsequent copulation, insemination, or fertilization.

SPIRACLE. Noun. (Latin, *spiraculum* = air hole. Pl., spiraculae, spiracles). A breathing pore or aperture. Typically paired lateral holes in the pleural wall of insect body segments through which air enters the tracheae.

Glossary

SPUR. Noun. (Anglo-Saxon, *spora* = spur). Any spinelike appendage of the integument that articulates with the body wall. Frequently found on the leg segments, such as the tibia.

STADIUM. Noun. (Latin, from Greek *stadion*, *stadios* = fixed, stable. Pl., stadia). The interval between the molts during larval development. See also Instar.

STAGE. Noun. (Old French, *estage* = dwelling, habitation). Any defined period in the development of an insect: the egg stage, larval stage, pupal stage, adult stage. Compare Instar.

STERNITE. Noun. (Greek, *sternon* = chest). 1. The ventral portion of a sclerotized ring or body segment. 2. A subdivision of a sternal sclerite or any one of the sclerotic components of a definitive sternum.

STERNUM. Noun. (Latin, *sternum* = breast bone. Pl., sterna). 1. The entire ventral division of any segment. 2. The ventral surface of the insect thorax between the coxal cavities.

STIGMA. Noun. (Greek, *stigma* = mark. Pl., stigmata). 1. A spiracle or pore of the respiratory system. 2. Hymenoptera: A pigmented spot along the costal margin of a wing, usually at the end of the radius. 3. Diptera: A colored wing spot near the tip of the auxiliary vein. 4. Lepidoptera: The specialized patch of black scales on the forewings of Hesperidae.

STING. Noun. (Anglo-Saxon, *stingan* = to sting). Aculeate Hymenoptera: A sclerotized, tapered, tubular shaft developed at the tip of the abdomen to inject venom into prey and for defense.

STIPES. Noun. (Latin, *stipes* = stalk. Pl., stipites). 1. The second segment of the maxilla in the insect head. Broadly attached to the cardo basad, bearing the movable palpus laterad, and attached to the galea and lacinia distad. Collectively, the cardo and stipes are called the *coxopodite* in the generalized mouthpart. Modified into a piercing structure in some Diptera and into a lever for flexing the proboscis in others. 2. Either of the pair of forceps in the male genitalia of aculeate Hymenoptera (sagittae). 3. The stalk of an elevated eye. 4. Distal portion of an embolus in spiders.

STRIDULATE. Verb. (Latin, *stridor* = grating). Insects: to make a creaking, grating, or hissing sound or noise by rubbing two ridged or roughened surfaces against each other. Example: Crickets.

STRIDULATING ORGANS. Parts of the insect structure that are used in making sounds. In general, one part is a filelike area and the opposing part is a scraper or rasp.

STRIGIL. Noun. (Latin, *strigilis* = comb, from *stringere* = to scrape, graze). 1. A tibial comb. 2. A scraper. 3. A structure adapted for cleaning the antenna. 4. Corixidae: The peculiar structure on the abdomen dorsally (sometimes shaped like a comb) opposed by a hair-fringed concavity at the base of the metatarsus. Hymenoptera, Lepidoptera: A curved, comblike movable spur on the distal end of the foretibia.

STRUCTURE. Noun. (Latin, *structura*, from *structum* = to build, to arrange). 1. The arrangement of constituent elements (parts, cells, tissues, organs) of a body. 2. Any organ, appendage, or part of an insect.

STYLE. Noun. (Greek, *stylos* = pillar; Latin, *stylus* = pricker. Pl., styli, styles). 1. A short, cylindrical appendage. 2. Aphididae: The slender tubular process at the end of the abdomen. 3. Coccidae: A long spinelike appendage at the end of the abdomen of the male; the genital spike. 4. Diptera: The ovipositor; the single immovable organ immediately below the forceps. Male Tipulidae: a thickened, jointed arista at or near the apex of the third antennal segment. 5. In the plural form (styli) applied to small, usually pointed, exarticulate appendages; most frequently found on the terminal segments of the abdomen. See also Stylus.

STYLET. Noun. (Latin, *stylus* = pricker). 1. A small style or stiff process. 2. Phthiraptera, some Hemiptera, Diptera: One of the mouthparts modified for piercing. 3. Hemiptera: A median dorsal element in the shaft of the ovipositor, formed of the united second valvulae. 4. Hymenoptera: First valvulae. Aculeata: Fused portions of second valvulae.

STYLIFORM. Adj. (Latin, *stylus* = pricker; *forma* = shape). 1. Shaped like a stylus; terminating in a long, slender point. Example: The antenna of some Diptera.

STYLUS. Noun. (Latin, *stylus* = pricker. Pl., styli). 1. A small, pointed, nonarticulated process. 2. Collembola: The dorsal, tapering process of the ventral abdominal appendages, freely movable at its base; a small appendage attached to the coxa of the middle and hind legs. 3. Coccidae: The outer sheath of the male genitalia. 4. Female Odonata: A small rod-shaped projection at the tip of the lateral gonapophyses of the ovipositor; in male nymphs, a short acute process on the ventral surface of the ninth abdominal segment. See also Style.

SUBCOXA. Noun. (Latin, *sub* = under; *coxa* = hip. Pl., subcoxae). The basal part of the limb base when divided from the coxa. The subcoxa usually is incorporated into the pleural wall of the body segment.

SUBFAMILY. Noun. (Latin, *sub* = under; *familia* = household, from *famulus* = servant). A division in zoological classification containing a group of closely related genera that differ from allied groups. Subfamily names end with "-inae."

Glossary

SUBGENITAL PLATE. An apical sternal sclerite or process that covers and protects the gonopore. The hypandrium sensu Snodgrass. Syn. Subgenital Shield.

SUBGENUAL ORGAN. A chordotonal organ in the proximal part of the tibia that is sensitive to vibrations.

SUBIMAGO. Noun. (Latin, *sub* = under; *imago* = likeness. Pl., subimagoes). 1. A stage between the pupa and the adult. 2. The stage in mayflies just after emergence from the pupa and before the final molt.

SUCTORIAL. Adj. (Latin, *sugere* = to suck). Mouthparts adapted for sucking. See also Haustellate.

SULCUS. Noun. (Latin, *sulcus* = furrow. Pl., sulci). A furrow or groove; typically a groove in the insect integument.

SUPERPARASITISM. Noun. (Latin, *super* = over; *para* = beside; *sitos* = food). 1. A biological phenomenon in which a female parasite oviposits more eggs on or in a host than can hatch and successfully develop to maturity. 2. Oviposition on or in a host previously parasitized by a conspecific female. 3. Parasitism of one host by more larvae than can survive to maturity.

SUTURE. Noun. (Latin, *sutura* = seam). 1. A seam or line of contact between two sclerites or hardened body parts that makes those body parts immovably connected. 2. The division of the distinct parts of the body wall. 3. The line of juncture of the elytra in Coleoptera or of the tegmina or hemelytra in other orders. Compare Sulcus. See also Apodeme, Phragma.

SYMBIONT. Noun. (Greek, *symbionai* = to live together). One of the partners in a symbiotic relationship. Example: An insect or other arthropod living in symbiosis with termites or ants. Alternate Symbiote.

SYMBIOSIS. Noun. (Greek, *symbionai* = to live together). A living together in more or less intimate association by organisms of different species. The association is not necessarily for mutual benefit. See also Commensalism, Parasitism.

SYMPATRIC. Adj. (Greek, *syn* = with; *patra* = homeland). Pertaining to organisms (usually species) that have overlapping geographical distributions.

SYMPHYTA. Suborder within the Hymenoptera, including sawflies and horntails. Adults possess an abdomen broadly sessile at its base and without a marked constriction.

Arthropod Collection and Indentification Techniques

SYNONYM. Noun. (Greek, *syn* = with; *onymia* = name). Two or more different names for the same thing. In zoological nomenclature, a different scientific name given to a species or genus previously named and described. See also Homonym.

SYNOPSIS. Noun. (Greek, *syn* = together; *opsis* = a view). A taxonomic publication that briefly summarizes current knowledge of a group. New information is not necessarily included in a synopsis.

SYNOPTIC COLLECTION. A group of taxa that has been authoritatively identified and that serves as the basis of comparison for the identification of other specimens.

SYSTEMATICS. (Greek, *systema* = a whole made of several parts). The practice of classifying organisms. See also Taxonomy.

SYSTEMATIST. Noun. (Greek, *systema* = a whole made of several parts). In zoology, a person who studies taxa and develops classifications of forms or groups according to biological relationships, phenetic affinities, or both. Compare Taxonomy.

– T –

TAGMA. Noun. (Greek *tagma* = corps. Pl., tagmata). A group of successive segments (somites) of a metamerically organized animal that forms a distinct region of the body (head, thorax, abdomen).

TARSUS. Noun. (Greek, *tarsos* = sole of the foot. Pl., tarsi). 1. The foot or jointed appendage attached at the apex of the tibia, bearing the claws and pulvilli. 2. The distal part of the insect leg, consisting of one to five segments.

TAXONOMIST. Noun. One who studies taxonomy. See Taxonomy.

TAXONOMY. Noun. (Greek, *taxis* = arrangement; *nomos* = law). 1. The system of arranging organisms into groups based on anatomical detail or biological relationship to each other. 2. The practice of identifying, describing, and naming organisms based on the rules of zoological nomenclature. Compare Systematics.

TEGULA. Noun. (Latin, *tegula* = tile. Pl., tegulae). Hymenoptera: Lepidoptera, Diptera: An articular sclerite on the mesothorax positioned at the base of the wing. Typically a small, scalelike sclerite and at the extreme base of the costa of the forewing. Misused for alula in Diptera. See also Patagium.

Glossary

TENACULUM. In Collembola, a minute organ with two divergent prongs, situated medially on the ventral surface of the third abdominal segment, serving to hold the furcula in place.

TENERAL. Adj. (Latin, *tener* = delicate). Describing the imago or adult shortly after emergence from the nymphal or pupal stage when the integument is not hardened or its color has not matured. Compare Callow.

TENTORIAL PITS. External depressions in the head at the base of the tentorial arms. The anterior tentorial pits are located in the subgenal suture or in the epistomal suture; the posterior tentorial pits are located in the lower ends of the postoccipital suture.

TENTORIUM. Noun. (Latin, *tentorium* = tent. Pl., tentoria). The endoskeleton of the insect head, consisting of two or three pairs of apodemes coalescing internally. The tentorium provides rigidity and strength to the head, supports the brain and foregut, and provides attachment surface for some muscles in the head.

TERGITE. Noun. (Latin, *tergum* = back). 1. A dorsal sclerite that is part of a body segment or sclerotized ring of a generalized body segment. 2. A sclerotized subdivision of a body tergum. Compare Tergum. Ant. Sternite.

TERGUM. Noun. (Latin *tergum* = back. Pl., terga). The upper or dorsal surface of any body segment of an insect, whether it consists of one or several sclerites. Compare Tergite. Ant. Sternum.

TERMINAL. Adj. (Latin, *terminus* = end). Pertaining to the apex or extreme tip of a structure or appendages. Ant. Basal. Syn. Apical.

TERMITOPHILOUS. Adj. (Latin, *termes* = woodworm; Greek, *philein* = to love). Termite-loving. A term applied to insects and other guests habitually living in a termite colony with and among the termites. Compare Myrmecophilous.

TERRESTRIAL. Adj. (Latin, *terra* = earth). Organisms living on or in the land. Compare Aquatic.

TERTIARY HYPERPARASITISM. A form of hyperparasitism in which hyperparasitic individuals attack one another. Conceptually divided into interspecific tertiary hyperparasitism (allohyperparasitism) and intraspecific tertiary hyperparasitism (autohyperparasitism). See also Hyperparasite.

TESTES. Plural noun. (Latin, *testis* = testicle. Sing., testis). The male reproductive gland that produces spermatozoa. Usually composed of many tubular structures called *follicles*.

Arthropod Collection and Indentification Techniques

THIGMOTACTIC. Adj. (Greek, *thigema* = touch; *taxis* = arrangement). Contact-loving or pertaining to species that tend to live in close proximity or in contact with a surface or in a crevice.

THORAX. Noun. (Greek, *thorax* = chest. Pl., thoraces). The second or middle region of the insect body. The thorax bears true legs and wings and is composed of three segments (pro-, meso-, and metathorax). When the prothorax is free (for example, Coleoptera, Orthoptera, and Hemiptera), the term *thorax* is sometimes used for that segment only. In Odonata (where the prothorax is small and not fused with the larger mesothorax and metathorax), the term *thorax* is commonly used for these segments, excluding the prothorax.

TIBIA. Noun. (Latin, *tibia* = shin. Pl., tibiae). The fourth division of the leg, articulated proximally with the femur and distally with the basitarsus. The tibia is typically long and slender. See also Leg.

TIMBAL. Noun. (French, *timbale* = kettledrum). The shell-like drum in cicadas, at the base of the abdomen, used to produce sound. Syn. Tympanum. Alternate: Tymbal.

TONGUE. Noun. (Anglo-Saxon, *tunge* = tongue). A term applied to the coiled mouth of Lepidoptera, the lapping organ of flies, the ligula of bees and wasps, and sometimes to the hypopharynx of other insects.

TOXOGNATH. Noun. (Greek, *toxikon* = poison; *gnathos* = jaw). A poison claw in centipedes. The modified first pair of legs in centipedes and not a fang as in Chelicerata.

TRACHEA. Noun. (Latin, *trachia* = windpipe. Pl., tracheae). An element of the respiratory system. Typically, tracheae are internal, elastic, spirally ringed air tubes in the terrestrial insect body. Compare Gill.

TRACHEATE. Adj. (Latin, *trachia* = windpipe). Supplied with tracheae. A general term applied to all arthropods that breathe by means of spiracular openings into a system of tubes that extend to all parts of the body.

TRANSECT. Noun. (Latin, *trans* = across; *secare* = to cut). A narrow area, of various widths, within which quantitative samples of physical and biological data are collected.

TRANSPIRATION. Noun. (Latin, *trans* = across; *spirare* = to breathe). The act or process of exhaling or passing off liquids as vapor. In plants, the passing of water vapor through the stomata.

TRANSVERSE. Adj. (Latin, *transversus* = across). 1. Pertaining to structures that are wider than long. 2. Structure extending across or cutting the longitudinal axis at right angles. Ant. Longitudinal.

Glossary

TRANSVERSE SUTURE. 1. A division across the body; more particularly across the dorsum of the thorax. 2. Diptera: The suture between the prescutum and the scutum of the mesothorax.

TRAP. Noun. (Middle English, *trappe*; Anglo-Saxon, *treppe* = trap). A device or machine designed or adapted for the collection of insects. Types: flight intercept trap, light trap, pit trap, Malaise trap.

TROCHANTER. Noun. (Greek, *trochanter* = runner). A segment of the insect leg. The trochanter is sometimes divided between the coxa and femur, sometimes fused with the femur. Typically the smallest segment of the insect leg.

TRUNCATION. Noun. (Latin, *truncatus* = cut off). A transverse, abrupt termination of a structure or appendage.

TUBERCLE. Noun. (Latin, *tuberculum* = a small hump. Pl., tuberculi). 1. A small, rounded protuberance. 2. Hymenoptera: A projection in adult Sphecoidea (wasps), consisting of rounded lobes on the dorsolateral margin of the pronotum. 3. Lepidoptera caterpillars: Body structures (of this character) sometimes bearing setae. Syn. Verruca.

TUBERCULATE. Adj. (Latin, *tuberculum* = a small hump). Covered or furnished with tubercles; formed like a tubercle. Alternate: Tuberculose, Tuberculous.

TUFT. Noun. (Middle English, from Old French, *tufe*; Greek, *zopf* = clump of hair, pigtail). Several closely arranged, elongate, flexible structures (setae, cuticular evaginations) whose bases are juxtaposed but whose apices are free and widely spaced.

TUMID. Adj. (Latin, *tumidus* = swollen). Pertaining to bodies or structures that are swollen or enlarged.

TURBINATE. Adj. (Latin, *turbo* = whirl). Top-shaped, nearly conical. Turbinate differs from pyriform in being shorter and more suddenly attenuated at the base.

TYMPANAL ORGAN. A complex auditory organ found in the wing, abdomen, and thorax of Hemiptera, Lepidoptera, and Neuroptera. Elements include sensory cells adjacent or attached to a thin, flexible, chitinous membrane (the tympanum). A chitinous frame surrounds the tympanum, and a trachea or air sac is adjacent to the inner surface of the tympanum. Compare Chordotonal Organ, Johnston's Organ, Subgenual Organ.

TYMPANUM. Noun. (Greek, *tympanon* = drum; Pl., tympana, tympanums). Any stretched membrane like the covering of a drum. Specifically applied to the membrane covering the auditory organs in Orthoptera and some Lepidoptera.

Arthropod Collection and Indentification Techniques

TYPE. Noun. (Latin, *typus* = pattern). Taxonomy: 1. The single specimen or member of a series of specimens from which a species is described. 2. The species upon which a genus is founded or that is selected as the type of a genus. 3. A holotype. See also Paratype.

TYPEGENUS. Families: The specific genus on which the family is founded. The typegenus is not necessarily the oldest genus. When the family takes its name from the typegenus, the family name follows nomenclatural changes.

TYPESPECIES. A species that is the type of a genus.

– U –

UNGUIS. Noun. (Latin, Pl. ungues). Clawlike structure of the maxilla.

UNISEXUAL. Noun. (Latin, *unus* = one; *sexus* = sex). 1. Parthenogenetic species in which only females exist. 2. Female-only generations in species that cyclically produce males. A biological feature widely distributed within the Insecta but best studied in aphids and cynipid wasps.

UNIVOLTINE. Adj. (Latin, *unus* = one; Italian, *volta* = time). Pertaining to populations or species with one generation per year or season. Compare Multivoltine.

UNSCLEROTIZED. Adj. (Middle English, *un* = not; Greek, *skleros* = hard). Pertaining to tissue not having sclerotin. In general, describes parts of the insect integument that are not hardened by the deposition of sclerotin.

UROGOMPHI. Plural noun. (sing., urogmphus). In larval Coleoptera, usually paried processes from the posterior end of the tergum of the ninth abodominal segment.

URTICATING HAIRS. (Latin, *urtica* = nettle). Some caterpillars and adult insects: Setae or chaetae connected with cutaneous poison glands, through which venom issues. Barbed setae that cause discomfort presumably induced by mechanical irritation.

– V –

VANNAL REGION. The wing area containing the vannal veins, or veins directly associated with the third axillary. When large, the vannal region is usually separated from the remigium by the plica vannalis, often forming an expanded fanlike area of the wing.

Glossary

VANNAL VEINS. The veins associated at their bases with the third axillary sclerite and occupying the vannal region of the wing. The "anal" veins except the postcubitus.

VANNUS. Noun. (Latin, *vannus* = fan). A large fanlike expansion of the posterior part of the insect wing.

VARIETY. Noun. (Latin, *varietas* = variety). A taxonomically ambiguous term "used to designate variants in size, structure, and color and varying ranks of all these," or to indicate a form for which the taxonomic status is uncertain. Varieties have no standing in zoological nomenclature.

VEIN. Noun. (Latin, *vena* = vein). Any chitinous, rodlike, or hollow tubelike structure supporting and stiffening the wings in insects, especially those extending longitudinally from the base of the wing to the outer margin.

VENATION. Noun. (Latin, *vena* = vein). The complete system of veins in an insect wing.

VENOM. Noun. (Latin, *venenum* = poison). Hymenoptera: Toxic fluid consisting of biologically active peptides, secreted from accessory glands and injected into prey or hosts by the sting or ovipositor. Venoms induce death or paralysis and cause pain, edematous reactions, and swelling of skin.

VENTER. Noun. (Latin, *venter* = belly). 1. The abdomen. 2. The ventral surface of the abdomen.

VENTRAD. Adv. (Latin, *venter* = belly; *ad* = to). Toward the venter; in the direction of the venter. Ant. Dorsad.

VENTRAL. Adj. (Latin, *venter* = belly). Pertaining to the undersurface of a structure. Ant. Dorsum.

VENTROLATERAL. Toward the venter on the side. Pertaining to the venter and the side.

VERRUCA. Noun. (Latin, *verruca* = wart. Pl., verrucae). Lepidopterous larvae: Somewhat elevated portions of the cuticle bearing tufts of setae (Imms). A wart or wartlike prominence.

VERTEX. Noun. (Latin, *vertex* = top). The top of the head between the eyes, front, and occiput.

VERTICAL BRISTLES. Diptera: Two pairs of bristles (an inner and an outer pair) more or less behind the upper and inner margins of the eyes; vertical cephalic bristles.

VESTIGIAL. Adj. (Latin, *vestigum* = trace). Small or degenerate. Descriptive of the remains of a previously functional part or organ, like remnants or vestiges. Descriptive of a structure in the process of disappearing.

VESTITURE. Noun (Latin, *vestitus* = garment). Clothing; the general surface covering of insects. Typically composed of scales, setae, or spines.

VISCERA. Plural noun. (Latin, *viscera* = bowels). The internal organs of the body.

VISCID. Adj. (Latin, *viscum* = mistletoe). Sticky; covered with a shiny, resinous, or greasy substance.

VIVIPARITY. Noun. (Latin, *vivus* = living; *parere* = to beget). A method of reproduction in which living young (not eggs) are extruded from the female body. Found in many groups of insects and varies in complexity. Offspring obtain nutrients directly from "milk glands" within the mother. Egg contains little or no yolk, lacks a chorion, and develops within an ovariole, and follicle cells supply some nutrients. Similar to ovoviviparity, in that there is a reduction in the number of offspring produced per female.

VIVIPAROUS. Adj. (Latin, *vivus* = living; *parere* = to beget). Pertaining to organisms that bear living young; opposed to organisms that lay eggs. Compare Oviparous, Ovoviviparous.

VULVA. Noun. (Latin, *vulva* = vulva). Insects: The orifice of the vagina in the female. Ticks: The female genitalia.

– W –

WAX GLAND. 1. Any gland that secretes a wax product in a scale, string, or powder. 2. Coccidae: The circumgenital and parastigmatic glands. 3. Hymenoptera: Apoidea: Abdominal glands that produce beeswax. 4. Neuroptera (Coniopterygidae): Glands on head, thorax, and abdomen that produce a meal-like wax. Wax glands have been identified on all tagmata and appendages.

WHORL. Noun. (Anglo-Saxon, *hweorfan* = to turn). 1. A ring or circle of setae surrounding a body or appendage segment. 2. A circular arrangement of structures that radiate outward from an imaginary center like the spokes of a wheel.

Glossary

WING. Noun. (Middle English, *winge*, from Old Norse, *vaengr*). Paired, often membranous and reticulated cuticular expansion of dorsolateral portion of the mesothorax and metathorax. Forewings (primaries) attached to mesothorax; hind wings (secondaries) attached to metathorax. Specifically adapted as organs of flight in insects or modified to protect the pair of wings involved in flapping flight. First detected in Carboniferous fossils, and present in most species.

WORKER. Noun. (Middle English, *werk*, *weore* = work). 1. The neuter or sterile individuals within the colony of a social species. 2. Isoptera: The sexually undeveloped forms, except the so-called soldiers. 3. Hymenoptera: Individuals anatomically female that lack the capacity to reproduce or lay unfertilized eggs that produce only males. Workers are responsible for nest building, brood care, and colony defense.

– X –

XENOBIOSIS. Noun. (Greek, *xenos* = guest, stranger; *biosis* = living). A commensalistic relationship in which one ant species lives as a guest in the nest of another species in mutual toleration.

XERIC. Adj. (Greek, *xeros* = dry). Pertaining to dry conditions or an absence of moisture.

XYLOPHAGOUS. Adj. (Greek, *xylon* = wood; *phagein* = to eat). Feeding in or upon woody tissue.

– Y –

– Z –

ZOOGEOGRAPHY. Noun. (Greek, *zoon* = animal; *ge* = earth; *graphein* = to write). The study of distribution, distributional patterns, and association or groupings of plants or animals in space, realm, or region.

ZOOPHAGOUS. Adj. (Greek, *zoon* = animal; *phagein* = to eat). Animal-eating; feeding on animals or animal products. Compare Herbivorous.

Arthropod Collection and Indentification Techniques

NOTES:

References

Arthropod Collection and Indentification Techniques

– A –

A'Brook, J. 1973. Observations on different methods of aphid trapping. Ann. Appl. Biol. 74: 253–267.

Acree, F., Jr., R. B. Turner, H. K. Gouck, M. Beroza & N. Smith. 1968. L-lactic acid: A mosquito attractant isolated from humans. Science (Washington, D.C.) 161: 1346–1347.

Acuff, V. R. 1976. Trap biases influencing mosquito collecting. Mosq. News 36: 173–196.

Agosti, D. 2001. Ants: Standard methods for measuring and monitoring biodiversity. Smithsonian Institution Press, Washington, DC. [Excellent example of specific procedures for surveying the diversity of one family of insects.]

Akar, H. & E. A. Osgood. 1987. Emergence trap and collecting apparatus for capture of insects emerging from soil. Entomol. News 98: 35–39.

Alderz, W. C. 1971. A reservoir-equipped Moericke trap for collecting aphids. J. Econ. Entomol. 64: 966–967.

Almand, L. K., W. L. Stering & C. L. Green. 1974. A collapsible truck-mounted aerial net for insect sampling. Tex. Agric. Exp. Stn. Misc. Publ. 1189: 1–4.

Anderson, G. S. 1999. Forensic entomology in death investigations. In: S. Fairgreave [ed.], Forensic anthropology case studies from Canada, Charles C Thomas, Springfield, IL.

Andreyev, S. V., B. K. Martin & V. A. Molchanova. 1970. Electric light traps in research on the protection of plants against insect pests. Entomol. Rev. 49: 290–297.

Apperson, C. S. & D. G. Yows. 1976. A light trap for collecting aquatic organisms. Mosq. News 36: 205–206.

Arensburger, P., T. R. Buckley, C. Simon, M. Moulds & K. E. Holsinger. 2004. Biogeography and phylogeny of the New Zealand cicada genera (Hemiptera: Cicadidae) based on nuclear and mitochondrial DNA data. J. Biogeography 31: 557-569.

Arnett, R. H. 1985. American insects. Van Nostrand Reinhold, New York.

Arnold, A. J. 1994. Insect suction sampling without nets, bags or filters. Crop Prot. 13: 73-77.

Atkins, M. D. 1957. An interesting attractant for *Priacma serrta* (LeC.) (Cupesidae: Coleoptera). Can. Entomol. 89: 214–219.

Azrang, M. 1976. A simple device for collecting insects. Entomol. Tidskr. 97: 92–94.

– B –

Bailey, W. 1991. Acoustic behavior of insects. Routledge, Chapman & Hall, London.

Baker, J. R. 1958. Principles of biological microtechnique. Methuen, London; John Wiley & Sons, New York.

Balogh, J. 1958. Lebensgemeinschaften der Landtiere; ihre Erforschung unter besonderer Berücksichtigung der zoozonologischen Arbeitsmethoden. Verlag Ungar. Akad. Wiss., Budapest; Akad. Verlag, Berlin. (In German.) [An extensive compilation, with 27 pp. of references.]

Banks, C. J. 1959. Experiments with suction traps to assess the abundance of Syrphidae (Diptera), with special reference to the aphidophagous species. Entomol. Exp. Appl. 2: 110–124.

References

Banks, N. 1909. Directions for preserving and collecting insects. U.S. Natl. Mus. Bull. 67. [Mostly of historical interest, but describes well the old methods and contains much general information about insects.]

Banks, W. A., C. S. Lofgren, D. P. Jouvenaz, C. E. Stringer, P. M. Bishop, D. F. Williams, D. P. Wojcik & R. M. Glancey. 1981. Techniques for collecting, rearing, and handling imported fire ants. U.S. Dep. Agric., Agric. Res. Serv., AAT-S-21.

Barber, H. S. 1931. Traps for cave-inhabiting insects. J. Elisha Mitchell Sci. Soc. 46: 259–266, pl. 23.

Barber, M. C. & R. W. Matthews. 1979. Utilization of trap nests by the pipe-organ muddauber, *Trypargilum politum* (Hymenoptera: Sphecidae). Ann. Entomol. Soc. Am. 72: 258–262.

Barnard, D. R. 1979. A vehicle-mounted insect trap. Can. Entomol. 111: 851–854.

Barnard, D. R. & M. S. Mulla. 1977. A non-attractive sampling device for collection of adult mosquitoes. Mosq. News 37: 142–144.

Barnes, H. F. 1941. Sampling for leather jackets with orthodichlorobenzene emulsion. Ann. Appl. Biol. 28: 23–28.

Barr, A. R., T. A. Smith, M. M. Boreham & K. E. White. 1963. Evaluation of some factors affecting the efficiency of light traps in collecting mosquitoes. J. Econ. Entomol. 56: 123-127.

Barrett, J. R., Jr., H. O. Deay & J. G. Hartsock. 1971. Striped and spotted cucumber beetle response to electric light traps. J. Econ. Entomol. 54: 413–416.

Bartelt, R. J., R. S. Vetter, D. G. Carlson & T. C. Baker. 1994. Influence of pheromones on *Carpophilus mutilatus* and *C. hemipterus* (Coleoptera: Nitidulidae) in a California date garden. J. Econ. Entomol. 87: 667–675.

Bartnett, R. E. & R. G. Stephenson. 1968. Effect of mechanical barrier mesh size on light trap collection in Harris County, Texas. Mosq. News 28: 108.

Batiste, W. C. & W. Joos. 1972. Codling moth: A new pheromone trap. Econ. Entomol. 65: 1741–1742.

Beatty, G. H. & A. F. Beatty. 1963. Efficiency in caring for large Odonata collections. Proc. North Cent. Branch Entomol. Soc. Am. 18: 149–153.

Beaudry, J. R. 1954. A simplification of Hubbell's method for trapping and preserving specimens of *Ceuthophilus* (Orthoptera, Gryllacrididae). Can. Entomol. 86: 121–122.

Beavers, J. B., T. P. McGovern, M. Beroza & R. A. Sutton. 1972. Synthetic attractants for some dipteran species. J. Econ. Entomol. 65: 1740–1741.

Becker, P., J. S. Mouve & F. J. A. Peralta. 1991. More about Euglossine bees in Amazonian forest fragments. Biotropica 23 (4 pt. B): 586–591.

Bedford, G. O. 1978. Biology and ecology of the Phasmatodea. Annu. Rev. Entomol. 23: 125–150.

Belding, M. J., S. A. Isard, A. D. Hewings & M. E. Irwin. 1991. Photovoltaic-powered suction trap for weakly flying insects. J. Econ. Entomol. 84: 306–310.

Belkin, J. N. 1962. The mosquitoes of the South Pacific (Diptera, Culicidae). 2 vols. University of California Press, Berkeley. [Vol. 1 includes introduction to methods.]

Bell, R. T. 1990. Insecta: Coleoptera Carabidae adults and larvae, pp. 1053–1092. In: D. L. Dindal, [ed.], Soil biology guide. John Wiley & Sons, New York.

Belton, P. 1962. Effects of sound on insect behavior. Proc. Entomol. Soc. Manit. 18: 22–30.

Belton, P. & R. H. Kempster. 1963. Some factors affecting the catches of Lepidoptera in light traps. Can. Entomol. 95: 832–837.

Belton, P. & A. Pucat. 1967. A comparison of different lights in traps for *Culicoides* (Diptera: Ceratopogonidae). Can. Entomol. 99: 267–272.

Bentley, J. W. 1992. An alcohol trap for capturing vespids and other Hymenoptera. Entomol. News 103: 86–88.

Beroza, M. 1970. Current usage and some recent developments with insect attractants and repellents in the USDA. In: M. Beroza [ed.], Chemicals controlling insect behavior, pp. 145–163, Academic Press, New York.

Beroza, M. 1972. Attractants and repellents for insect control. In: Pest control: Strategies for the future, pp. 226–253. Wash. Acad. Sci., Washington, DC.

Beroza, M. & N. Green. 1963. Materials tested as insect attractants. U.S. Dep. Agric., Agric. Handb. 239.

Beroza, M., E. C. Paszek, E. R. Mitchell, B. A. Bierl, J. R. McLaughlin & D. L. Chambers. 1974. Tests of a 3-layer laminated plastic bait dispenser for controlled emission of attractants from insect traps. Environ. Entomol. 3: 926–928.

Berte, S. B. 1979. An improved method for preserving color patterns in pinned insects. Entomol. News 90: 147–148.

Bidlingmayer, W. L. 1967. A comparison of trapping methods for adult mosquitoes: Species response and environmental influence. J. Med. Entomol. 4: 200–220.

Birch, M. [ed.]. 1974. Pheromones. North-Holland, Amsterdam. 495 pp.

Blakeslee, T. E. et al. 1959. *Aedes vexans* and *Culex salinarius* light trap collection at five elevations. Mosq. News 19: 283.

Bland, R. C. & H. E. Jacques. 1978. How to know the insects, 3rd ed. Wm. C. Brown, Dubuque, Iowa. [Successor to Jacques, H. E., 1947, 2nd ed.]

Blum, M. S. & J. P. Woodring. 1963. Preservation of insect larvae by vacuum dehydration. J. Kans. Entomol. Soc. 36: 96–101.

Blume, R. R., J. A. Miller, J. L. Eschle, J. J. Matter & M. O. Pickens. 1972. Trapping tabanids with modified Malaise traps baited with CO_2. Mosq. News 32: 90–95.

Borgmeier, T. 1964. How to prepare minuten-pin double mounts of small Diptera. Stud. Entomol. 7: 489.

Borror, D. J. & R. W. White. 1970. A field guide to the insects of America north of Mexico. Houghton Mifflin, Boston.

Borror, D. J., C. A. Triplehorn & N. F. Johnson. 1989. An introduction to the study of insects, 6th ed. Saunders College Publishing, Philadelphia.

Bosic, J. J. 1997. Common names of insects & related organisms. Entomological Society of America, College Park, MD.

Boudreaux, H. B. 1979. Arthropod phylogeny with special reference to insects. John Wiley & Sons, New York.

References

Bowles, D. E., K. Stephan & M. L. Mathis. 1990. A new method for collecting adult phryganeid caddisflies. Entomol. News. 101: 222–224.

Bradbury, W. C. & P. E. Morrison. 1975. A portable electric aspirator for collecting large insects. Can. Entomol. 107: 107–108.

Bram, R. A. 1978. Surveillance and collection of arthropods of veterinary importance. U.S. Dep. Agric., Agric. Handb. 518.

Brandenburg, R. L. & M. G. Villani [eds.]. 1995. Handbook of turfgrass insect pests. Entomological Society of America, Landham, MD.

Breyev, K. A. 1963. The effect of various light sources on the numbers and species of blood-sucking mosquitoes (Diptera, Culicidae) collected in light traps. Entomol. Obozr. 42: 280–303. (In Russian; translated into English in Entomol. Rev. 42: 155–168.)

Briggs, J. B. 1971. A comparison of pitfall trapping and soil sampling in assessing populations of two species of ground beetles (Col.: Carabidae). Rep. E. Malling Res. Stn. 1970: 108–112.

Brindle, A. 1963. Terrestrial Diptera larvae. Entomol. Rec. J. Var. 75: 47–62. [Includes collecting methods.]

Broadbent, K., J. P. Doucaster, R. Hull & M. A. Watson. 1948. Equipment used in trapping and identifying alate aphids. Proc. R. Entomol. Soc. Lond., Ser. A, Gen. Entomol. 23: 57.

Broadbent, L. 1949. Aphid migration and the efficiency of the trapping method. Ann. Appl. Biol. 35: 379–394.

Broce, A. B., J. L. Goodenough & J. R. Coppedge. 1977. A wind-oriented trap for screwworm flies. J. Econ. Entomol. 70: 413–416.

Brockway, P. B. et al. 1962. A wind directional trap for mosquitoes. Mosq. News 22: 404–405.

Brown, A. W. A. 1954. Studies on the responses of the female *Aedes* mosquito. Pt. IV. The attractiveness of coloured clothes to Canadian species. Bull. Entomol. Res. 45: 67–68.

Brown, R. D. 1973. Funnel for extraction of leaf litter organisms. Ann. Entomol. Soc. Am. 66: 485–486.

Brues, C. T., A. L. Melander & F. M. Carpenter. 1954. Classification of insects. Bull. Mus. Comp. Zool. 73: 1–917.

Burbutis, P. P. & J. A. Stewart. 1979. Blacklight trap collection of parasitic Hymenoptera. Entomol. News 90: 17–22.

Burditt, A. K. 1982. *Anastrepha suspensa* (Loew) (Diptera: Tephritidae) McPhail traps for survey and detection. Fla. Entomol. 65: 367–373.

Burditt, A. K. 1988. Western cherry fruit fly (Diptera: Tephritidae): Efficacy of homemade and commercial traps. J. Entomol. Soc. B. C. 85: 53–57.

Buriff, C. R. 1973. Recapture of released apple maggot flies in sticky board traps. Environ. Entomol. 2: 757–758.

Burrells, W. 1978. Microscope technique: A comprehensive handbook for general and applied microscopy. Fountain Press, London; Halsted Press, New York; Intl. Publ. Serv., New York.

Byers, G. W. 1959. A rapid method for making temporary insect labels in the field. J. Lepid. Soc. 13: 96–98.

– C –

Cade, W. 1975. Acoustically orienting parasitoids: Fly phonotaxis to cricket song. Science (Washington, DC) 190: 1312–1313.

Callahan, P. S., A. N. Sparks, J. W. Snow & W. W. Copeland. 1972. Corn earworm moth: Vertical distribution in nocturnal flight. Environ. Entomol. 1: 497–504.

Campion, D. G. 1972. Some observations on the use of pheromone traps as a survey tool for *Spodoptera littoralis*. Cent. Overseas Pest Res. Rep. 4.

Cantrall, I. J. 1939–1940. Notes on collecting and studying Orthoptera. Ward's Combined Entomol. Nat. Sci. Bull. 13(3): 1–5; (4): 1–6; (5): 4,5; (6): 5–7; (7): 4,5 (all pp. unnumbered). [Much of the same material is in Cantrall, I. J., 1941.]

Cantrall, I. J. 1941. Notes on collecting and preserving Orthoptera. Compendium of entomological methods, pt. II. Ward's Combined Entomol. Nat. Sci. Bull. 13(4) :1–9.

Carayon, J. 1969. Emploi du noir chlorazol en anatomie microscopique des insectes. Ann. Soc. Entomol. Fr. (n.s.) 5: 179-193. [On the use of the dye chlorazol black to color cuticle in microscopical preparations.]

Carlberg, U. 1986. *Phasmida*: A biological review (Insecta). Zool. Anz. 216: 1–18.

Carlson, D. 1971. A method of sampling larval and emerging insects using an aquatic black light trap. Can. Entomol. 103: 1365–1369.

Carroll, J. F. 1988. Short-duration dry ice sampling for American dog ticks (Acari: Ixodidae) in Maryland: A comparison with dragging. J. Entomol. Sci. 23: 131–135.

Castner, E. & J. Byrd. [eds]. 2001. Forensic entomology: The utility of arthropods in legal investigations. CRC Press, Boca Raton, FL.

Catts, E. P. 1970. A canopy trap for collecting Tabanidae. Mosq. News 30: 472–474.

Chamberlin, J. C. 1940. A mechanical trap for the sampling of aerial insect populations. U.S. Dept. Agric., Bur. Entomol. Plant Quar., ET-163.

Chandler, D. S. 1990. Insecta: Coleoptera Pselaphidae. In: D. L. Dindal [ed.], Soil biology guide, pp. 1175–1190. John Wiley & Sons, New York.

Chapman, J. A. & J. M. Kinghorn. 1955. Window flight traps of insects. Can. Entomol. 87: 46–47.

Chapman, T. W., A. P. Willsie, P. G. Kevan & D. S. Willis. 1990. Fiberglass resin for determining nest architecture of ground-nesting bees. J. Kans. Entomol. Soc. 63: 641–643.

Cheng, L. 1975. A simple emergence trap for small insects. Pan-Pac. Entomol. 50: 305–307.

Chiang, H. C. 1973. A simple trap for certain minute flying insects. Ann. Entomol. Soc. Am. 66: 704.

Christiansen, K. A. 1990. Insecta: Collembola. In: D. L. Dindal [ed.], Soil biology guide, pp. 965–995. John Wiley & Sons, New York.

Chu, H. F. 1949. How to know the immature insects. An illustrated key for identifying the orders and families of many immature insects with suggestions for collecting, rearing, and studying them. Wm. C. Brown, Dubuque, IA.

Clark, J. D. & C. E. Curtis. 1973. A battery-powered light trap giving two years' continuous operation. J. Econ. Entomol. 66: 393–396.

References

Clark, W. H. & P. E. Blom. 1979. Use of a hand sprayer as a collecting technique. Entomol. News 99: 247–248.

Clark, W. H. & R. E. Gregg. 1986. Housing arthropods and other invertebrates stored in alcohol. Entomol. News 97:237–240.

Clarke, J.F.G. 1941. The preparation of slides of the genitalia of Lepidoptera. Bull. Brooklyn Entomol. Soc. 36: 149–161.

Clifford, C. W., R. M. Roe & J. P. Woodring. 1977. Rearing methods for obtaining house crickets, *Acheta domesticus*, of known age, sex, and instar. Ann. Entomol. Soc. Am. 70: 69–74.

Clinch, P. G. 1971. A battery-operated vacuum device for collecting insects unharmed. N. Z. Entomol. 5: 2830.

Coffey, M. D. 1966. Studies on the association of flies (Diptera) with dung in southeastern Washington. Ann. Entomol. Soc. Am. 59: 207–218.

Cogan, B. H. & K. V. G. Smith. 1974. Insects: instructions for collectors No. 4a. 5th ed., rev. Brit. Mus. (Nat. Hist.) Publ. 705.

Colless, D. H. 1959. Notes on the culicine mosquitoes of Singapore. VI. Observations on catches made with baited and unbaited trap-nets. Ann. Trop. Med. Parasitol. 53: 251–258.

Coluzzi, M. & V. Petrarca. 1973. Aspirator with paper cup for collecting mosquitoes and other insects. Mosq. News 33: 249–250.

Comstock, J. H. 1940. An introduction to entomology, 9th ed. Comstock, Ithaca, NY.

Cook, E. F. 1954. A modification of Hopkin's technique for collecting ectoparasites from mammal skins. Entomol. News 15: 35–37.

Cooke, J. A. L. 1969. Notes on some useful arachnological techniques. Bull. Br. Arachnol. Soc. 1: 42–44.

Coon, B. F. & J. O. Pepper. 1968. Alate aphids captured in air traps at different heights. J. Econ. Entomol. 61: 1473-1474.

Copeland, T. P. & G. Imadate. 1990. Insecta: Protura. In: D. L. Dindal [ed.], Soil biology guide, pp. 911–933. John Wiley & Sons, New York.

Corbet, P. S. 1965. An insect emergence trap for quantitative studies in shallow ponds. Can. Entomol. 97: 845–848.

Corbet, P. S., C. Longfield & N. W. Moore. 1960. Dragonflies. Collins, London.

Coulson, R. N., R. T. Franklin & D. A Crosley, Jr. 1970. A self-maintaining window trap for collecting flying insects. Entomol. News 81: 164.

Cushing, C. E. 1964. An apparatus for sampling drifting organisms in streams. J. Wildl. Manage. 28: 592–594.

– D –

Dales, R. P. 1953. A simple trap for tipulids (Dipt.). Entomol. Mon. Mag. 89: 304.

Darling, D. C. & L. Packer. 1988. Effectiveness of Malaise traps in collecting Hymenoptera: The influence of trap design, mesh size, and location. Can. Entomol. 120: 787–796.

Darling, D. C. & R. C. Plowright. 1990. HPLABEL: A program and microfont for the generation of date/locality labels using a laser printer. Entomol. News 101: 143–146.

Davidson, J. & D. C. Swan. 1933. A method for obtaining samples of the population of Collembola (Symphyleona) in pastures. Bull. Entomol. Res. 24 (pt. 3): 351–352.

Davies, J. B. 1971. A small mosquito trap for use with animal or carbon dioxide baits. Mosq. News 31: 441–443.

Davis, E. W. & B. J. Landis. 1949. An improved trap for collecting aphids. U. S. Dep. Agric., Bur. Entomol. Plant Quar., ET-278.

DeBarro, P. J. 1991. A cheap lightweight efficient vacuum sampler. J. Aust. Entomol. Soc. 30: 207–208.

Debolt, J. W., D. L. Jay & R. W. Ost. 1975. Light traps: Effect of modifications on catches of several species of Noctuidae and Arctiidae. J. Econ. Entomol. 68: 186–188.

DeFoliart, G. R. 1972. A modified dry-ice-baited trap for collecting haematophagous Diptera. J. Med. Entomol. 9: 107–108.

DeJong, D. J. 1967. Some problems connected with the use of light traps. Entomophaga 3: 25–32.

Dethier, V. G. 1955. Mode of action of sugar-baited fly traps. J. Econ. Entomol. 48: 235–239.

Dethier, V. G. 1992. Crickets and katydids, concerts and solos. Harvard University Press, Cambridge, MA.

Dietrick, E. J. 1961. An improved backpack motor fan for suction sampling of insect populations. J. Econ. Entomol. 54: 394–395.

Dietrick, E. J., E. I. Schlinger & R. Van den Bosch. 1959. A new method for sampling arthropods using a suction collecting machine and modified Berlese funnel separator. J. Econ. Entomol. 52: 1085–1091.

Dindal, D. L. [ed.]. 1990. Soil biology guide. John Wiley & Sons, New York.

Doane, J. F. 1961. Movement on the soil surface of adult *Ctenicera aeripennis destructor* (Brown) and *Hypolithus bicolor* Esch. (Coleoptera: Elateridae) as indicated by pitfall traps, with notes on captures of other arthropods. Can. Entomol. 93: 636–644.

Dodge, H. R. 1960. An effective, economical flytrap. J. Econ. Entomol. 53: 1131–1132.

Dodge, H. R. & J. M. Seago. 1954. Sarcophagidae and other Diptera taken by trap and net on Georgia mountain summits in 1952. Ecology 35: 50–59.

Dominick, R. B. 1972. Practical freeze-drying and vacuum dehydration of caterpillars. J. Lepid. Soc. 26: 69–79.

Dresner, E. 1970. A sticky trap for Mediterranean fruit fly survey. J. Econ. Entomol. 63: 1813–1816.

Dryden, M. W. & A. B. Broce. 1993. Development of a trap for collecting newly emerged *Ctenocephalides felis* (Siphonaptera: Pulicidae). 30: 901–906.

Dunn, G. A. & R. M. Reeves. 1980. A modified collection net for catching insects under cloth bands on trees. Entomol. News 91: 7–9.

Dybas, H. S. 1990. Insecta: Coleoptera Ptiliidae. In: D. L. Dindal [ed.], Soil biology guide, pp. 1093–1112. John Wiley & Sons, New York.

References

– E –

Eads, R. B, G. C. Smith & G. O. Maupin. 1982. A CO_2 platform trap for taking adult *Dermacentor andersoni* (Acari: Ixodidae). Proc. Entomol. Soc. Wash. 84: 342–348.

Eastop, V. 1955. Selection of aphid species by different kinds of insect traps. Nature (Lond.) 176: 936.

Edmondson, W. T. & G. G. Winberg [eds.]. 1971. Manual for estimating secondary production in fresh waters. Blackwell Sci. Publ. (IBP Handb. 17), Oxford, England.

Edmunds, G. F., Jr., & W. P. McCafferty. 1978. A new J. G. Needham device for collecting adult mayflies (and other out-of-reach insects). Entomol. News 80: 193–194.

Edmunds, G. F., Jr., S. L. Jensen & L. Berner. 1976. The mayflies of North and Central America. University of Minnesota Press, Minneapolis.pp. 8–26. [Methods of collecting and preservation.].

Edwards, D. & N. Leppla. 1987. Arthropod species in culture in the United States and other countries. Entomological Society of America, Lanham, MD.

Edwards, J. G. 1963. Spreading blocks for butterfly wings. Turtox News. 41: 16–19.

Edwards, S. R., B. M. Bell & M. E. King. 1981. Pest control in museums: A status report (1980). Assoc. Syst. Collect. vii, 34, a-iii, A-30, b-iii, B-20, c-i, C-5, d-v, D-35, e-iii, E-14, f-iii, F-29, g-iii, G-10 pp. [Includes insecticides and their properties, regulations for their use, identification of pests, list of agencies, etc.]

Ellington, J. J., K. Kiser, M. Cardenas, J. Duttle & Y. Lopez. 1984a. The Insectavac: A high-clearance, high-volume arthropod vacuuming platform for agricultural ecosystems. Environ. Entomol. 13: 259–265.

Ellington, J. J., K. Kiser, G. Ferguson & M. Cardenas. 1984b. A comparison of sweepnet, absolute and insectavac sampling methods in cotton ecosystems. J. Econ. Entomol. 77: 599–605.

Elliott, J. M. 1970. Methods of sampling invertebrate drift in running water. Ann. Limnol. 6: 133–159.

Ellis, K. A., G. A. Surgeoner & C. R. Ellis. 1985. Versatile program for generating insect labels on an IBM-PC microcomputer. Can. Entomol. 117: 1447-1448.

Emden, H. F. van [ed.]. 1972. Aphid technology. Academic Press, London. [Collection and preservation on pp. 1–10, also much data on sampling, extraction, trapping, etc., in other chapters.]

Erzinclioglu, Y. Z. 1983. The application of entomology to forensic medicine. Med. Sci. Law, 23, 57-63.

Essig, E. O. 1942. College entomology. Macmillan, New York.

Essig, E. O. 1958. Insects and mites of western North America. Macmillan, New York.

Ettinger, W. S. 1979. A collapsible insect emergence trap for use in shallow standing water. Entomol. News 90: 114–117.

Evans, G. O. 1992. Principles of acarology. CAB International, Wallingford, CT.

Evans, G. O., J. G. Sheals & D. MacFarlane. 1964. The terrestrial Acari of the British Isles: An introduction to their morphology, biology and classification. Vol.1. Introduction and biology. Brit. Mus. (Nat. Hist.), London. [Techniques treated on pp. 61–88.]

Evans, L. J. 1975. An improved aspirator (pooter) for collecting small insects. Proc. Br. Entomol. Nat. Hist. Soc. Trans. 8: 8–11.

Everett, R. & J. L. Lancaster, Jr. 1968. A comparison of animal- and dry-ice-baited traps for the collection of tabanids. J. Econ. Entomol. 61: 863–864.

– F –

Fahy, E. 1972. An automatic separator for the removal of aquatic insects from detritus. J. Appl. Ecol. 6: 655–658.

Fallis, A. M. & S. M. Smith. 1964. Ether extracts from birds and CO_2 as attractants for some ornithophilic simuliids. Can. J. Zool. 42: 723–730.

Ferguson, L. M. 1990. Insecta: Microcoryphia and Thysanura. In: D. L. Dindal [ed.], Soil biology guide, pp. 935–949. John Wiley & Sons, New York.

Fessenden, G. R. 1949. Preservation of agricultural specimens in plastics. U.S. Dept. Agric. Misc. Publ. 679.

Finch, S. & G. Skinner. 1974. Some factors affecting the efficiency of water-traps for capturing cabbage root flies. Ann. Appl. Biol. 77: 213–226.

Fincher, G. T. & T. B. Stewart. 1968. An automatic trap for dung beetles. J. Ga. Entomol. Soc. 3: 11-12.

Fischer, R. W. & F. Jursic. 1958. Rearing houseflies and roaches for physiological research. Can. Entomol. 90: 1–7.

Flaschka, H. & J. Floyd. 1969. A simplified method of freeze-drying caterpillars. J. Lepid. Soc. 23: 43–48.

Fleming, W. E., et al. 1940. The use of traps against the Japanese beetle. U.S. Dep. Agric. Circ. 594.

Foote, R. H. 1948. A synthetic resin imbedding technique. U.S. Public Health Serv. CDC Bull. 1948 (July-Sept.): 58–59.

Ford, R. L. E. 1973. Studying insects, a practical guide. Warne, London & New York.

Fredeen, F.J.H. 1961. A trap for studying the attacking behavior of black flies *Simulium arcticum* Mall. Can. Entomol. 93: 73–78.

Freeman, T. J. 1972. Laboratory manual—slide mounting techniques. Calif. Dep. Agric., Lab., Serv. Entomol., Sacramento.

Frost, S. W. 1952. Light traps for insect collections, survey, and control. Pa. Agric. Exp. Stn. 550.

Frost, S. W. 1957. The Pennsylvania insect light trap. J. Econ. Entomol. 50: 287–292.

Frost, S. W. 1958. Insects attracted to light traps placed at different heights. J. Econ. Entomol. 51: 550–551.

Frost, S. W. 1964. Killing agents and containers for use with insect light traps. Entomol. News 75: 163–166.

Frost, S. W. 1970. A trap to test the response of insects to various light intensities. J. Econ. Entomol. 63: 1344–1346.

Furumizo, R. T. 1975. Collection and isolation of mites from housedust samples. Calif. Vector Views 22: 19–27.

References

– G –

Gary, N. E. & J. M. Marston. 1976. A vacuum apparatus for collecting honeybees and other insects in trees. Ann. Entomol. Soc. Am. 69: 287–288.

Gerberich, J. B. 1945. Rearing tree-hole organisms in the laboratory. Am. Biol. Teach. 7: 83–85.

Gerking, S. D. 1957. A method of sampling the littoral macrofauna and its application. Ecology 38: 219–226.

Gier, H. T. 1949. Differential stains for insect exoskeleton. J. Kans. Entomol. Soc. 22: 79–80.

Gillies, M. T. 1969. The ramp-trap, an unbaited device for flight studies of mosquitoes. Mosq. News 29: 189–193.

Gillies, M. T. & W. F. Snow. 1967. A CO_2-baited sticky trap for mosquitoes. Mosq. News 29: 189–193.

Gist, C. S. & D. A. Crossley. 1973. A method for quantifying pitfall trapping. Environ. Entomol. 2: 951–952.

Glasgow, J. P. & B. J. Duffy. 1961. Traps in field studies of *Glossina pallidipes* Austen. Bull. Entomol. Res. 52: 795–814.

Glen, D. M. 1976. An emergence trap for bark-dwelling insects; its efficiency and effects on temperature. Ecol. Entomol. 1: 91–94.

Glick, P. A. 1939. The distribution of insects, spiders, and mites in the air. U.S. Dep. Agric. Tech. Bull. 673.

Glick, P. A. 1957. Collecting insects by airplane in southern Texas. U.S. Dep. Agric. Tech. Bull. 1158.

Gojmerac, W. L. & E. C. Davenport. 1971. Tabanidae (Diptera) of Kegonsa State Park, Madison, Wisconsin: Distribution and seasonal occurrence as determined by trapping and netting. Mosq. News 31: 572–575.

Golding, F. D. 1941. Two new methods of trapping the cacao moths (*Ephestia cautella*). Bull. Entomol. Res. 32: 123–132.

Golding, F. D. 1946. A new method of trapping flies. Bull. Entomol. Res. 37: 143–154.

Goma, L. K. H. 1965. The flight activity of some East African mosquitoes (Diptera, Culicidae). I. Studies on a high steel tower in Zika forest, Uganda. Bull. Entomol. Res. 56: 17–35.

Goodenough, J. L. & J. W. Snow. 1973. Increased collection of tobacco budworm by electric grid traps as compared with blacklight and sticky traps. J. Econ. Entomol. 66: 450–453.

Goonewardene, H. F., B. G. Townshend, R. G. Bingham & R. Borton. 1973. Improved technique for field use of female Japanese beetles as lures. J. Econ. Entomol. 66: 396–397.

Gordh, G. & J. C. Hall. 1979. A critical point drier used as a method of mounting insects from alcohol. Entomol. News 90: 57–59.

Gordon, W. M. & E. J. Gerberg. 1945. A directional mosquito barrier trap. J. Natl. Malar. Soc. 4: 123–125.

Governatori, M., C. Bulgarini, F. Rivasi & S. Pampiglione. 1993. A new portable aspirator for Culicidae and other winged insects. J. Am. Mosq. Control Assoc. 9: 460–462.

Graham, H. M., P. A. Glick & J. P. Hollingsworth. 1961. Effective range of argon glow lamp survey traps for pink bollworm adults. J. Econ. Entomol. 54: 788–789.

Grandjean, F. 1949. Observation et conservation des tres petits arthropodes. Bull. Mus. Natl. Hist. Nat., ser. 2, 21: 363–370.

Granger, C. A. 1970. Trap design and color as factors in trapping the salt marsh greenhead fly. J. Econ. Entomol. 63: 1670–1672.

Gray, J. S. 1985. A carbon dioxide trap for prolonged sampling of *Ixodes ricinus* L. populations. Exp. Appl. Acarol. 1335–1344.

Gray, T. G. & A. I. Ibaraki. 1994. An economical cage for confining and collecting emerging insects and their parasites. Can. Entomol. 126: 447–448.

Gray, T. G., R. F. Shepherd, D. L. Struble, J. B. Byers & T. F. Maher. 1991. Selection of pheromone trap and attractant dispenser load to monitor black army cutworm *Actebia fennica*. J. Chem. Ecol. 17: 309–316.

Green, C. H. 1994. Bait methods for tsetse fly control. Adv. Parasitol. 34: 229–291.

Greenberg, B. & J. C. Kunich. 2002. Entomology and the law: Flies as forensic indicators. Cambridge University Press, London.

Greenslade, P. 1973. Sampling ants with pitfall traps: Digging-in effects. Insectes Soc. 20: 343–353.

Greenslade, P. & P. J. M. Greenslade. 1971. The use of baits and preservatives in pitfall traps. J. Aust. Entomol. Soc. 10: 253–260.

Greenslade, P. J. M. 1964. Pitfall trapping as a method for studying populations of Carabidae. J. Anim. Ecol. 33: 301–310.

Gressitt, J. L., J. Sedlacek, K. A. J. Wise & C. M. Yoshimoto. 1961. A high-speed airplane trap for airborne organisms. Pac. Insects 3: 549–555.

Grigarick, A. A. 1959. A floating pan trap for insects associated with the water surface. J. Econ. Entomol. 52: 348–349.

Grimstone, A. 1972. A guidebook to microscopical methods. Cambridge University Press, Cambridge, England.

Gruber, P. & C. A. Prieto. 1976. A collection chamber suitable for recovery of insects from large quantities of host plant material. Environ. Entomol. 5: 343–344.

Gui, H. L., L. C. Porter & G. F. Prideax. 1942. Response of insects to color intensity and distribution of light. Agric. Eng. 23: 51–58.

Gullan, P. J. & P. S. Cranston [eds.]. 2004. The insects: An outline of entomology, 3rd ed. Blackwell Publishing, Oxford England. [Example of general textbook, including collection, preservation and identification techniques.]

Gulliksen, B. & K. M. Deras. 1975. A diver-operated suction sampler for fauna on rocky bottoms. Oikos 26: 246–249.

Gurney, A. B., J. P. Kramer & G. C. Steyskal. 1964. Some techniques for the preparation, study, and storage in microvials of insect genitalia. Ann. Entomol. Soc. Am. 57: 240–242.

– H –

Hafraoui, A., E. J. Harris & A. Chakir. 1980. Plastic traps for detection and survey of the Mediterranean fruit fly, *Ceratitis capitata* (Diptera: Tephritidae) in Morocco. Proc. Hawaii. Entomol. Soc. 23: 199–203.

References

Haglund, W. D. and M. Sorg [eds.]. 2002. Advances in forensic taphonomy: Method, theory and archeological perspectives. CRC Press, Boca Raton, FL.

Halstead, J. A. & R. D. Haines. 1987. Flume collecting: A rediscovered collecting method with notes on insect extracting techniques. Pan-Pac. Entomol. 63: 383–431.

Hammond, H. E. 1960. The preservation of lepidopterous larvae using the inflation and heat-drying technique. J. Lepid. Soc. 14: 67–78.

Hanec, W. & G. K. Bracken. 1964. Seasonal and geographical distribution of Tabanidae (Diptera) in Manitoba, based on females captured in traps. Can. Entomol. 96: 1362–1369.

Hansens, E. J., E. M. Bosler & J. W. Robinson. 1971. Use of traps for study and control of saltmarsh greenhead flies. J. Econ. Entomol. 64: 1481–1486.

Hardwick, D. F. 1950. Preparation of slide mounts of lepidopterous genitalia. Can. Entomol. 82: 231-235.

Hardwick, D. F. 1968. A brief review of the principles of light trap design with a description of an efficient trap for collecting noctuid moths. J. Lepid. Soc. 22: 65–75.

Hargrove, J. W. 1977. Some advances in the trapping of tsetse (*Glossina* spp.) and other flies. Ecol. Entomol. 2: 123–137.

Harris, E. J., S. Nakagawa & T. Urago. 1971. Sticky traps for detection and survey of three tephritids. J. Econ. Entomol. 64: 62–65.

Harris, R. H. 1964. Vacuum dehydration and freeze-drying of entire biological specimens. Ann. Mag. Nat. Hist. 7(13): 65–74.

Harris, T. L. & W. P. McCafferty. 1977. Assessing aquatic insect flight behavior with sticky traps. Gr. Lakes Entomol. 10: 233–239.

Hartstack, A. W., J. P. Hollingsworth & D. A. Lindquist. 1968. A technique for measuring trapping efficiency of electric insect traps. J. Econ. Entomol. 61: 546–552.

Harwood, J. & S. Areekul. 1957. A rearing trap for producing pomace flies for bioassay of insecticides. J. Econ. Entomol. 50: 512–513.

Harwood, R. F. 1961. A mobile trap for studying the behavior of flying bloodsucking insects. Mosq. News 21: 35–39.

Hathaway, D. O. 1981. Codling moth: Field evaluations of blacklight and sex attractant traps. U.S. Dep. Agric., Agric. Res. Serv., AAT-W-19.

Hazeltine, W. R. 1962. A new insect clearing technique. J. Kans. Entomol. Soc. 35: 165–166.

Heathcote, G. D. 1957a. The comparison of yellow cylindrical, flat, and water traps to Johnson suction traps, for sampling aphids. Ann. Appl. Biol. 45: 133–139.

Heathcote, G. D. 1957b. The optimum size of sticky aphid traps. Plant Pathol. 6: 104–107.

Heathcote, G. D., J. M. P. Palmer & L. R. Taylor. 1969. Sampling for aphids by traps and by crop inspection. Ann. Appl. Biol. 63: 155–166.

Hendrix, W. H., III, & W. B. Showers. 1990. Evaluation of differently colored bucket traps for black cutworm and armyworm (Lepidoptera: Noctuidae). J. Econ. Entomol. 83: 596–598.

Herting, B. 1969. Tent window traps used for collecting tachinids (Dipt.) at Delémont, Switzerland. Commonw. Inst. Biol. Control Tech. Bull.

Hienton, T. E. 1974. Summary of investigations of electric insect traps. U.S. Dep. Agric. Tech. Bull. 1498.

Hill, M. N. 1971. A bicycle-mounted trap for collecting adult mosquitoes. J. Med. Entomol. 8: 108–109.

Hillis, D. M., C. Moritz, & B. K. Maple. 1966. Molecular Systematics, 2nd ed. Sinauer Associates, Sunderland, MA.

Hinton, H. E. 1951. Myrmecophilous Lycaenidae and other Lepidoptera — a summary. Proc. So. London Entomol. Nat. Hist. Soc. 1949-1950: 111–175.

Hocking, B. 1953. Plastic embedding of insects — a simplified technique. Can. Entomol. 85: 14–18.

Hocking, B. 1963. The use of attractants and repellents in vector control. Bull. W.H.O. 29 (suppl.): 121–126.

Hocking, B. & G. E. Hudson. 1974. Insect wind traps: Improvements and problems. Queensl. Entomol. 10: 275–284.

Hodgson, C. E. 1940. Collection and laboratory maintenance of Dytiscidae (Coleoptera). Entomol. News 64: 36–37.

Hoffard, W. H. 1980. How to collect and prepare forest insect and disease organisms and plant specimens for identification. U.S. Dep. Agric., Forest Serv., Southeast Area, State and Private Forestry, Gen. Rep. SA-GR13.

Holbrook, J. E. R. 1927. Apparatus and method used to remove pins from insect specimens. J. Econ. Entomol. 20: 642–643.

Holbrook, R. F. & M. Beroza. 1960. Gypsy moth (*Porthetria dispar*) detection with the natural female sex lure. J. Econ. Entomol. 53: 751–756.

Hollingsworth, J. P. & A. W. Hartstack, Jr. 1972. Effect of components on insect light trap performance. Trans. Am. Soc. Agric. Eng. 15: 924–927.

Hollingsworth, J. P., C. P. Briggs, P. A. Glick & H. M. Graham. 1961. Some factors influencing light trap collections. J. Econ. Entomol. 54: 305–308.

Hollingsworth, J. P., J. G. Hartsock & J. M. Stanley. 1963. Electrical insect traps for survey purposes. U.S. Dep. Agric., Agric. Res. Serv. 42-3-1.

Hollis, D. [ed.]. 1980. Animal identification: A reference guide. Vol. 3, Insects. Brit. Mus. Nat. Hist., London; John Wiley & Sons, New York.

Holtcamp, R. H. & J. I. Thompson. 1985. A lightweight, self-contained insect suction sampler. J. Aust. Entomol. Soc. 24: 310–302.

Holzapfel, E. P., H. B. Clagg & M. L. Goff. 1978. Trapping of air-borne insects on ships on the Pacific. Pt. 9. Pac. Insects 19: 65–90.

Hood, J. D. 1947. Microscopical whole-mounts of insects, 3rd ed. Cornell University, Ithaca, NY. [Mimeographed; out of date, but of considerable interest.]

Hopkins, G. H. E. 1949. The host associations of the lice of mammals. Proc. Zool. Soc. (London), 119 (pt. II): 387–604. [Methods on pp. 395–401.]

Hottes, F. C. 1951. A method for taking aphids in flight. Pan-Pac. Entomol. 27: 190.

Houseweart, M. W., D. T. Jennings & J. C. Rea. 1979. Large-capacity pitfall trap. Entomol. News 90: 51–54.

References

Howell, J. F. 1980. Codling moth: Blacklight trapping and comparisons with fermenting molasses bait and pheromone traps. U.S. Dep. Agric., Agric. Res. Serv., ARR-W-22.

Howell, J. R., Jr., M. Cheikh & E. J. Harris. 1975. Comparison of the efficiency of three traps for the Mediterranean fruit fly baited with minimum amounts of Trimedlure. J. Econ. Entomol. 68: 277–279.

Hower, R. O. 1979. Freeze-drying biological specimens: A laboratory manual. Smithsonian Institution Press, Washington, DC.

Howland, A. F, J. W. Debolt, W. W. Wolf, H. H. Toba, T. Gibson & A. N. Kishaba. 1969. Field and laboratory studies of attraction of the synthetic sex pheromone to male cabbage looper moths. J. Econ. Entomol. 62: 117–122.

Hoy, J. B. 1970. Trapping the stable fly by using CO_2 or CO as attractants. J. Econ. Entomol. 63: 792–795.

Hubbell, T. H. 1956. A new collecting method: The oatmeal trail. Entomol. News 67: 49–51.

Huber, J. T. 1986. Systematics, biology, and hosts of the Mymaridae and Mymarommatidae (Insecta: Hymenoptera): 1758–1984. Entomography 4: 185–243.

Huffacker, C. B. & R. C. Back. 1943. A study of methods of sampling mosquito populations. J. Econ. Entomol. 36: 561–569.

Humason, G. L. 1979. Animal tissue techniques, 4th ed. W. H. Freeman, San Francisco.

Hurd, P. D. 1954. Myiasis resulting from the use of the aspirator method in the collection of insects. Science (Washington, DC) 119: 814–815.

– I –

Inouye, D. W. 1991. Quick and easy insect labels. J. Kans. Entomol. Soc. 64: 242–243.

– J –

Jacobson, M. 1972. Insect sex pheromones. Academic Press. New York.

Jacobson, M. & M. Beroza. 1964. Insect attractants. Sci. Am. 211 (No. 2, Aug.): 20–27. [Includes excellent figures of Steiner, McPhail, and USDA beetle traps.]

Jacques, H. E. 1947. How to know the insects, 2nd ed. Wm. C. Brown, Dubuque, IA. (See also Bland & Jacques 1978.)

James, S. H. and J. J. Nordby [eds.]. 2005. Forensic science: An introduction to scientific and investigative techniques. 2nd ed. Taylor & Francis, New York.

Jenkins, J. 1991. A simple device to clean insect specimens for museums and scanning electron microscopy. Proc. Entomol. Soc. Wash. 93: 204–205.

Jeppson, L. R. & H. H. Kiefer. 1975. Mites injurious to economic plants. University of California Press, Berkeley. [Includes data on collection and preservation.]

Johnson, C. G. 1950. The comparison of suction trap, sticky trap, and tow-net for the quantitative sampling of small airborne insects. Ann. Appl. Biol. 37: 268–285.

Johnson, C. G. & L. R. Taylor. 1955. Development of large suction traps for airborne insects. Ann. Appl. Biol. 43: 51–62.

Johnson, C. G., T. R. E. Southwood & H. M. Entwistle. 1957. A new method of extracting arthropods and molluscs from grassland and herbage with a suction apparatus. Bull. Entomol. Res. 48: 211–218.

Jonasson, P. M. 1954. An improved funnel trap for capturing emerging aquatic insects, with some preliminary results. Oikos 5: 179–188.

Joossee, E. N. G. 1975. Pitfall-trapping as a method for studying surface-dwelling Collembola. Z. Morphol. Oekol. Tiere 55: 587–596.

– K –

Kasprzak, K. 1993. Methods for fixing and preserving soil animals. In: M. Gorny & L. Grum [eds.], Methods on soil zoology, pp. 321–345. Elsevier, Amsterdam.

Kempson, D., M. Lloyd & R. Ghelardi. 1962. A new extractor for woodland litter. Pedobiologia 3: 1–21.

Kevan, D. K. McE. [ed.]. 1955. Soil zoology. Butterworth, London.

Kevan, D. K. McE. 1962. Soil animals. Witherby, London. [Sampling and extraction on pp. 102–125.]

Kieckhefer, R. W., D. A. Dickmann & E. L. Miller. 1976. Color responses of cereal aphids. Ann. Entomol. Soc. Am. 69: 721–724.

Kim, K. C., H. D. Pratt & C. J. Stojanovich. 1986. The sucking lice of North America. Pennsylvania State University Press, University Park.

Kimerle, R. A. & N. H. Anderson. 1967. Evaluation of aquatic insect emergence traps. J. Econ. Entomol. 60: 1255–1259.

Kissinger, D. G. 1982. Insect label production using a personal computer. Proc. Entomol. Soc. Wash. 84: 855–857.

Klein, M. G., K. O. Lawrence & T. L. Ladd, Jr. 1973. Japanese beetles: Shielded traps to increase captures. J. Econ. Entomol. 66: 562–563.

Klots, A. B. 1932. Directions for collecting and preserving insects. Ward's Nat. Sci. Estab., Rochester, NY.

Knodel, J. J. & A. M. Agnello. 1990. Field comparison of nonsticky and sticky pheromone traps for monitoring fruit pests in western New York. J. Econ. Entomol. 63: 197–204.

Knox, P. C. & K. L. Hays. 1972. Attraction of *Tabanus* spp. (Diptera: Tabanidae) to traps baited with carbon dioxide and other chemicals. Environ. Entomol 1: 323–326.

Knudsen, J. W. 1966. Biological techniques. Harper & Row, New York.

Knudsen, J. W. 1972. Collecting and preserving plants and animals. Harper & Row, New York. [Insects on pp. 128–176.]

Kogan, M. & D. C. Herzog [eds.]. 1980. Sampling methods in soybean entomology. Springer Verlag, New York.

Kosztarab, M. 1966. How to build a herbarium of insect and mite damage. Turtox News 44: 290–294.

Kovrov, B. G. & A. S. Monchadskii. 1963. The possibility of using polarized light to attract insects. Entomol. Obozr. 42: 49–55. (In Russian; transl. into English in Entomol. Rev. 42: 25–28.)

Krantz, G. W. 1978. A manual of acarology, 2nd ed. Oregon State University Book Stores, Corvallis. [Collection and preservation on pp. 77–98.]

Kring, J. B. 1970. Red spheres and yellow panels combined to attract apple maggot flies. J. Econ. Entomol. 63: 466–469.

Krombein, K. V. 1967. Trap-nesting wasps and bees: Life histories, nests, and associates. Smithsonian Institution Press, Washington, DC. [Trap-nesting technique described on pp. 8–14.]

Kronblad, W. & S. Lundberg. 1978. Bilhåvning-en intressant fangstmetod för skalbaggar och andra insekter. Entomol. Tidskr. 99: 115–118. [Describes net attached to roof of automobile.]

Kühnelt, W. 1976. Soil biology, with special reference to the animal kingdom. Michigan State University Press, East Lansing. [Observation and collecting techniques, pp. 35–65; bibliography, pp. 385–466.]

– L –

LaGasa, E. H. & S. D. Smith. 1978. Improved insect emergence trap for stream community population sampling: Environmental impact of insecticides. U.S. Dep. Agric. Res. Note PSW, U.S. Pac. Southwest. Forest and Range Exp. Stn. 328.

Laird, N. 1981. Blackflies: The future biological methods in integrated control. Academic Press, London. [Includes chapters on trapping and sampling methods.]

Lambert, H. L. & R. T. Franklin. 1967. Tanglefoot traps for detection of the balsam woolly aphid. J. Econ. Entomol. 60: 1525–1529.

Lammers, R. 1977. Sampling insects with a wetland emergence trap: Design and evaluation of the trap with preliminary results. Am. Midl. Nat. 97: 381–389.

Landlin, J. 1976. Methods of sampling aquatic beetles in the transitional habitats at water margin. Freshwater Biol. 6: 81-87.

Lane, J. 1965. The preservation and mounting of insects of medical importance. WHO/Vector Control 152.65. Mimeographed.

Lane, R. S. & J. R. Anderson. 1976. Extracting larvae of *Chrysops hirsuticallus* (Diptera: Tabanidae) from soil: Efficiency of two methods. Ann. Entomol. Soc. Am. 69: 854–856.

Langford, T. E. & J. R. Daffern. 1975. The emergence of insects from a British river warmed by power station cooling water. Pt. 1. The use and performance of insect emergence traps. Hydrobiologia 46: 71–114.

Lawson, D. L. & R. W. Merritt. 1979. A modified Ladell apparatus for the extraction of wetland macroinvertebrates. Can. Entomol. 111: 1389–1393.

Lee, W. L., B. M. Bell & J. F. Sutton [eds.]. 1982. Guidelines for acquisition and management of biological specimens. Assoc. Syst. Collect., Lawrence, KS.

Leech, H. B. 1955. Cheesecloth flight traps for insects. Can. Entomol. 87: 200.

Lehker, G. E & H. O. Deay. 1969. How to collect, preserve, and identify insects. Purdue Univ., Coop. Ext. Serv., Ext. Circ. 509.

LeSage, L. 1991. An improved technique for collecting large samples of arthropods. Entomol. News 102: 97.

LeSage, L. & A. D. Harrison. 1979. Improved traps and techniques for the study of emerging aquatic insects. Entomol. News 90: 65–78.

Levin, M. D. 1957. Artificial nesting burrows for *Osmia ligniaria* Say. J. Econ. Entomol. 50: 506–507.

Lewis, G. G. 1965. A new technique for spreading minute moths. J. Lepid. Soc. 19: 115–116.

Lewis, T. & L. R. Taylor. 1965. Diurnal periodicity of flight by insects. Trans. R. Entomol. Soc. Lond. 116: 393–469.

Lincoln, R. J. & J. G. Sheals. 1979. Invertebrate animals. Collection and preservation. Brit. Mus. (Nat. Hist.), Cambridge University Press, Cambridge, England.

Lindeberg, B. 1958. A new trap for collecting emerging insects from small rockpools, with some examples of the results obtained. Ann. Entomol. Fenn. 24: 186–191.

Lindroth, C. H. 1957. The best method for killing and preserving beetles. Coleopt. Bull. 11: 95-96.

Lipovsky, L. J. 1951. A washing method of ectoparasite recovery with particular reference to chiggers. J. Kans. Entomol. Soc. 24: 151–156.

Lipovsky, L. J. 1953. Polyvinyl alcohol with lacto-phenol, a mounting and clearing medium for chigger mites. Entomol. News 64: 42–44.

Loschiavo, S. R. 1974. The detection of insects by traps in grain-filled boxcars during transit. Proc. Int. Working Conf. Stored-Product Entomol. 1: 639–650.

Lowe, R. B. & L. G. Putnam. 1964. Some simple and useful technological improvements in light traps. Can. Entomol. 96: 129.

Luff, M. L. 1968. Some effects of formalin on the numbers of Coleoptera caught in pitfall traps. Entomol. Mon. Mag. 104: 115–116.

Luff, M. L. 1975. Some factors influencing the efficiency of pitfall traps. Oecologia (Berl.) 19: 345–357.

Lumsden, W. H. R. 1958. A trap for insects biting small vertebrates. Nature (Lond.) 181: 819–820.

– M –

Macan, T. T. 1964. Emergence traps and the investigation of stream faunas. Rev. Hydrobiol. 3: 75–82.

MacFadyen, A. 1962. Soil arthropod sampling. Adv. Ecol. Res. 1: 1–34.

Macleod, J. & J. Donnelly. 1956. Methods for the study of blowfly populations. I. Bait trapping: significance of limits for comparative sampling. Ann. Appl. Biol. 44: 80–104.

MacSwain, J. W. 1949. A method for collecting male Stylops. Pan-Pac. Entomol. 25: 89–90.

Martin, J. E. H. 1977. The insects and arachnids of Canada. Pt. 1. Collecting, preparing, and preserving insects, mites, and spiders. Can. Dep. Agric., Biosyst. Res. Inst. Publ. 1643.

Masner, L. & G. A. P. Gibson. 1979. The separation bag — a new device to aid in collecting insects. Can. Entomol. 111: 1197–1198.

Mason, H. C. 1963. Baited traps for sampling *Drosophila* populations in tomato field plots. J. Econ. Entomol. 56: 897–899.

References

Mason, W. T. & J. E. Sublette. 1971. Collecting Ohio River basin Chironomidae (Diptera) with a floating sticky trap. Can. Entomol. 103: 397–404.

Masteller, E. C. 1977. An aquatic insect emergence trap on a shale stream in western Pennsylvania. Melsheimer Entomol. Ser. 23: 10–15.

Matheson, R. 1951. Entomology for introductory courses, 2nd ed. Comstock, Ithaca, NY.

Maxwell, C. W. 1965. Tanglefoot traps as indicators of apple maggot activities. Can. Entomol. 97: 110.

McCafferty, W. P. 1981. Aquatic entomology. Jones & Bartlett, Boston, MA.

McCauley, V. J. E. 1976. Efficiency of a trap for catching and retaining insects emerging from standing water. Oikos 27: 339–346.

McClain, D. C., G. C. Rock & J. B. Woolley. 1990. Influence of trap color and San Jose scale (Homoptera: Diaspidiidae) pheromone on sticky trap catches of 10 aphelinid parasitoids (Hymenoptera). Environ. Entomol. 19: 926–931.

McClung, C. E. (Jones, R. McC., ed.). 1964. Handbook of microscopical technique for workers in animal and plant tissues, 3rd ed. Hafner, New York.

McDaniel, B. 1979. How to know the mites and ticks. Wm. C. Brown, Dubuque, IA.

McDonald, J. L. 1970. A simple, inexpensive alcohol light trap for collecting Culicoides. Mosq. News 30: 652–654.

McKillop, W. B. & W. B. Preston. 1981. Vapor degreasing: A new technique for degreasing and cleaning entomological specimens. Can. Entomol. 113: 251–253.

McNutt, D. N. 1976. Insect collecting in the tropics. Center for Overseas Pest Research, London.

Menzies, D. R. & E. A. C. Hagley. 1977. A mechanical trap for sampling populations of small, active insects. Can. Entomol. 109: 1405–1407.

Merrill, W. & J. M. Skelly. 1968. A window trap for collection of insects above the forest canopy. J. Econ. Entomol. 61: 1461–1462.

Merritt, R. W. & J. H. Poorbaugh, Jr. 1975. A laboratory collection and rearing container for insects emerging from cattle droppings. Calif. Vector Views 22: 43–46.

Merritt, R. W., K. W. Cummins & V. H. Resh. 1984. Collecting, sampling, and rearing methods for aquatic insects. In: R. W. Merritt & K. W. Cummins [eds.], An introduction to the aquatic insects of North America, 2nd ed., pp 11–26. Kendall/Hunt, Dubuque, IA.

Meyerdirk, D. E., W. G. Hart & J. Burnside. 1979. Evaluation of a trap for the citrus blackfly, *Aleurocanthus woglumi* (Homoptera: Aleyrodidae). Can. Entomol. 111: 1127–1129.

Meyers, E. G. 1959. Mosquito collections by light traps at various heights above the ground. Proc. Calif. Mosq. Control Assoc. 27: 61–63.

Miller, R. S., S. Passoa, R. D. Waltz & V. Mastro. 1993. Insect removal from sticky traps using a citrus oil solvent. Entomol. News. 104: 209–213.

Miller, T. A., R. G. Stryker, R. N. Wilkinson & S. Esah. 1970. The influence of moonlight and other environmental factors on the abundance of certain mosquito species in light-trap collections in Thailand. J. Med. Entomol. 7: 555–561.

Minter, D. M. 1961. A modified Lumsden suction trap for biting insects. Bull. Entomol. Res. 52: 233–238.

Mitchell, E. R., J. C. Webb, A. H. Baumhover, R. W. Hines, J. W. Stanley, R. G. Endris, D. A. Lindquist, & S. Masuda. 1972. Evaluation of cylindrical electric grids as pheromone traps for loopers and tobacco hornworms. Environ. Entomol. 1: 365–368.

Mitchell, E. R., J. C. Webb, J. C. Benner, Jr. & J. M. Stanley. 1973. A portable cylindrical electric grid trap. J. Econ. Entomol. 66: 1232–1233.

Mitchell, E. R., J. M. Stanley, J. C. Webb & A. H. Baumhover. 1974. Cylindrical electric grid traps: The influence of elevation, size and electrode spacing on captures of male cabbage loopers and tobacco hornworms. Environ. Entomol. 3: 49–50.

Mitchell, R. D. & D. R. Cook. 1952. The preservation and mounting of water-mites. Turtox News 30: 169–172.

Montgomery, S. L. 1982. Biogeography of the moth genus *Eupithecia* in Oceana and the evolution of ambush predation in Hawaiian caterpillars (Lepidoptera: Geometridae). Entomol. Gen. 8: 549–556.

Montgomery, S. L. 1983. Carnivorous caterpillars: the behavior, biogeography and conservation of *Eupithecia* (Lepidoptera: Geometridae) in the Hawaiian Islands. GeoJournal 7.6: 549-556.

Moore, R, R. T. Clarke & S. Creer. 1993. An insect sorting device to be used in conjunction with insect suction samplers. Bull. Entomol. Res. 83: 113–120.

Moreland, C. R. 1955. A wind frame for trapping insects in flight. J. Econ. Entomol. 47: 944.

Morgan, C. V. G. & N. H. Anderson. 1958. Techniques for biological studies of tetranychid mites. Can. Entomol. 90: 212–215.

Morgan, N. C., A. W. Waddell & W. B. Hall. 1963. A comparison of the catches of emerging aquatic insects in floating box and submerged funnel traps. J. Anim. Ecol. 32: 203–219.

Morgan, N. O. & E. C. Uebel. 1974. Efficacy of the Assateague insect trap in collecting mosquitoes and biting flies in a Maryland salt marsh. Mosq. News 34: 196–199.

Morrill, W. L. 1975. Plastic pitfall trap. Environ. Entomol. 4: 596.

Morrill, W. L. & W. H. Whitcomb. 1972. A trap for alate imported fire ants. J. Econ. Entomol. 65: 1194–1195.

Morris, C. D. & G. R. DeFoliart. 1969. A comparison of mosquito catches with miniature light traps and CO_2-baited traps. Mosq. News 29: 424–426.

Morris, K. R. S. 1961. Effectiveness of traps in tsetse surveys in the Liberian rain forest. Am. J. Trop. Med. Hyg. 10: 905–913.

Mulhern, T. D. 1942. New Jersey mechanical trap for mosquito surveys. N.J. Agric. Exp. Stn. Circ. 421: 1–8.

Mullen, M. A. 1992. Development of a pheromone trap for monitoring *Tribolium castaneum*. J. Stored Prod. Res. 28: 245–249.

Mullen, M. A., H. A. Highland, R. E. Taggart & B. W. Lingren. 1992. Insect monitoring system. U.S. Dep. Agric. patent 5,090,153.

Muma, M. H. 1975. Long-term can trapping for population analyses of ground-surface, arid-land arachnids. Fla. Entomol. 58: 257–270. [Includes bibliography of can and pitfall trapping.]

Mundie, J. H. 1956. Emergence traps for aquatic insects. Mitt. Int. Ver. Theor. Angew. Limnol. 7: 113.

Mundie, J. H. 1964. A sampler for catching emerging insects and drifting materials in streams. Limnol. Oceanogr. 9: 456–459.
Mundie, J. H. 1966. Sampling emerging insects and drifting materials in deep flowing water. Gewaess. Abwaess. 41/42: 159–162.
Mundie, J. H. 1971. Insect emergence traps. In: W. T. Edmondson & G. G. Winberg [eds.], Manual for estimating secondary production in fresh waters, pp. 80–93. Blackwell Scientific (IBP Handb. 17), Oxford, England. [References on pp. 80–93.]
Murphy, D. H. & H. Gisin. 1959. The preservation and microscopic preparation of anopheline eggs in a lacto-glycerol medium. Proc. R. Entomol. Soc. Lond. Ser. A, Gen. Entomol. 34 (10–12): 171–174.
Murphy, P. W. [ed.]. 1962. Progress in soil zoology. Butterworth, London. [Includes reference on sticky traps, pp. 226–227.]
Murphy, W. L. 1985. Procedures for removal of insect specimens from sticky-trap material. Ann. Entomol. Soc. Am. 78: 881.
Murray, T. D. & W. N. Charles. 1975. A pneumatic grab for obtaining large, undisturbed mud samples: its construction and some applications for measuring the growth of larvae and emergence of adult Chironomidae. Freshwater Biol. 5: 205–210.

– N –

Nakagawa, S., D. L. Chambers, T. Urago & R. T. Cunningham. 1971. Trap-lure combinations for surveys of Mediterranean fruit flies in Hawaii. J. Econ. Entomol. 64: 1211–1213.
Nakagawa, S., D. Suda, T. Urago & E. J. Harris. 1975. Gallon plastic tub, a substitute for the McPhail trap. J. Econ. Entomol. 68: 405–406.
Nantung Institute of Agriculture. 1975. Effects of light traps equipped with two lamps on capture of insects. Acta Entomol. Sin. 18: 289–294. (In Chinese, with English summary.)
Neal, J. W., Jr. [ed.]. 1979. Pheromones of the Sesiidae (formerly Aegeriidae). U.S. Dep. Agric., Agric. Res. Serv., ARR-NE-6. [Proceedings of a symposium, including trapping and perspective of pheromone research.]
Needham, J. G. [ed.]. 1937. Culture methods for invertebrate animals. Comstock, Ithaca, NY. [Includes methods of collecting and rearing some insects.]
Newell, I. M. 1955. An autosegregator for use in collecting soil-inhabiting arthropods. Trans. Am. Microsc. Soc. 74: 389–392.
Newhouse, V. T., R. W. Chamberland, J. G. Johnston & W. D. Sudia. 1966. Use of dry ice to increase mosquito catches of the CDC miniature light trap. Mosq. News 26: 30–35.
Newton, A. F., Jr. 1990. Insecta: Coleoptera Staphylinidae adults and larvae. In: D. L. Dindal [ed.], Soil biology guide, pp. 1137–1174. John Wiley & Sons, New York.
Newton, A. F., Jr., & S. B. Peck. 1975. Baited pitfall traps for beetles. Coleopt. Bull. 29: 45–46.
Nicholls, C. F. 1960. A portable mechanical insect trap. Can. Entomol. 92: 48–51.
Nicholls, C. F. 1970. Some entomological equipment, 2nd ed. Can. Dep. Agric. Inf. Bull. 2.

Nichols, S. W. & R. T. Schuh [eds.]. 1989. The Torre-Bueno glossary of entomology. New York Entomol. Soc.

Nielsen, B. O. 1974. Registrering af insektakivitet på bøgestammer ved hjaelp af fangtragte. Entomol. Meddel. 42: 1–18. [Describes catches of insects in funnel traps on trunks of beech trees.]

Nielsen, E. T. 1960. A note on stationary nets. Ecology 41: 375–376.

Nijholt, W. W. & J. A. Chapman. 1968. A flight trap for collecting living insects. Can. Entomol. 100: 1151–1153.

Norris, K. R. 1966. The collection and preservation of insects. Aust. Entomol. Soc. Handb. 1.

– O –

Oldroyd, H. 1958. Collecting, preserving and studying insects. Hutchinson, London; Macmillan, New York.

Olmi, M. 1984a. A Revision of the Dryinidae (Hymenoptera). Mem. Am. Entomol. Inst. (Gainesville) 37(1): 1–946.

Olmi, M. 1984b. A revision of the Dryinidae (Hymenoptera). Mem. Am. Entomol. Inst. (Gainesville) 37(2): 947–1913.

Onsager, J. A. 1976. Influence of weather on capture of adult southern potato wireworm in blacklight traps. U.S. Dep. Agric. Tech. Bull. 1527.

– P –

Parker, S. P. [ed.]. 1982. Synopsis and classification of living organisms. Vol. 2. McGraw-Hill, New York.

Parkman, J. P. & J. H. Frank. 1993. Use of a sound trap to inoculate *Steinernema scapterisci* (Rhabditida: Steinernematidae) into pest mole cricket populations (Orthoptera: Gryllotalpidae). Florida Entmol. 65: 105-110.

Parman, D. C. 1931. Construction of the box-type trap for eye gnats and blow flies. U.S. Dep. Agric. Bull. E-299.

Parman, D. C. 1932. A box-type trap to aid in the control of eye gnats and blowflies. U.S. Dep. Agric. Circ. 247.

Peacock, J. W. & R. A. Cuthbert. 1975. Pheromone-baited traps for detecting the smaller European elm bark beetle. Coop. Econ. Insect Rep. 25: 497–500.

Pearce, M. J. 1990. A new trap for collecting termites and assessing their foraging activity. Trop. Pest Manag. 36: 310–311.

Peck, S. B. 1990. Insecta: Coleoptera Silphidae and the associated families Agrytidae and Leiodidae. In: D. L. Dindal [ed.], Soil biology guide, pp. 1113-1136. John Wiley & Sons, New York.

Peck, S. B. & A. E. Davies. 1980. Collecting small beetles with large-area "window" traps. Coleopt. Bull 34: 237–239.

Pennak, R. W. 1978. Fresh-water invertebrates of the United States, 2nd ed. John Wiley & Sons, New York. [Apparatus, formulas, etc. on pp. 769–782.]

Pennington, N. E. 1967. Comparison of DDVP and cyanide as killing agents in mosquito light traps. J. Med. Entomol. 4: 518.

Pérez Pérez, R. & S. D. Hensley. 1973. A comparison of pheromone and blacklight traps for attracting sugar borer [*Diatraea saccharalis* (F.)] adults from a natural population. P. R. J. Agric. 57: 320–329.

Peterson, A. 1948. Larvae of insects. Pt. 1. Lepidoptera and Hymenoptera. Pt. 2. Coleoptera, Diptera, Neuroptera, Siphonaptera, Mecoptera. Edwards, Ann Arbor, MI.

Peterson, A. 1964. Entomological techniques, 10th ed. Edwards, Ann Arbor, MI.

Peterson, B. V., J. W. McWade & E. F. Bord. 1961. A simple method for preparing uniform minuten-pin double mounts. Bull. Brooklyn Entomol. Soc. 56: 19–21.

Pickens, L. G., N. O. Morgan & R. W. Miller. 1972. Comparison of traps and other methods for surveying density of populations of flies in dairy barns. J. Econ. Entomol. 65: 144–145.

Pieczynski, E. 1961. The trap method of capturing water mites (Hydracarina). Ekol. Pol. B7: 111–115.

Pinniger, D. B. 1975. The use of bait traps for assessment of stored-product insect populations. Coop. Econ. Insect Rep. 25: 907909.

Powers, W. J. 1969. A light-trap bag for collecting live insects. J. Econ. Entomol. 62: 735–736.

Pratt, H. D. 1944. Studies on the comparative attractiveness of 25-, 50-, and 100-watt bulbs for Puerto Rican *Anopheles*. Mosq. News 4: 17–18.

Preiss, F. J., H. L. Barefoot, R. G. Stryker & W. W. Young. 1970. Effectiveness of DDVP as a killing agent in mosquito killing jars. Mosq. News 30: 417–419.

Pritchard, M. H. & G. O. W. Kruse. 1982. The collection and preservation of animal parasites. Univ. Nebr. Tech. Bull. 1.

Prokopy, R. J. 1968. Sticky spheres for estimating apple maggot adult abundance. J. Econ. Entomol. 61: 1082–1085.

Prokopy, R. J. 1973. Dark enamel spheres capture as many apple maggot flies as fluorescent spheres. Environ. Entomol. 2: 953–954.

– R –

Race, S. R. 1960. A comparison of two sampling techniques for lygus bugs and stink bugs on cotton. J. Econ. Entomol. 53: 689–690.

Reeves, R. M. 1980. Use of barriers with pitfall traps. Entomol. News 91: 10–12.

Reeves, W. C. 1951. Field studies on carbon dioxide as possible host stimulant to mosquitoes. Proc. Soc. Exp. Biol. Med. 77: 64–66.

Reeves, W. C. 1953. Quantitative field studies on a carbon dioxide chemotropism of mosquitoes. Am. J. Trop. Med. Hyg. 2: 325–331.

Reierson, D. A. & R. E. Wagner. 1975. Trapping yellowjackets with a new standard plastic wet trap. J. Econ. Entomol. 68: 395–398.

Rennison, B. D. & D. H. Robertson. 1959. The use of carbon dioxide as an attractant for catching tsetse. Rep. E. Afr. Trypan. Res. Organ. 1958: 26.

Richards, O. W. & R. G. Davies. 1977. Imm's general textbook of entomology, 10th ed. Vol. 1, Structure, physiology, and development; 2, Classification and biology. Chapman & Hall, London; John Wiley & Sons, New York.

Richards, W. R. 1964. A short method for making balsam mounts of aphids and scale insects. Can. Entomol. 96: 963–966.

Riley, C. V. 1892. Directions for collecting and preserving insects. Bull. U.S. Natl. Mus., part F 39.

Riley, G. B. 1957. A modified sweep-net for small insects. Bull. Brooklyn Entomol. Soc. 52: 95–96.

Roberts, R. H. 1972. Relative attractiveness of CO_2 and a steer to Tabanidae, Culicidae, and Stomoxys calcitrans (L.). Mosq. News 32: 208–211.

Roe, R. M. & C. W. Clifford. 1976. Freeze-drying of spiders and immature insects using commercial equipment. Ann. Entomol. Soc. Am. 69: 497–499.

Rogers, D. J. & D. T. Smith. 1977. A new electric trap for tsetse flies. Bull. Entomol. Res. 67: 153–159.

Rogoff, W. M. 1978. Methods for collecting eye gnats (Diptera: Chloropidae). U.S. Dep. Agric., Agric. Res. Serv., ARM-W-2 (iii). [Includes extensive bibliography.]

Rohlf, F. J. 1957. A new technique in the preserving of soft-bodied insects and spiders. Turtox News 35: 226–229.

Roling, M. P. & W. H. Kearby. 1975. Seasonal flight and vertical distribution of Scolytidae attracted to ethanol in an oak-hickory forest in Missouri. Can. Entomol. 107: 1315–1320.

Ross, H. H., C. A. Ross, & J. R. P. Ross. 1982. A textbook of entomology, 4th ed. John Wiley & Sons, New York.

Rudd, W. G. & R. L. Jensen. 1977. Sweep net and ground cloth sampling for insects in soybeans. J. Econ. Entomol. 70: 301–304.

– S –

Sabrosky, C. W. 1966. Mounting insects from alcohol. Bull. Entomol. Soc. Am. 12: 349.

Sabrosky, C. W. 1971. Packing and shipping of pinned insects. Bull. Entomol. Soc. Am. 17: 6–8.

Saferstein, R. 2004. Criminalistics: An introduction to forensic science. Prentice Hall, Upper Saddle River, NJ.

Salmon, J. T. 1946. A portable apparatus for the extraction from leaf mould of Collembola and other minute organisms. Dom. Mus. Rec. Entomol. (Wellington) 1: 13–18.

Salt, G. & F.S.J. Hollick. 1944. Studies of wireworm populations. 1. A census of wireworms in pasture. Ann. Appl. Biol. 31: 53–64.

Sanders, D. P. & R. C. Dobson. 1966. The insect complex associated with bovine manure in Indiana. Ann. Entomol. Soc. Am. 59: 955–959.

Sauer, R. J. 1976. Rearing insects in the classroom. Am. Biol. Teach. 1976 (Apr.): 216–221.

Schlee, D. 1966. Präparation und Ermittlung von Messwerton an Chironomiden (Diptera). Gewaess. Abwaess. 41/42: 169–192. [Describes preparation of Chironomidae for study.]

Schmid, J. M., J. C. Mitchell & M. H. Schroeder. 1973. Bark beetle emergence cages modified for use as pitfall traps. U.S. Dep. Agric., For. Serv. RM-244.

References

Scott, J. A. 1986. The butterflies of North America: A natural history and field guide. Stanford University Press, Stanford, CA.

Seber, G.A.F. 1973. The estimation of animal abundance. Griffin, London.

Service, M. W. 1976. Mosquito ecology: Field sampling methods. Applied Science, London. [Includes extensive bibliographies.]

Shipley, A. E. 1904. The orders of insects. Zool. Anz. 28: 259–262.

Sholdt, L. L. & P. Neri. 1974. Mouth aspirator with holding cage for collecting mosquitoes and other insects. Mosq. News 34: 236.

Shorey, H. H. 1973. Behavioral responses to insect pheromones. Annu. Rev. Entomol. 18: 349–380.

Shorey, H. H. & J. J. McKelvey, Jr. 1977. Chemical control of insect behavior: Theory and application. John Wiley & Sons, New York.

Shubeck, P. P. 1976. An alternative to pitfall traps in carrion beetle studies (Coleoptera). Entomol. News 87: 176–178.

Singer, G. 1964. A simple aspirator for collecting small arthropods directly into alcohol. Ann. Entomol. Soc. Am. 57: 796–798.

Singer, G. 1967. A comparison between different mounting techniques commonly employed in acarology. Acarologia 9: 475-484.

Sladeckova, A. 1962. Limnological investigation methods for the periphyton ("aufwuchs") community. Bot. Rev. 28: 286–350.

Smith, B. J. 1976. A new application in the pitfall trapping of insects, to investigate migratory patterns. Trans. Ky. Acad. Sci. 37: 94–97.

Smith, J. G., A. C. Pereira, B. S. Correa & A. R. Panizzi. 1977. Confecção de aparelhos de baix custo para coleta e criaçã de insetos. An. Soc. Entomol. Bras. 6: 132-135. (In Portuguese). [Describes low-cost carrying cages, pitfall traps, and rearing cages.]

Smith, J. S., Jr. 1974. S-1 black-light insect-survey trap plans and specifications. U.S. Dep. Agric., Agric. Res. Serv., ARS-S-31.

Smith, K. V. G. 1974. Rearing the Hymenoptera parasitica. Leaf. Amat. Entomol. Soc. 35.

Snoddy, E. L. & K. L. Hays. 1966. Carbon dioxide trap for Simuliidae. J. Econ. Entomol. 59: 242–243.

Sommerman, K. M. 1967. Modified car-top insect trap functional to 45 mph. Ann. Entomol. Soc. Am. 50: 857.

Southwood, T. R. E. 1979. Ecological methods with particular reference to the study of insect populations, 2nd ed. John Wiley & Sons, New York. [Includes extensive bibliographies.]

Sparks, A. N., J. R. Raulston, P. D. Lingren, J. E. Carpenter, J. A. Klun & B. G. Mullinix. 1980. Field response of male *Heliothis virescens* to pheromonal stimuli and traps. Bull. Entomol. Soc. Am. 25: 268–274. [Includes pictures of several types of traps used.]

Stanley, J. M. & C. B. Dominick. 1970. Funnel size and lamp wattage influence on light-trap performance. J. Econ. Entomol. 63: 1423–1426.

Stanley, J. M., J. C. Webb, W. W. Wolf & E. R. Mitchell. 1977. Electrocutor grid insect traps for research purposes. Trans. Am. Soc. Agric. Eng. 20: 175–178.

Steck, W. & B. K. Bailey. 1978. Pheromone traps for moths: Evaluation of cone trap design and design parameters. Environ. Entomol. 7: 419–455.

Stehr, F. W. 1987. Immature insects. Vol. 1. Kendall/Hunt. Dubuque, IA.

Stehr, F. W. 1991. Immature insects. Vol. 2. Kendall/Hunt, Dubuque, IA.

Stein, J. D. 1976. Insects: A guide to their collection, identification, preservation and shipment. U.S. Dep. Agric., For. Serv., RM-311.

Stein R., W. V. Eisenberg & P. M. Brickey, Jr. 1968. Extraneous materials. Staining technique to differentiate insect fragments, bird feathers, and rodent hairs from plant tissue. J. Assoc. Off. Anal. Chem. 51: 513–518.

Stewart, J. W. & T. L. Payne. 1971. Light trap screening for collecting small soft-bodied insects. Entomol. News 82: 309–311.

Stewart, P. A. & J. J. Lam. 1968. Catch of insects at different heights in traps equipped with blacklight lamps. J. Econ. Entomol. 61: 1227–1230.

Steyskal, G. C. 1957. The relative abundance of flies (Diptera) collected at human feces. Z. Angew. Zool. 44: 79-83.

Steyskal, G. C. 1977. History and use of the McPhail trap. Fla. Entomol. 60: 11–16.

Steyskal, G. C. 1981. Bibliography of the Malaise trap. Proc. Entomol. Soc. Wash. 83: 225–229.

Steyskal, G. C., W. L. Murphy and E. M. Hoover. 1986. Insects and mites: Techniques for collection and preservation. U.S. Dept. of Ag. Misc. Pub. No. 1443.

Still, G. W. 1960. An improved trap for deciduous tree fruit flies. J. Econ. Entomol. 53: 967.

Stoetzel, M. B. 1985. Pupiform larvae in the Phylloxeridae (Homoptera: Aphidoidea). Proc. Entomol. Soc. Wash. 87: 535–537.

Stoetzel, M. B. 1989. Common names of insects and related organisms. Entomological Society of America, College Park, MD.

Strandtmann, R. W. & G. W. Wharton. 1958. Manual of mesostigmatid mites parasitic on vertebrates. Univ. Md. Inst. Acarol. Contrib. 4: 1–330.

Strenzke, K. 1966. Empfohlene Methodenzur Aufzucht und Präparation terrestrischer Chironomiden. Gewaess. Abwaess. 41/42: 163–168. [Methods for rearing and preparation of terrestrial Chironomidae (Diptera).]

Stryker, R. G. & W. W. Young. 1970. Effectiveness of carbon dioxide and L (+) lactic acid in mosquito light trap with and without light. Mosq. News 30: 388–393.

Stubbs, A. & P. Chandler [eds.]. 1978. A dipterist's handbook. Vol. 2, The amateur entomologist. J. Amateur Entomol. Soc., Hanworth, England. [Collecting and recording, pp. 1–37.]

Stys, P. & I. Kerzhner. 1975. The rank and nomenclature of higher taxa in recent Heteroptera. Acta Entomol. Bohemoslov. 72: 65–79.

Summers, C. G. S., R. E. Garrett & F. G. Zalom. 1984. New suction device for sampling arthropod populations. J. Econ. Entomol. 77: 817–823.

References

– T –

Tachikawa, T. 1981. Hosts of encyrtid genera in the world (Hymenoptera: Chalcidoidea). Mem. Coll. Agric. Ehime Univ. 25: 85–110.

Tagestad, A. D. 1974. A technique for mounting Microlepidoptera. J. Kans. Entomol. Soc. 47: 26–30.

Tagestad, A. D. 1977. Easily constructed individual mounting blocks for macro- and microlepidoptera. J. Kans. Entomol. Soc. 50: 27–30.

Tarshis, I. B. 1968a. Use of fabrics in streams to collect blackfly larvae. Ann. Entomol. Soc. Am. 61: 260–261.

Tarshis, I. B. 1968b. Collecting and rearing blackflies. Ann. Entomol. Soc. Am. 61: 1072–1083.

Tashiro, H. 1990. Insecta: Coleoptera Sacarabaeidae larvae. In: D. L. Dindal [ed.], Soil biology guide, pp. 1191–1209. John Wiley & Sons, New York.

Taylor, L. R. 1962a. The absolute efficiency of insect suction traps. Ann. Appl. Biol. 50: 405–421.

Taylor, L. R. 1962b. The efficiency of cylindrical sticky insect traps and suspended nets. Ann. Appl. Biol. 50: 681–865.

Tedders, W. L. & G. W. Edwards. 1972. Effects of blacklight trap design and placement on catch of adult hickory shuckworms. J. Econ. Entomol. 65: 1624–1627.

Teskey, H. J. 1962. A method and apparatus for collecting larvae of tabanid (Diptera) and other inhabitants of wetland. Proc. Entomol. Soc. Ont. 92: 204–206.

Thomas, D. B. & E. L. Sleeper. 1977. Use of pit-fall traps for estimating the abundance of arthropods, with special reference to the Tenebrionidae (Coleoptera). Ann. Entomol. Soc. Am. 70: 242–248.

Thompson, P. H. 1969. Collecting methods for Tabanidae (Diptera). Ann. Entomol. Soc. Am. 62: 50–57.

Thompson, P. H. & E. J. Gregg. 1974. Structural modifications and performance of the modified animal trap and the modified Manitoba trap for collection of Tabanidae (Diptera). Proc. Entomol. Soc. Wash. 76: 119–122.

Thompson, S. R. & R. L. Brandenburg. 2004. A modified pool design for collecting adult mole crickets (Orthoptera: Gryllotalpidae). Florida Entomol. 87:4 582-584.

Thorsteinson, A. J., B. G. Bracken & W. Hanec. 1965. The orientation behavior of horse flies and deer flies (Tabanidae: Diptera). III. The use of traps in the study of orientation of tabanids in the field. Entomol. Exp. Appl. 8: 189192.

Tindale, N. B. 1962. The chlorocresol method for field collecting. J. Lepid. Soc. 15: 195–197.

Townes, H. 1962. Design for a Malaise trap. Proc. Entomol. Soc. Wash. 83: 225–229.

Townes, H. 1972. A light-weight Malaise trap. Entomol. News 83: 239–247.

Traver, J. R. 1940. Compendium of entomological methods. pt. I. Collecting mayflies (Ephemeroptera). Ward's Nat. Sci. Estab., Rochester, NY.

Tretzel, E. 1955. Technik und Bedeutung des Fallenfanges für ökologische Untersuchungen. Zool. Anz. 155: 276–287. [Technique and significance of the pitfall trap for ecological research.]

Turnbull, A. L. & C. F. Nicholls. 1966. Quick trap for area sampling of arthropods in grassland communities. J. Econ. Entomol. 59: 1100–1104.

Turnock, W. J. 1957. A trap for insects emerging from the soil. Can. Entomol. 89: 455–456.

– U –

U.S. Department of Agriculture, Agricultural Research Service. 1961. Response of insects to induced light. Presentation papers. U.S. Dep. Agric., Agric. Res. Serv., ARS 20-10. [Includes 11 pp. of bibliography on insects and light.]

U.S. Department of Agriculture, Extension Service. 1970. 4-H Clubs entomology publications (loose leaf, in 4 pts.): 1, How to make an insect collection; 2, Key to orders, rearing cages, experimental activities; 3, Teen and junior leader's guide; 4, Reference material.

U.S. Department of Agriculture, Plant Pest Control Division. 1966–1970. Survey methods; selected references 1942–1970. Pts. 1–35. USDA Coop. Econ. Insect Rep. 17: 97–112, 326–336, 862–868, 977–982, 1071–1076; 18: 70–72, 369–376, 564–568, 746–750, 911–914, 969–974, 990–996, 1054–1058, 1103–1106; 19: 88–96, 335–337, 353-354, 505–506, 785–786, 797-798, 838–842, 865–870; 20: 11–12, 34–36, 43–44, 151–152, 207–208, 239–242, 345–346, 527–528, 727–732, 763–767; 21: 35–40, 87–94; 22: 285–298. [References categorized into population measurement, forecasting, rearing, equipment and technology, traps, attractants, and pictorial keys. The various parts are not in chronological order.]

Urquhart, F. A. 1965. Introducing the insect. Frederick Warne, London. [Elementary directions for making insect collections, pp. 1–19.]

Usinger, R. L. [ed.]. 1956. Aquatic insects of California with keys to North American genera and California species. University of California Press, Berkeley.

– V –

Van Cleave, J. & J. A. Ross. 1947. A method of reclaiming dried zoological specimens. Science (Washington, DC) 105: 318.

Vick, K. W., R. W. Mankin, R. R. Cogburn, M. Mullen, J. E. Throne, V. F. Wright & L. D. Cline. 1990. Review of pheromone-baited sticky traps for detection of stored-product insects. J. Kans. Entomol. Soc. 63: 526–532.

Viggiani, G. 1984. Bionomics of the Aphelinidae. Annu. Rev. Entomol. 29: 257–276.

Viggiani, G. 1990. Hyperparasites. In: D. Rosen [ed.], Armored-scale insects, their biology, natural enemies and control, Vol. B. Elsevier Science, Amsterdam.

Vite, J. P. & E. Baader. 1990. Present and future use of semiochemicals in pest management of bark beetles. J. Chem. Ecol. 16: 3031–3041.

von Hagen, K. B. & J. Kadereit. 2001. The phylogeny of Gentianella (Gentianaceae) and its colonization of the southern hemisphere are revealed by nuclear and chloroplast DNA sequence variation. Organisms Diversity and Evolution, 1, 61-79.

References

– W –

Wagstaffe, R. & J. H. Fidler. 1955. The preservation of natural history specimens. Vol. 1, Invertebrates. Philosophical Library, New York.

Walker, A. R. & P. F. L. Boreham. 1976. Saline, as a collecting medium for culicoides (diptera: ceratopogonidae) in blood-feeding and other studies. Mosq. News 36: 18–20.

Walker, T. J. 1982. Sound traps for sampling mole cricket flights (Orthoptera: Gryllotalpidae: Scapteriscus). Florida Entomol. 65(1): 105-110.

Walsh, G. B. 1933. Studies in the British necrophagous Coleoptera. II. The attractive powers of various natural baits. Entomol. Mon. Mag. 69: 28–32.

Waltz, R. D. & W. P. McCafferty. 1984. Indication of mounting media information. Entomol. News 95: 31–32.

Waterhouse, D. [ed.]. 1971. The insects of Australia. CSIRO, Melbourne University Press, Melbourne.

Waters, T. F. 1969. Subsampler for dividing large samples of stream invertebrate drift. Limnol. Oceanogr. 14: 813–815.

Watson, G. E. & A. B. Amerson, Jr. 1967. Instructions for collecting bird parasites. Smithson. Inst., Mus. Nat. Hist., Inf. Leafl. 477.

Watson, J. A. L. & G. B. Smith. 1991. Archaeognatha (microcoryphia), In: I. D. Nauman [ed.], The insects of Australia: A textbook for students and research workers, pp. 272-274. Melbourne University Press, Melbourne, Australia.

Weatherston, J. 1976. A new insect trap for use with lepidopteran sex pheromones. Can. Dep. Fish. For. Bi-Mon. Res. Notes 32: 9–10.

Weaver, J. S., III, & T. R. White. 1980. A rapid, steam bath method for relaxing dry insects. Entomol. News 91: 122–124.

Weber, R. G. 1987. An underwater light trap for collecting bottom-dwelling aquatic insects. Entomol. News. 98: 246–252.

Welch, P. S. 1948. Limnological methods. McGraw-Hill, New York.

Wells, J. D. et al. 2001. Human and insect mitochondrial DNA analysis from maggots. J. Forensic Sci., 46 (3) 685-687.

Wells, J. D. & Sperling, F. A. 2001. DNA-based identification of forensically important Chrysomyinae (Diptera: Calliphoridae). Forensic Sci. Int., 120, 110-115.

Wellso, S. G. & R. L. Fischer. 1971. Insects taken at Japanese beetle traps baited with anethole-eugenol in southern Michigan in 1968. Mich. Entomol. 4: 105–108.

Weseloh, R. M. 1974. Relationships between different sampling procedures for the gypsy moth, *Porthetria dispar* (Lepidoptera, Lymantriidae), and its natural enemies. Can. Entomol. 106: 225-231.

White, E. G. 1964. A design for the effective killing of insects caught in light traps. N. Z. Entomol. 3: 25–27.

Whitsel, R. H. & R. G. Schoeppner. 1965. The attractiveness of carbon dioxide to female *Leptoconops torrens* Tns. and L. kerteszi Kieff. Mosq. News 25: 403–410.

Whittaker, R. H. 1952. A study of summer foliage insect communities in the Great Smoky Mountains. Ecol. Monogr. 22 (1): 1–44.

Wiens, J. E. & L. Burgess. 1972. An aspirator for collecting insects from dusty habitats. Can. Entomol. 104: 1557–1558.

Wilkinson, R. S. 1969. A blacklight box trap for nocturnal insects. Newsl. Mich. Entomol. Soc. 14(3/4): 3.

Willey, R. L. 1971. Microtechniques: A laboratory guide. Macmillan, New York.

Williams, C. B. 1948. The Rothamsted light trap. Proc. R. Entomol. Soc. Lond., Ser. A, Gen. Entomol. 23: 80–85.

Williams, C. B. 1951. Comparing the efficiency of insect traps. Bull. Entomol. Res. 42: 513-517.

Williams, C. B. & P. S. Milne. 1935. A mechanical insect trap. Bull. Entomol. Res. 26: 543–551.

Williams, D. F. 1973. Sticky traps for sampling populations of *Stomoxys calcitrans*. J. Econ. Entomol. 66: 1279–1280.

Williams, L., III, & L. E. O'Keeffe. 1990. An ultrasonic procedure for cleaning sticky-trapped arthropods. J. Kans. Entomol. Soc. 63(3): 461–462.

Williams, R. W. 1954. A study of the filth flies in New York City — 1953. J. Econ. Entomol. 47: 556–562.

Williamson, K. 1954. The Fair Isle apparatus for collecting bird ectoparasites. Br. Birds 47: 234–235.

Wilson, B. H., N. O. Tugwell & E. C. Burns. 1966. Attraction of tabanids to traps baited with dry ice under field conditions in Louisiana. J. Med. Entomol 3: 148–149.

Wilson, J. G., D. R. Kinzer, J. R. Sauer & J. A. Hair. 1972. Chemo-attraction in the lone star tick (Acarina: Ixodidae): I. Response of different developmental stages to carbon dioxide administered via traps. J. Med. Entomol. 9: 245–252. [A modification of this method is described in Eads et al. 1982.]

Wilson, L. F. 1969. The window-pane insect trap. Newsl. Mich. Entomol. Soc. 14(3/4): 1–3.

Wilson, S. W., J. L. Smith & A. H. Purcell, III. 1993. An inexpensive vacuum collector for insect sampling. Entomol. News. 104: 203–208.

Wilton, D. P. 1963. Dog excrement as a factor in community fly problems. Proc. Hawaii. Entomol. Soc. 118: 311–317.

Wirth, W. W. & N. Marston. 1968. A method for mounting small insects on microscope slides in Canada balsam. Ann. Entomol. Soc. Am. 61: 783–784.

Woke, P. A. 1955. Aspirator-cage combinations for delicate and infected arthropods. Ann. Entomol. Soc. Am. 48: 485–488.

Wood, D. M. & D. M. Davies. 1966. Some methods of rearing and collecting black flies. Proc. Entomol. Soc. Ont. 96: 81–90.

Wood, D. M., P. T. Danf & R. A. Ellis. 1979. The insects and arachnids of Canada. Pt. 6. The mosquitoes of Canada: Diptera: Culicidae. Can. Dep. Agric., Biosyst. Res. Inst., Publ. 1686 [Methods of collecting, preservation, and rearing on pp. 45–53.]

Woodring, J. P. 1968. Automatic collecting device for tyroglyphid (Acaridae) mites. Ann. Entomol. Soc. Am. 61: 1030–1031.

Woodring, J. P. & M. S. Blum. 1963. Freeze-drying of spiders and immature insects. Ann. Entomol. Soc. Am. 56: 138–141.

Woolley, T. A. 1988. Acarology: Mites and human welfare. John Wiley & Sons, New York.

– Y –

Yates, M. 1974. An emergence trap for sampling adult tree-hole mosquitoes. Entomol. Mon. Mag. 109: 99101.

– Z –

Zimmerman, E. C. 1978. Microlepidoptera. Insects of Hawaii 9: 1–1903.

Zolnerowich, G., J. M. Heraty & J. B. Wooley. 1990. Separation of insect and plant material from screen-sweep samples. Entomol. News 101: 301–306.

Zycherman, L. A. & J. R. Schrock [eds.]. 1988. A guide to museum pest control. Associaton of Systematic Collections. [Excellent summary of pests, their biology and identification, and control methods.]

NOTES:

Appendices

APPENDIX I – Liquid Preservation Formulas

Distilled water should be used in these formulas if available, but rainwater or bottled drinking water is satisfactory. "Parts" is by volume.

AGA (Alcohol–Glycerin–Acetic Acid) Solution
- Commercial ethanol (ethyl alcohol) .... 8 parts
- Water .... 5 parts
- Glycerin .... 1 part
- Glacial acetic acid .... 1 part

Barber's Fluid
- Commercial ethanol (ethyl alcohol) .... 53 parts
- Water .... 49 parts
- Ethyl acetate (acetic ether) .... 19 parts
- Benzene (benzol) .... 7 parts

Embalming Fluid
- Toluene (or xylene) .... 60 ml
- Tertiary butyl alcohol (TBA) .... 25 ml
- Ethyl alcohol .... 15 ml
- Phenol .... 5 g
- Naphthalene .... 20 g
- Canada balsam in xylene .... 10 drops

    Note: *Embalming fluid contains many toxicants (toluene, phenol, TBA). The original formula calls for paradichlorobenze (PDB). This carcinogen has been replaced with naphthalene here. PBD kills and repels insects; naphthalene repels insects. Embalming fluid is introduced into the specimen with a hypodermic syringe inserted between sclerites or into the anus. The fluid should be injected into each body region.*

Essig's Aphid Fluid
- Lactic acid .... 20 parts
- Glacial acetic acid .... 4 parts
- Phenol (saturated H_2O solution) .... 2 parts
- Distilled H_2O .... 1 part

Hoyer's Medium
- Chloral hydrate .... 200 g
- Distilled water .... 50 ml
- Gum arabic (granules) .... 30 g
- Glycerin .... 20 ml

    There are many variants of Hoyer's Medium. We cite the original. Hoyer's Medium is a temporary mountant that breaks down with time. Thus, it should not be considered as suitable for mounting valuable specimens or important dissections. Chloral hydrate is a controlled substance in the United States and may be difficult to obtain. To prepare Hoyer's Medium, dissolve gum arabic in water at room temperature. Add chloral hydrate and allow the mixture to stand for a day or two until all solids have dissolved. Add glycerin. Filter through glass wool. Store in a glass-stoppered bottle.

KAA (Kerosene–Acetic Acid) Solution
- Commercial ethanol (ethyl alcohol) .... 10 parts
- Glacial acetic acid .... 2 parts
- Kerosene .... 1 part

    Mix in the order given. For very soft-bodied larvae, use half as much kerosene or less. Formerly, this solution was called KAAD, and one part of dioxane was included.

APPENDIX II – Guidelines for Mounting Small and Soft-Bodied Specimens (SEL)

Aphids, scale insects, and whiteflies cannot be pinned because of their small size and their tendency to shrivel. The following procedures for mounting certain insects and mites for scientific study are preferred by the Systematic Entomology Laboratory (ARS, USDA). A few of the chemicals indicated by an asterisk (*) in these procedures are hazardous. Carefully investigate their properties to ensure their safe use.

1. Place specimen in 10% potassium hydroxide (KOH)* solution and heat gently until body contents are softened, or leave in KOH solution at room temperature for up to 48 hours.
2. Remove specimen from KOH and place in 70% ethanol for five minutes. (Note for aleyrodids: If black specimens have not turned brown at this point, bleach them in peroxide-ammonia solution [one drop ammonia to six drops hydrogen peroxide] until brown. Next place in 95% ethanol for five minutes; then proceed to step 7.)
3. Remove specimen from 70% ethanol and place in Essig's aphid fluid. Make an incision halfway across the body between the second and third pairs of legs. Then squeeze a few times to remove and flush out body contents. If feasible, one or two well-formed embryos should be left in bodies of aphids. Excess wax may be removed by placing specimens in tetrahydrofuran*. Note: This is a hazardous chemical and should be used under an exhaust hood and in a very well-ventilated place to avoid inhalation of fumes.
4. Remove from Essig's fluid and place in 95% ethanol for five minutes.
5. Remove from the ethanol and place in acid fuchsin stain for about five minutes or until properly stained, then place in 70% ethanol for five or more minutes to remove excess stain.
6. Remove from 70% ethanol and place in 95% ethanol for five minutes.
7. Remove from 95% ethanol and place in clove oil; leave for five minutes or until specimen appears nonshiny, dull, and flat.
8. Remove from clove oil and place specimen dorsum upward on a slide in one to two drops of Canada balsam. If mounting three or more specimens on one slide, place them in a row right to left with one specimen ventral side up. Keep all specimens neatly horizontal with heads pointed in the same direction.
9. Put coverslip in place and attach labels, preferably so that they may be read with heads of specimens toward you.

Mounting Thrips. For detailed study, mount thrips in Canada balsam as described here. Place each specimen by itself centrally on a slide with wings, legs, and antennae spread for easy observation of structures. Most specimens should be cleared for optimal appearance of surface detail, but a few specimens should be left in their natural color by omitting steps two and three. Rapid identifications may be made from temporary mounts in glycerin or Hoyer's Medium, but they usually cause distortion. Excess or used fluids may be removed at each step with a pipette.

1. Soak specimen for 24 hours in clean 60% ethanol to remove collecting fluid.
2. Macerate in cold 5% sodium hydroxide* solution for 30 minutes or up to four hours for especially dark specimens.
3. Wash briefly in 50% ethanol and then leave in 60% ethanol for 24 hours.
4. Dehydrate through a series of ethanol solutions: 70% for one hour, 80% for two hours, and 100% for ten minutes (change alcohol once). Place specimen in clove oil until clear (30 seconds to 10 minutes). Spread appendages carefully at each stage. Dehydration and clearing may be promoted by puncturing thoracic and abdominal membranes in one or two places with a fine needle.
5. Place ventral side uppermost on a 13-mm coverslip in Canada balsam, then lower slide onto coverslip. This method is easier to control than the usual method of lowering coverslip onto slide with forceps.
6. Use two labels on slide, one at each side of specimen, with host, locality, altitude, date, and collector's name on right-hand label and determination and sex data on left-hand label.
7. Cure in oven at 40°C within a few minutes of preparation, or leave for up to 6 weeks for complete curing.

Mounting Mites. Mites are most easily mounted if collected in AGA solution. Mount specimens collected in 70–80% ethanol on slides as soon as possible. The following procedures do not apply to mites of the family Eriophyidae:

1. Place drop of Hoyer's Medium in center of a clean 2.5 by 7.6 cm glass microscope slide.

2. Remove mite directly from host, or pour specimens from collecting vial into small casserole, watchglass, or petri dish. Avoid pouring too much fluid from vial into dish; the less fluid, the easier it is to pick out the mite. The mites may be removed from the host or from the fluid by dipping a needle into the Hoyer's Medium on the slide and then quickly touching the mite with it.

3. Place one specimen in the medium on a slide. Press the specimen to the surface of the slide and spread all legs laterally. Most mites should be mounted dorsoventrally, but males of many species, such as those of the Acaridae and Tetranychidae, should be mounted laterally to allow examination in profile of the specifically characteristic aedeagus. Some mites may require a small body puncture to eliminate the contents. Heavily pigmented mites may be cleared in a solution of lactophenol before mounting. The solution may be heated to hasten clearing.

4. With forceps, carefully place a clean, small (13 mm or smaller) coverslip on the mite in Hoyer's Medium. Gently press the coverslip with forceps to hold the mite in position.

5. Place the slide on a hotplate set at 60°C and remove it rapidly when a single bubble forms in the Hoyer's Medium. Avoid more bubbling or mite will be displaced.

6. Turn the slide so that the anterior part of the mite is directed toward you.

7. Place label on slide to right of coverslip; host, locality, collector, date, and serial number data should be shown on this label.

8. Place slide in oven at 45–50°C for 24 hours or longer to cure and solidify the Hoyer's Medium. The slide must be kept horizontal until the medium is firm and there is no danger of the coverslip moving.

9. After removing slide from oven, ringing coverslip with additional Hoyer's Medium helps to prevent the mount from drying out.

APPENDIX III – DIRECTORY
State Extension Service Directors and Administrators, August 2005

ALABAMA
Dr. W. Gaines Smith, Interim Director
Alabama Cooperative Extension System
109D Duncan Hall
Auburn University, AL 36849-5612
334/844-4444; (FAX) 334/844-5544
wsmith@acesag.auburn.edu

Dr. Virginia Caples, Administrator
1890 Extension Programs
Alabama A&M University
P.O. Box 967
Normal, AL 35762
256/851-5943; (FAX) 256/851-5840
vcaples@asnaam.aamu.edu

Dr. Velma L. Blackwell, Administrator
1890 Extension Programs
Tuskegee University
Kellogg Conference Center For Continuing Education
Tuskegee, AL 36088
334/727-8601; (FAX) 334/724-4199
blackwell@tusk.edu

Dr. Thomas S. Harris
Associate Administrator
1890 Extension Programs
Tuskegee University
Rm. 102 Morrison /Mayberry Hall
Tuskegee, AL 36088
334/724-4441; (FAX) 334/727-8812
tharris@acesag.auburn.edu

ALASKA
Dr. Anthony T. Nakazawa, Director
Cooperative Extension Service
University of Alaska Fairbanks
P.O. Box 756180
Fairbanks, AK 99775-6180
907/474-7246; (FAX) 907/474-6971
fnatn@uaf.edu

ARIZONA
Dr. James A. Christenson, Director
Cooperative Extension Service
College of Agriculture
University of Arizona
Forbes Bldg., Rm. 301
Tucson, AZ 85721
520/621-7209; (FAX) 520/621-1314
jimc@ag.arizona.edu

ARKANSAS
Dr. Ivory W. Lyles, Director
Cooperative Extension Service
University of Arkansas
2301 South University Avenue
Little Rock, AR 72204
501/ 671-2001; (FAX) 501/ 671-2107
ilyles@uaex.edu

Dr. Jacqueline W. McCray, Administrator
1890 Agricultural Programs
University of Arkansas
Box 4990
1200 N. University Drive
Pine Bluff, AR 71611
870/575-7199; (FAX) 870/543-8033
mccray_ j@vx4500.uapb.edu

CALIFORNIA
Dr. W. R. Gomes, Director
Cooperative Extension Service
University of California
300 Lakeside Dr., 6th Fl.
Oakland, CA 94612-3560
510/987-0060; (FAX) 510/451-2317
wr.gomes@ucop.edu

COLORADO
Mr. Marc A. Johnson, Interim Director
Cooperative Extension Service
Colorado State University
1 Administration Building
Fort Collins, CO 80523-4040
970/491-6281; (FAX) 970/491-6208
Marc.Johnson@colostate.edu

CONNECTICUT
Dr. Kirklyn M. Kerr, Director
Cooperative Extension System
College of Agriculture & Natural Resources
University of Connecticut
Young Building, Rm. 218
1376 Storrs Road, Uuit 4066
Storrs, CT 06269-4066
860/486-2917; (FAX) 860/486-5113
kirklyn.kerr@uconn.edu

Dr. Nancy H. Bull, Associate Director
Cooperative Extension System
College of Agriculture & Natural Resources
University of Connecticut
Young Building, Rm 232
1376 Storrs Road, Unit 4134
Storrs, CT 06269-4134
860/486-1987; (FAX) 860/486-0264
nancy.bull@uconn.edu

DELAWARE
Dr. Janice A. Seitz, Director
Cooperative Extension Service
University of Delaware
113 Townsend Hall
Newark, DE 19716
302/831-2501; (FAX) 302/831-6758
No e-mail address

Dr. Kenneth W. Bell, Administrator
1890 Extension Programs
Delaware State University
1200 North DuPont Highway
Dover, DE 19901
302/857-6400; (FAX) 302/739-4997
kbell@dsc.edu

DISTRICT OF COLUMBIA
Dr. Alford H. Ottley, Acting Director
Cooperative Extension Service
University of the District of Columbia
4200 Connecticut Avenue, NW
Washington, D.C. 20008
202/274-7100; (FAX) 202/274-7016
aottley@udc.edu

FLORIDA
Dr. Larry R. Arrington, Interim Director
Cooperative Extension Service
University of Florida
Box 110210
Gainesville, FL 32611-0210
352/392-1761; (FAX) 352/846-0458
lrarrington@mail.ifas.ufl.edu

Dr. Lawrence Carter, Administrator
1890 Extension Programs
Florida A&M University
Perry-Paige Building, Room 215
Tallahassee, FL 32307
850/599-3546; (FAX) 850/561-2151
lca@gnv.ifas.ufl.edu

GEORGIA
Dr. Gale A. Buchanan , Director
Cooperative Extension Service
College of Agricultural & Environmental Sciences
The University of Georgia
Room 101, Conner Hall
Athens, GA 30602-7501
706/542-3924; (FAX) 706/542-0803
agdean@uga.cc.uga.edu

Dr. Mack C. Nelson, Administrator
1890 Extension Programs
Ft. Valley State University
124 Stallworth Building
Ft. Valley, GA 31030
478-825-6827/6320 ; (FAX) 478-825-6376
nelsonm@fvsu.edu

HAWAII
Dr. Andrew G. Hashimoto, Director
Cooperative Extension Service
University of Hawaii at Manoa College of Tropical Agriculture
 & Human Resources
3050 Maile Way, Gilmore 202
Honolulu, HI 96822
808/956-8234; (FAX) 808/956-9105
aghashim@hawaii.edu

APPENDIX

HAWAII (cont'd)
Dr. Barry M. Brennan, Associate Director
Cooperative Extension Service
University of Hawaii at Manoa College of Tropical Agriculture & Human Resources
3050 Maile Way, Gilmore 203
Honolulu, HI 96822
808/956-8397; (FAX) 808/956-9105
barryb@hawaii.edu

IDAHO
Dr. Charlotte V. Eberlein, Interim Director
Cooperative Extension System
University of Idaho
Twin Falls R&E Center
P.O. Box 1827
315 Falls Avenue
Twin Falls, ID 83303-1827
208/736-3603; (FAX) 208/736-0843
ceberl@uidaho.edu

ILLINOIS
Dr. Dennis R Campion, Director
Cooperative Extension Service
University of Illinois
123 Mumford Hall
1301 West Gregory Drive
Urbana, IL 61801
217/333-5900; (FAX) 217/244-5403
dcampion@uiuc.edu

INDIANA
Dr. David C. Petritz, Director
Cooperative Extension Service
Purdue University
1140 Agriculture Admin. Bldg.
West Lafayette, IN 47907-114
765/494-8489; (FAX) 765/494-5876
petritz@ces.purdue.edu

IOWA
Dr. Stanley R. Johnson, Director
Cooperative Extension Service
Iowa State University
315 Beardshear
Ames, IA 50011-2020
515/294-6192; (FAX) 515/294-9781
vpforext@exnet.iastate.edu

IOWA (cont'd)
Dr. Michael D. Ouart
Associate Vice Provost/Associate Director
Cooperative Extension Service
Iowa State University
Extension Administration
2150 Beardshear Hall
Ames, IA 50011-2046
515/294-1482; (FAX) 515/294-4715
michaelo@iastate.edu

KANSAS
Dr. Fred A. Cholick, Director
Cooperative Extension Service
Kansas State University
College of Agriculture
115 Waters Hall
Manhattan, KS 66506
785/532-7137; (FAX) 785/532-6563
fcholick@oznet.ksu.edu

Dr. Daryl D. Buchholz, Associate Director
Cooperative Extension Service
Kansas State University
Research and Extension
123 Umberger Hall
Manhattan, KS 66506
785/532-5820; (FAX) 785/532-6290
dbuchhol@oznet.ksu.edu

KENTUCKY
Dr. Larry W. Turner, Director
Cooperative Extension Service
University of Kentucky
S-107 Ag. Science Bldg.-North
Lexington, KY 40546-0091
859/257-4302; (FAX) 859/257-3501
larry.turner@uky.edu

Dr. Harold R. Benson, Administrator
1890 Extension Programs
Kentucky State University
Cooperative Extension Program Facility
400 E. Main Street
Frankfort, KY 40601
502/227-6310; (FAX) 502/227-5933
hbenson@gwmail.kysu.edu

LOUISIANA
Dr. Paul D. Coreil, Director
Cooperative Extension Service
Louisiana State University
P.O. Box 25100
Baton Rouge, LA 70894-5100
225/578-6083; (FAX) 225/578-4225
pcoreil@agcenter.lsu.edu

Dr. Leodrey Williams, Administrator
1890 Extension Programs
Southern University & A&M College
P.O. Box 10010
Baton Rouge, LA 70813
225/771-2242; (FAX) 225/771-2861
lwilliams2@aol.com

MAINE
Dr. Lavon L. Bartel, Director
Cooperative Extension Service
University of Maine
Cooperative Extension
5741 Libby Hall, Rm. 102
Orono, ME 04469-5741
207/581-2811; (FAX) 207/581-1387
lbartel@umext.maine.edu

MARYLAND
Dr. Bruce Gardner, Interim Director
Cooperative Extension Service
University of Maryland
College of Agriculture & Natural Resources
1296 Symons Hall
College Park, MD 20742-5551
301/405-2072; (FAX) 301/314-9146
bruceg@arec.umd.edu

Dr. Henry M. Brooks, Administrator
1890 Extension Programs
University of Maryland
Eastern Shore
Princess Anne, MD 21853
410/651-6206; (FAX) 410/651-6207
hmbrooks@mail.umes.edu

MASSACHUSETTS
Mr. Stephen J. Demski, Interim Director
Cooperative Extension Service
University of Massachusetts
212 Stockbridge Hall
Amherst, MA 01003
413/545-4800; (FAX) 413/545-6555
No e-mail address

MICHIGAN
Dr. Thomas Coon, Director
Cooperative Extension Service
Michigan State University
108 Agriculture Hall
East Lansing, MI 48824-1039
517/355-2308; (FAX) 517/355-6473
coont@msu.edu

MINNESOTA
Dr. Charles H. Casey, Interim Director
Minnesota Extension Service
University of Minnesota
Room 240 Coffey Hall
1420 Eckles Avenue
St. Paul, MN 55108-6070
612/624-2703; (FAX) 612/625-6227
casey002@umn.edu

MISSISSIPPI
Dr. Vance H. Watson, Interim Director
Cooperative Extension Service
Mississippi State University
Box 9601
Bost Extension Center, Rm 201
Mississippi State, MS 39762
662/325-3036; (FAX) 662/325-8407
vwatson@dafvm.msstate.edu

Dr. Leroy Davis, Administrator
1890 Extension Programs
Alcorn State University
1000 ASU Drive, #690
Lorman, MS 39096
601/877-6136/6137; (FAX) 601/877-6219
ldavis@lorman.alcorn.edu

APPENDIX

MISSOURI
Dr. Thomas A. Henderson, Interim Director
Cooperative Extension Service
University of Missouri
108 Whitten Hall
Columbia, MO 65211
573/882- 7754; (FAX) 573/882-1955
hendersont@umsystem.edu

Dr. Dyremple Marsh, Interim Administrator
1890 Extension Programs
Lincoln University
820 Chestnut, P. O. Box 29
Jefferson City, MO 65102
573/681-5547; (FAX) 573/681-5881
marshd@lincolnu.edu

MONTANA
Dr. Douglas L. Steele, Director
Cooperative Extension Service
Montana State University
205 Culbertson Hall
P.O. Box 172230
Bozeman, MT 59717-2230
406/994-1750; (FAX) 406/994-1756
dsteele@montana.edu

NEBRASKA
Dr. Elbert C. Dickey, Director
Cooperative Extension
University of Nebraska
211 Agriculture Hall
Lincoln, NE 68583-0703
402/472-2966; (FAX) 402/472-5557
edickey1@unl.edu

NEVADA
Ms. Karen L. Hinton, Director
Nevada Cooperative Extension
University of Nevada Cooperative Extension
Mail Stop 404
Reno, NV 89557
775/784-7070; (FAX) 775/784-4881
hintonk@unce.unr.edu

NEW HAMPSHIRE
Dr. John E. Pike, Director
Cooperative Extension Service
University of New Hampshire
59 College Road
103A Taylor Hall
Durham, NH 03824-3587
603/862-1520; (FAX) 603/862-1585
john.pike@unh.edu

NEW JERSEY
Dr. Karyn Malinowski, Director
Cooperative Extension Service
Rutgers University
Martin Hall Room 309
88 Lippman Drive
New Brunswick, NJ 08901-8525
732/932-9000 X581; (FAX) 732/932-6633
malinowski@rce.rutgers.edu

NEW MEXICO
Dr. Jerry G. Schickedanz, Director
Cooperative Extension Service
New Mexico State University
P.O. Box 30003
Department 3AE
Las Cruces, NM 88003
505/646-3016; (FAX) 505/646-5975
adean@nmsu.edu

Dr. Paul Gutierrez, Associate Director
Cooperative Extension Service
New Mexico State University
P.O. Box 3003, MSC 3AE
Las Cruces, NM 88003
505/646-3015; (FAX) 505/646-5975
pgutierrez@nmsu.edu

NEW YORK
Dr. Helene Roberts Dillard, Director
Cooperative Extension Service
Cornell University
366 Roberts Hall
Ithaca, NY 14853
607/255-2237; (FAX) 607/255-0788
hrd1@cornell.edu

NORTH CAROLINA
Dr. Jon F. Ort, Director
Cooperative Extension Service
North Carolina State University
Box 7602
Raleigh, NC 27695-7602
919/515-2811; (FAX) 919/515-3135
jon_ort@ncsu.edu

Dr. Marion Ray McKinnie, Administrator
1890 Extension Programs
North Carolina A&T State University
205 Coltrane Hall
Greensboro, NC 27420-1928
336/334-7691; (FAX) 336/334-7077
mckinnie@ncat.edu

NORTH DAKOTA
Mr. Duane Hauck, Interim Director
NDSU Extension Service
North Dakota State University
315 Morrill Hall, P.O. Box 5437
Fargo, ND 58105-5437
701/231-8944; (FAX) 701/231-8520
ext-dir@ndsuext.nodak.edu

OHIO
Dr. Keith L. Smith, Director
Ohio State University Extension
Ohio State University
3 Agricultural Admin. Bldg.
2120 Fyffe Road
Columbus, OH 43210
614/292-4067; (FAX) 614/688-3807
smith.150@osu.edu

OKLAHOMA
Dr. Sam E. Curl, Director
Cooperative Extension Servic
Oklahoma State University
139 Agriculture Hall
Stillwater, OK 74078-6019
405/744-5398; (FAX) 405/744-5339
securl@okstate.edu

OKLAHOMA (cont'd)
Dr. Marvin Burns, Administrator
1890 Extension Programs
Langston University
Agricultural Research Building, Room 100-S
P.O. Box 730
Langston, OK 73050
405/466-3836; (FAX) 405/466-3138
mburns@luresext.edu

Mr. Sherman Lewis, Associate Administrator
Langston University
P.O. Box 1258
Langston, OK 73050
405/466-6023; (FAX) 405/466-9937
sllewis@lunet.edu

OREGON
Dr. L. J. Knoong, Interim Director
Cooperative Extension Service
Oregon State University
Extension Administration
101 Ballard Extension Hall
Corvallis, OR 97331-3606
541/737-2713; (FAX) 541/737-4423
l.j.koong@oregonstate.edu

PENNSYLVANIA
Dr. Daney G. Jackson, Director
Cooperative Extension Service
The Pennsylvania State University
217 Ag. Admin. Bldg.
University Park, PA 16802
814/863-3438;(FAX) 814/863-7905
djackson@psu.edu

RHODE ISLAND
Dr. Jeffrey Seemann, Director
Cooperative Extension Service
University of Rhode Island
College of the Environment and Life Sciences
12 Woodward Hall
9 East Alumni Avenue
Kingston, RI 02881
401/874-5493; (FAX) 401/874-4017
Jseemann@uri.edu

SOUTH CAROLINA
Dr. John W. Kelly, Director
Cooperative Extension Service
Clemson University
130 Lehotsky Hall
Clemson, SC 29634
864/656-3642; (FAX) 864/656-3608
jkelly@clemson.edu

Dr. James H. Walker, Jr, Administrator
1890 Extension Programs
South Carolina State University
P.O. Box 7453
Orangeburg, SC 29117
803/536-8969; (FAX) 803/536-7102
jhwalker@scsu.edu

SOUTH DAKOTA
Dr. Gerald W. Warmann, Director
Cooperative Extension Service
South Dakota State University
Ag Hall 154, Box 2207D Brookings, SD 57007
605-688-4792; (FAX) 605-688-6347
warmann.gerald@ces.sdstate.edu

TENNESSEE
Dr. Charles L. Norman, Director
Tennessee Agricultural Ext. Service
The University of Tennessee
121 Morgan Hall
Knoxville, TN 37996-4530
865/974-7114; (FAX) 865/974-1068
clnorman@utk.edu

Dr. Clyde E. Chesney, Administrator
1890 Extension Program
Tennessee State University
3500 John A. Merritt Building
Nashville, TN 37209-1561
615/963-5526; (FAX) 615/963-5394
cchesney@picard.tnstate.edu

TEXAS
Dr. Ed Smith, Director
Texas Cooperative Extension
Texas A&M University
112 Jack K. Williams Administration Bldg.
7101 TAMU
College Station, TX 77843-7101
979/845-7967; (FAX) 979/845-9542
egsmith@tamu.edu

Dr. Linda Williams Willis, Administrator
1890 Extension Programs
Prairie View A&M University
P.O. Box 3059
Prairie View, TX 77446-3059
936/857-2023; (FAX) 409/857-2004
lw-willis@tamu.edu

UTAH
Dr. Jack M. Payne, Director
Cooperative Extension Service
Utah State University
4900 Old Main Hill
Logan, UT 84322-4900
435/797-2201; (FAX)435/797-3268
jackp@ext.usu.edu

VERMONT
Dr. Douglas Lantagne, Interim Director
Cooperative Extension Service
University of Vermont Extension Systems
601 Main Street
Burlington, VT 05405
802/656-8679; (FAX) 802/656-8642
doug.lantagne@uvm.edu

VIRGINIA
Dr. Patricia M. Sobrero, Director
Cooperative Extension Service
Virginia Polytechnic Institute and State University
Office of the Director
104 Hutcheson Hall, MC 402
Blacksburg, VA 24061
540/231-4152; (FAX) 540/231-4163
psobrero@vt.edu

Arthropod Collection and Indentification Techniques

VIRGINIA (cont'd)
Dr. Charles Stallings, Associate Director
Cooperative Extension Service
Virginia Polytechnic Institute
  and State University
106B Hutcheson Hall
Blacksburg, VA 24061
540/231-6371; (FAX) 540/231-5545
cstallin@vt.edu

Dr. Lorenza W. Lyons, Administrator
1890 Extension Programs
Virginia State University
P.O. Box 9081
Petersburg, VA 23806
804/524-5961; (FAX) 804/524-5967
llyons@vsu.edu

WASHINGTON
Dr. Linda Kirk Fox, Director
Cooperative Extension Service
Washington State University
P.O. Box 641046
411 Hulbert Hall
Pullman, WA 99164-6230
509/335-2933; (FAX) 509/335-2926
lkfax@wsu.edu

WEST VIRGINIA
Lawrence S. Cote, Director
Cooperative Extension Service
West Virginia University
P.O. Box 6031
Morgantown, WV 26506-6031
304/293-5691; (FAX) 304/293-7163
lcote@wvu.edu

WISCONSIN
Dr. Arlen G. Leholm, Director
Cooperative Extension Service
University of Wisconsin
432 N. Lake Street
Room 601, Extension Building
Madison, WI 53706
608/263-2775; (FAX) 608/265-4545
arlen.leholm@ces.uwex.edu

WYOMING
Dr. Glen Whipple, Director
Cooperative Extension Service
University of Wyoming
College of Agriculture
Room 103, P.O. Box 3354
Laramie, WY 82071-3354
307/766-5124; (FAX) 307/766-3998
glen@uwyo.edu

TERRITORIES:

AMERICAN SAMOA
Dr. Carol S. Whitaker, Director
Land Grant Programs
American Samoa Community College
Agriculture, Human and Natural Resources
P.O. Box 5319
Pago Pago, AS 96799
684/699-1394; (FAX) 684/699-5011
whitakerc@rocketmail.com

GUAM
Dr. Lee S. Yudin, Acting Director
Cooperative Extension Service
University of Guam
College of Agriculture & Life Sciences
UOG Station
Mangilao, Guam 96923
671/735-2000/1; (FAX) 671/734-6842
Lyudin@uog9.edu

MICRONESIA
Dr. Singeru Singeo, Acting Director
College of Micronesia
AES/CES P.O. Box 1179
Kolonia, Pohnpei 96941
691/320-2728; (FAX) 9011-691/320-2726
No e-mail addess

NORTHERN MARIANAS
Mr. Anthony C. Benavente, Acting Director
Cooperative Extension Service
Northern Marianas College
Box 501250
Saipan, MP 96950
670/234-5498; (FAX) 670-234-0500
Anthonyb@nmcnet.edu

PUERTO RICO
Dr. John Fernandez-Van Cleve
Acting Director
Cooperative Extension Service
University of Puerto Rico at Mayaguez
College of Agricultural Sciences
P.O. Box 9030
Mayaguez, PR 00681-9030
787/833-3486; (FAX) 787/833-4220
john@uprmedu

VIRGIN ISLANDS
Mr. Kwame N. Garcia, Sr., Director
Cooperative Extension Service
University of the Virgin Islands
RR02 Box 10,000
Kingshill, St. Croix, VI 00850
809-692-4091/4903
(FAX) 809-692-4085
kgarcia@uvi.edu

APPENDIX IV – Submitting Specimens for Identification (SEL)

Communications & Taxonomic Services Unit (CTSU)

The CTSU is the unit in the Systematic Entomology Laboratory that receives specimens, assigns/distributes them to identifiers, and reports the identifications to submitters. With the exception of some common species encountered in offices, residences/yards, CTSU does not directly provide identifications of insects or mites.

Instructions for Completing Identification Request Form, ARS 748

Dates
Use MM/DD/YY format for dates. If the Date Needed is less than two months from the Date Submitted, explain under Remarks.

Specimen Disposition
The expense of returning specimens is a major budgetary concern for SEL. When you do not need the submitted specimens returned to you (e.g. you have retained specimens in your collection that you are certain are duplicates), it will help us if you can indicate the Keep or Discard option for Specimen Disposition. When it is important that specimens be returned, be certain to specify Return for Specimen Disposition so that we will be sure to return any specimens that are not needed for the National Collection. Please note that the identifier will often feel compelled to retain certain specimens for the National Collection, particularly when they represent species that are undescribed or poorly represented in the National Collection, when they voucher new host records, or when they require special preparation prior to identification.

Level of Identification Requested
For various reasons, it is not always possible to ID specimens to the species level, but if that is the level you need, so indicate; if other ID levels (family or generic) suffice, please indicate that. If, instead of (or in addition to) a name, you need an indication of whether the species is predatory, parasitic, phytophagous, saprophagous, please indicate that in Remarks.

Tentative Identification
If you have some assessment of the kind (e.g. family, genus) of insect or mite being submitted, please indicate. List tentative identifications on separate sheet if multiple species are being submitted.

Host
If specimens are reared from a known host (plant, insect, etc.), indicate the name of the host. Family, genus, and species are preferable to common names.

Project Description
Include 1) the project title, 2) the name of the project leader, and 3) citations of any prior correspondence pertaining to this submittal. Information you provide regarding your project will help us in determining the priority to be assigned to your lot.

Remarks
Include reasons for needing report sooner than two months from date of submittal (if you have so indicated above), and Reason for Submittal if you have selected Other under Reasons for Submittal.

Arthropod Collection and Indentification Techniques

Preparation of Specimens Submitted for Identification

Because of a lack of sufficient technical assistance to prepare submitted specimens, we are including the following instructions to help you to prepare suitable specimens. If you are uncertain how to prepare specimens, please contact CTSU. Improperly prepared specimens will not be accepted.

Pinned

Most specimens should be pinned (while the specimens are fresh); those too small or fragile for direct pinning (e.g. tiny moths or flies) should be double-mounted on minuten nadeln or carefully glued to paper points. Glue the point to the right side of the specimen, using care that the glue does not conceal critical characters. DO NOT pin specimens that have dried; glue them to points. DO NOT glue tiny moths or flies to points; use minuten nadeln. Specimens should be pinned while fresh. Moths should be submitted with wings spread. Examination of properly prepared genitalia is necessary to identify many insect species to genus or species level. Specific instructions for preparing genitalia will be supplied upon request. Puparia, pupal skins, cocoons, etc., should be placed in a gelatin capsule or glued to a card either on separate pins or the same pin. Specifically, adults of the following must be pinned or point-mounted: Coleoptera, Lepidoptera, Heteroptera, and most Diptera, Nematocera and Acalyptrata (latter may be sent in alchohol or pinned).

Slide Mounts

Those skillful at preparing slides should submit mites (except Eriophyidae), thrips, aphids, most scale insects, mosquito larvae, and some Mymaridae on microscope slides. This will enable us to identify material more quickly. Larval ticks are accepted as slide mounts if they are not engorged. Specific instructions for slide-mounting the above groups will be supplied upon request.

Alcohol

Submit specimens of the following in alcohol: most scale insects, ichneumonid wasps (95% ethanol), mayflies, all soft-bodied insects (including all larvae and pupae), and most insects under 2mm in length, except as indicated below. Also, specimens in the groups listed under Slide Mounts should be preserved in 70% alcohol if not mounted. NEVER submit adults of the following groups in alcohol: beetles, moths, true bugs, or flies other than Acalyptrata--which may be in alcohol or pinned--and Nematocera. Place only one kind of insect in each vial. Screw-capped vials with polyethylene-cone seals are excellent, but other kinds of screw caps are problematic and should be avoided in favor of shell vials with neoprene or silicone stoppers. Do not wax or tape vial caps; instead, use parafilm if necessary. Vials should be of clear glass and be large enough to allow the use of forceps or an eye dropper for specimen removal. However, use of vials that are too large can make it very difficult to find and remove tiny specimens. Do not use methanol or formalin solutions; use 70, 80, or 95% ethanol if possible. Isopropanol ("rubbing alcohol") is adequate only for temporary storage. To prevent dilution of the alcohol and subsequent decomposition of specimens, fresh alcohol must be placed in vials within 24 hours after initial immersion of specimens. All vials should be filled to the top, and it is especially important to eliminate air bubbles in vials containing soft-bodied insects.

Preparation of Specific Forms

Kill larvae by placing them in boiling water or in an alcohol/glacial acetic acid mixture; then transfer them to 70% alcohol. Ichneumonid wasps and mayflies should be preserved in 95% ethanol. Thrips should be killed and placed in AGA (9 parts 70% ethanol, 1 part glycerine, 1 part glacial acetic acid). Nymphal, adult, and engorged ticks should be preserved in 70-80% ethanol.

Dry, Unmounted

If whiteflies or diaspidid and asterolecaniid scales are not mounted on slides, they should be submitted on parts of the host plant placed between pieces of dry paper toweling, blotters, or other absorbent materials. Do not place specimens of these families in plastic bags. Galls containing insects should be submitted with plant material in 70, 80, or 95% ethanol. Submit eriophyid mites dry on the host plant.

APPENDIX

Pill Boxes

Pill boxes and match boxes are not acceptable containers for submitting insects, but they may be used to submit associated plant samples, galls, or similar material. Soft tissue paper or cellucotton (not cotton) should be used in such boxes.

Identification Capability

Except as listed below, adults and immatures of all groups of insects, mites, and ticks are accepted for identification. On a case-by-case basis, depending upon the justification, material of some of the following groups may be accepted for identification or we may provide names and addresses of other experts.

Protura, Diplura, Collembola, Thysanura

Coleoptera: Staphylinoidea, Hydrophyloidea, Eucinetoidea, Dascilloidea, Scarabaeoidea, Byrrhoidea, Dryopoidea, Elateroidea (Elateridae accepted), Cucujoidea, Coccinellidae (Coccinellini accepted)

Diptera: Tanyderidae, Blephariceridae, Deuterophlebiidae, Nymphomyiidae, Axymiidae, Pachyneuridae, Psychodidae, Trichoceridae, Anisopodidae, Scatopsidae, Dixidae, Thaualeidae, Chironomidae, Agromyzidae, Sciomyzoidea, Lauxanioidea, Sphaeroceroidea, Ephydroidea, Periscelidae, Asteiidae, Carnidae, Canacidae, Chloropidae, Muscidae

Homoptera: Aleyrodidae

Hymenoptera: Stephanoidea, Ceraphronoidea, Proctotrupoidea, Cynipoidea, Trichogrammatidae, Evaniioidea, Chrysidoidea, Sphecoidea, Scolioidea, Pompiloidea, Vespoidea. Groups of Hymenoptera Parasitica not included in the foregoing list can be accepted for identification, but they must be reared from hosts that are known at least to family. Specialists may be willing to accept non-reared material in special cases, but please do not submit such material without inquiring in advance.

Lepidoptera: Gelechioidea, Copromorphoidea, Yponomeutidae, Cossoidea, Hesperoidea, Geometroidea, Bombycoidea, Sphingoidea. Papilionoidea (true butterflies) from outside America north of Mexico, can only be accepted if 1) voucher specimens may be retained 2) the specimens have been reared or are parasitized, and 3) all specimens are fully labeled, including food plant data.

Instructions for Submittal of Specimens

1. Use this form for submitting each lot, not for each individual specimen.

   Place a copy of the form with the specimens, and mail them to:

   USDA, ARS, Systematic Entomology Laboratory
   Communications & Taxonomic Services Unit
   Bldg. 005, Room 137, BARC-West
   10300 Baltimore Avenue
   Beltsville, Maryland 20705-2350

   If you wish to e-mail us copies of the form and covering correspondence in Advance. Our e-mail address is:

   IDservice@sel.barc.usda.gov

Arthropod Collection and Indentification Techniques

2. All specimens whether pinned, on slides, or in vials must be labeled with complete collection data, including locality (country, state, or other political subdivision, plus city or other landmark designation), latitude and longitude coordinates (if known), date of collection, and name (initials plus last name) of the collector(s). For specimens submitted in alcohol, please include one locality label per specimen (i.e. if there are 5 specimens in a vial, send 5 labels, not 1).

3. When the host is known, provide a label identifying the host as precisely as possible (i.e. by genus and species, not common name). For reared specimens, the emergence date should be indicated.

4. There should be a separate label with a brief, unique reference number (RN) to facilitate reporting of identifications. Assign the RNs after sorting material to order, family, and morphospecies. If specimens have already been labeled with voucher numbers relating to sites or other collecting or rearing circumstances, we still want separate RN labels placed on specimens following the sort to lookalikes. Submittals for which there are no RNs and/or that have not been sorted will be assigned a lower priority than they would otherwise rate.

5. Specimens will likely be retained for the National Collection, so if you will need to know the localities or other information for such specimens, you should keep a record of the correspondence between the RN labels and the data on other labels.

6. Specimens that are inadequately prepared and labeled will be returned to the submitter unidentified.

Inquiries About ID Status

1. Please reference the CTSU lot number when you know it.
2. Please use, mail, e-mail (address cited above) or FAX: 301-504-6482.
3. Because voice phone inquiries are disruptive, whenever possible, please avoid using our voice number (301-504-7041) for making inquiries about the status of a lot you have submitted for identification.

Credit for Identifications Provided

Publications and reports citing identifications provided via SEL/CTSU should cite the SEL taxonomists or extramural collaborators in the format indicated in the identification report. If the name of the identifier cannot be cited in the tables or text where the insect or mite names are cited, footnotes or some analogous means of reference should be used.

Analogous formats should be used for cooperating identifiers with other organizations or locations.

Whenever reprints of publications citing SEL-provided identifications can be spared, we would greatly appreciate receiving copies.

Notices Concerning Reports of Identifications

E-mail is our preferred method for sending reports of identifications. When an e-mail address is supplied, we will do our best to report identifications as they are reported by specialists. Because sending partial reports via the Postal Service requires more of our time than e-mailing them, we will report on larger submittals only after several or more identifiers have completed their assignments.

Use of FAX for reporting of identifications will be restricted to lots for which WE assign URGENT priority and then only when the submitter does not have an e-mail address.

For the sake of consistency in the SEL Identification System (SELIS), parentheses are no longer used in the author field to report names of authors of species or subspecies when the species or subspecies is placed in a genus other than that in which it was originally described.

APPENDIX

303

IDENTIFICATION REQUEST

Priority:	Lot Number:
Date Submitted:	Number of Specimens:
Date Needed:	Specimen Disposition: ☐ Return ☐ Keep/Discard

Name:

Address:

Telephone: FAX:

E-mail:

Submitter's Reference Number: Tentative Identification:

Level of Identification Requested:
☐ Family ☐ Genus ☐ Species

Host:

Reason for Identification:
☐ A – Biological Control
☐ B – Damaging Crop/Plants
☐ C – Suspected Pest of Regulatory Concern
☐ D – Stored Product Pest
☐ E – Livestock, Wildlife, or Domestic Animal Pest
☐ F – Danger to Human Health
☐ G – Household Pest
☐ H – Possible Immigrant
☐ I – Reference Collection
☐ J – Survey
☐ K – Thesis Problem
☐ L – Other (elaborate below)

Affiliation:
☐ APHIS/PPQ ☐ Private Individual
☐ ARS ☐ Other Federal (US)
☐ CICP ☐ Other State Agency
☐ Commercial Organization ☐ Private University
☐ US Department of Defense ☐ State Agriculture Agency
☐ Foreign ☐ State University
☐ US Forest Service

Project Support:
☐ APHIS/PPQ ☐ ARS ☐ DOI ☐ EPA ☐ FAO ☐ FS ☐ Hatch ☐ NIH ☐ NRCS ☐ NSF ☐ USAID ☐ Other

Collecting Permits:
☐ Required ☐ Not Required   If required, please submit copies with specimens.

Project Description:

Remarks:

USDA

Communications & Taxonomic Services Unit – Systematic Entomology Laboratory
Building 005 – Room 137 – BARC-West
10300 Baltimore Avenue – Beltsville – Maryland – 20705

ars

OMB 0518-0032-sel-1 (2/2006)

NOTES:

Index

A

abbreviations 86
Acari 110, 111, 114, 115
Acaridae 290
Acariformes 115
acarology 198
acetone 14, 43, 67, 68
aculeate 197
adhesive 16, 42, 51-52, 54-59, 61, 93-94
adult mosquito 190
aerial net 10-12
AGA (alcohol-glycerin-acetic acid) 44, 288, 290, 300
alcohol 8, 13, 21-25, 30, 43-45, 51, 55, 57, 67, 68, 71-73, 75, 78-79, 83, 89, 92
alderflies 183
Aleyrodidae 177
Amblycera 123
Amblypygi 112
ambush bug 176
ametabolous 122
ammonia water 14
anamorphic 157
angelwing 186
animal kingdom 105
Anisoptera 160
Anoplura 122-123, 140, 174-175
antennules 117
antlions 124, 183, 184
ants 121, 124, 157, 163, 167, 177, 180, 184, 186, 193-196
aphids 123, 164, 167, 172, 177-178, 289
APHIS quarantine inspectors 95
Apocrita 193
apotele 116
Apterygota 119
aquarium 37
aquatic canvas net 33

aquatic dip net 33
aquatic insects 5, 33
aquatic net 10, 12
Arachnida 107, 110, 114
Araneida 110
Archaeognatha 119, 121, 158-159
arctiid 186
armored scales 178
artificial diets 41
artificial refuge traps 29
aspirators 8, 17-19
assassin bug 176
association 36, 87
Astigmata 115
attractants 29

B

backswimmer 176
Baetidae 160
bags 9
baits 29-30
Barber's fluid 288
barklice 172
bark beetles 124, 180
barriers 22
basonym 105
basswood 60-61
battery 26-27
bat flies 189
beating sheets 19
bed bugs 123, 140, 177
bees 124, 154, 182, 193-195, 197
beetles 124, 179, 180-181, 188
behavior 5, 6
Berlese funnel 21
binomen 104-105
biological supply houses 8-10, 60, 72, 76-77, 84, 88-90, 95
bird lice 123
Birge-Ekman dredge 33
bite 6

Bittacidae 182
black fly 190
Blattaria 163-164
Blattodea 122-123, 130, 150, 163, 181
bleach 72, 81
bleaching 72
body lice 123, 174
booklice 123, 172
Boreidae 183
botanist's vasculum 9
bottle flies 189
Brachycera 189
bristletails 121, 158-159
brush 9, 19-20, 41, 55, 81, 92
butterflies 124, 185, 186-188

C

caddisflies 124, 184-185
calcium cyanide 15
camel's hair 9
Campodeidae 157
Canada balsam 288-289
carbon dioxide 32
carbon tetrachloride 14, 48, 50, 91
card point 56, 58-59, 87
care of the collection 89, 96
carrion beetle 180
case-making larvae 184
castes 168
caterpillars 186
cat flea 192
Cellosolve 28, 51-54
cellulose 167
centipedes 108, 117
cerci 130, 138, 148-150, 157, 189
Chelisochidae 167
Chelonethida 113
chewing lice 123, 173-174

Chilopoda 108, 117
chloroform 14, 48, 91
Chrysopidae 183
cicadas 123, 178
Cimicidae 177
classification 103, 183
click beetle 179
cockroaches 123, 162-164, 166
Coleoptera 122-124, 127, 142-143, 149, 152, 154-155, 179-181
collecting nets 10
collection 7, 10, 18, 23, 25-26, 33, 34, 40, 44, 45, 47, 51- 52, 59-60, 65-68, 73, 76, 83, 85-92, 95
collection data 45, 47, 51, 65, 85, 87, 89
collection information 34, 66, 88
collector 86
collector's bag 7
Collembola 119, 121, 144, 158
color traps 28
common name 105, 158-159, 166-167, 170, 172, 175
common scorpionflies 183
computer generated labels 84
containers 8, 13, 36, 48, 92
corpse 6
Corrodentia 122-123, 172
Corydalidae 183
coverslips 76
CO_2 Trap 32
crab louse 174
crepuscular 161
crickets 122, 162, 166
criminal investigations 6
critical-point drying 67-69, 82
crochets 186
Crustacea 107
Cryptostigmata 115
cryptozoic 166
ctenidia 192
Culicidae 190
curing 77
curved forceps 7

D

damselflies 122, 160-161
date 86
DDVP 16
deep freeze 50
deer flies 124
degreasing 50-51
dehydration 71
Dermaptera 122-123, 127, 147, 166-167
diagnosis 103
diapause 20, 39-40
dichlorvos 16, 91
dichotomy 119
Dictyoptera 122
Diplopoda 108, 117
Diplura 119, 121, 144, 157
diplurans 121
Diptera 105, 122, 124, 131, 140, 143, 152-154, 156, 181, 189-191, 195
direct pinning 53, 58-59, 87
diseases 6
dish trap 23
diurnal 161, 170, 185-186
diving beetle 181
DNA 42
dobsonflies 124, 183
dogface 185
dog flea 192
dormancy 39
dorsal 111-113, 115-116
double mounts 51, 53, 55, 56, 58, 59, 87
dragonflies 122, 160-161
drag cloth 19
drawing pens 84
dredge 33
drying 67, 96
dry preservation 45

E

earwigs 123, 166-167
Echinophthiridae 175
Ectognatha 119-121, 158
ectoparasites 34
electrical grid traps 29
embalming fluid 288
embedding 69
Embiidina 122-123, 138, 151, 169
Embioptera 123, 169
emergence traps 24
Endopterygota 122-124
Entognatha 118-121
Entomobryidae 158
entomology 198
envelopes 8, 45-49
environmental disturbance 5
Ephemeridae 160
Ephemeroptera 118, 122, 131, 134, 142, 159-160
ephydrid 190
eradication 5
Eriophyidae 292
Essig's fluid 288-289
ether 14, 48, 50-51
ethyl acetate 14-15, 28, 48, 91
ethyl alcohol 43
euparal 75-76, 79-83
Euplexoptera 123
evidence 6
Exopterygota 118, 122
exoskeleton 118
Extension service 291-298
extractors 20-21
exuviae 36

F

families 105, 132, 140, 157-158, 160, 162, 167, 173, 174, 179, 183-185, 189-192
feces 31
firebrats 121, 159

INDEX

fishflies 183
flag 19
flagellate protozoans 167
flashlight 9, 20, 30-31
flatid bug 178
flat headed wood borer 180
fleas 121, 124, 158, 192-193
flesh fly 190
flowerpot cage 38-39
flushing agents 6
food 34, 37, 40-41, 48, 83, 96
footspinners 169
forceps 7
forensic 35, 87, 101
Forficulidae 167
form, ARS 748 299, 303
formalin 44
free-living caddisfly larva 185
freeze-drying 44, 67-69, 82
freezer 13, 42, 45, 50, 68
fringe-winged insects 175
fumigation 91
fungivorous 173, 175
furcula 158

G

gall wasps 196
gardener's trowel 9
genitalia 73, 77
genus 103-105, 161, 196
giant water bug 176
glaciers 166
glassine envelopes 8, 46-47, 62
glue 34, 54-55, 57-59, 62
glue traps 34
glycerine 44
glyptol 74, 76
gnats 124, 189
gonopore 157
gonostyli 157
grasshoppers 122, 162
ground beetles 124, 180
ground litter 20

Grylloblattaria 166
Grylloblattodea 123, 150, 166

H

habitat 5-6, 37-38, 86-87
Haematopinidae 175
hairstreak 185
Haller's organ 114
halteres 131, 189
hand lens 9
hangingflies 124, 182
headlamp 9, 30-31
head lice 123, 174
heliconid line 186
hellgrammites 124
hemimetabolous 122
Hemiptera 122-123, 136, 140, 153, 175-177, 181
Hemoerobiidae 183
Heptageniidae 160
herbarium sheets 57, 84
Herkol 16
Heteroptera 123
Hexapoda (Insecta) 108, 117, 119, 124-125, 157
hister beetle 180
historical baseline 5
hobby 101
holometabolous insects 122, 124, 189
Holothyrina 116
Homoptera 122-123, 128, 131, 136, 154, 177-179
honey 104, 163, 194-195
honeydew 178
honey bee 194
hoppers 123
horse flies 189
hospitals 6
host 6
host plant 38, 41, 86, 96
house flies 124, 189-190
housing the collection 89

Hoyer's medium 72-74, 90, 288-290
Hymenoptera 122, 124, 137, 146-147, 154-155, 177, 179, 181, 193-197
hypermetamorphic 181, 194
hypodermic syringe 13
hypostome 114, 116

I

ichneumon wasp 124, 195
identification 6, 100-102, 299-302
identification labels 88
imago 118
indicators 5
inflation of larvae 67
injection 13, 43, 80
ink 83-84, 87-88
Insecta 117, 121-122, 125, 157, 172
insect audits 5
insect drawer 90
insect pin 52-53, 55-56, 59, 62, 65, 80, 90
inspections 6
instar 118, 170, 175, 181, 194
interception nets 22
international code 105
invasion 5
Ischnocera 123
Isopoda 117
isopropyl alcohol 43
Isoptera 122-123, 129, 138, 150-151, 167-168
Ixodida 116

J

Japygidae 157

K

KAA (KAAD) 288
katydids 122, 162
keys 103
kick screen 33
killing container 13

killing jar 8, 11-12, 14-17, 20, 22, 24-25, 30-31, 43, 49
king 168-169
knife 9, 56, 65

L

labeling 4, 6, 42, 77, 83, 87-88
   microscope slides 88
   vials 87
labels 84-85, 87-88, 96
Labiduridae 167
Labiidae 167
labium 142, 161, 175
lacewings 124, 183-184
lanyard 9
larva 118, 124, 154, 184-185, 189-190, 192, 194
leaf-cutter bee 194
leafhoppers 123, 177-179
leaf beetles 124
leaf insects 123, 165
Lepidoptera 122, 124, 132, 135, 143, 146, 151-152, 154, 156, 182, 185-188, 194-195
Lepismatidae 159
life stages 101
light 22, 25-27, 40
light sheets 27
light traps 25
liquid-preserved specimens 51
liquid killing agents 14
litigation 6
loading cartons 94
lobster trap 24
locality 85
locusts 122
longhorned beetle 179
lures 29-30
Lycaenidae 186

M

maceration 70-71, 289
Machilidae 158
maggot 190
Malacostraca 117
Malaise traps 22
Mallophaga 119, 122-123, 139, 173-175
Manitoba trap 28
mantidflies 124
mantids 123, 162, 164-165, 178
mantispid 184
Mantodea 122-123, 130, 149, 164, 181
mayflies 122, 131, 159-160
mealybugs 177, 178, 179
Mecoptera 122, 124-125, 135, 137, 147-148, 152, 182-183
medico-criminal investigations 35
Megaloptera 134, 142, 183
Meinertellidae 158
Mengeidae 181
mercury vapor lamp 26
Mesostigmata 116
metallic wasp 195
metamorphosis 118, 122, 124, 158-159, 161-163, 165, 167, 170, 172-175, 177, 182
radical 122
metazoan animals 198
Microcoryphia 119, 121, 158-159
microlepidoptera 61, 65, 71
microscope slides 88, 90, 95
Microthelyphonida 112
microvial 73, 78-80, 83
midges 124, 189
millipedes 108, 117
minuten 52, 55-56, 59, 62, 65, 73, 78, 80
mites 103-105, 107, 110-111, 114, 116, 197-198

moisture 38
mold 91-92
molecular research 42
molts 117-118, 157, 181
monarch butterfly 187
monarch caterpillar 187
monitoring 6
monocondylic 159
mosquito 124, 154, 189
   larva (wiggler) 190
   pupa (tumbler) 190
moths 124, 132, 185, 187-188
mound-building ants 196
mounting 47-48, 60, 69, 72-73, 75-76, 81-82, 289, 290
   boards 60
   larvae 82
   medium 70-75, 77, 81, 88
   mites 290
   small specimens 47, 69, 76
   soft body specimens 289
   thrips 289
   wings 81
multivoltine 175
museum pests 91, 94
myiasis 190
Myriapoda 117
myrmecophilous 195
Myrmeleontidae 183
Myrmeleontids 184

N

naiads 118, 170, 171
names 104-105, 114, 119-121, 124, 169, 172, 183, 189
necrophilous 173
Nematocera 189
Neoptera 122, 192
neotenics 168
Neuroptera 122, 124-125, 134, 136-137, 141-142, 152, 155, 183-184

INDEX

New Jersey trap 25
no-pest strips 16
no-see-ums 189
nocturnal 113, 159, 161-163, 167, 185, 187
Nogos 16
notebook 8
Notoptera 123
nuptial chamber 168
Nuvan 16, 91
nymphs 118, 122, 159-165, 167, 169-170, 172, 174

O

oatmeal 31
Odonata 122, 134, 142, 160-161
ootheca 163-165
Opilioacarida 116
Opiliones 114
orders 157
ordinal classification 119
Orthoptera 122, 129, 149, 162-164, 166, 181
owlflies 183
oystershell scales 178

P

packaging and shipping specimens 92
packing materials 92
Paleoptera 122, 159
Palpigradi 112
Panorpidae 183
paper 8, 13-16, 20, 25, 31, 36, 45-49, 51-52, 55-57, 62, 64-65, 83-84, 94-96
papering 46
para-dichlorobenzene (PDB) 91
paralyzing a butterfly 11
Parasitiformes 115
parthenogenesis 158, 175, 195
parthenogenetic 163

paurometabolous 122
Pauropoda 109, 117
Pediculidae 174
Pediculus 174
Pedipalpida 112
pencil 42, 84, 87-88
permanent mounting media 75
pest control industry 101
pest insects and mites 101
pest management audits 34
petiole 137, 147, 194
Phalangida 114
Phasmatodea 123, 151, 165
Phasmida 123, 130, 165
pheromones 31, 32
Philopteridae 173
photographers 101
photographs 6
phragmotic 168
Phthiraptera 123
Phylloxeridae 177
phylogeneticists 119
physogastry 168
picture-wing fly 189
pinned specimens 93
pinning block 59, 87
pitfall trap 23
placing labels 87
plaited-winged insects 170
Planipennia 136-137, 152
plant bugs 123, 176-177
plant clippers 9
plant press 9
plasticine 69
Plecoptera 122-123, 133, 143, 170-171
Poduridae 158
point punch 56
poison 15, 16
pollination 195
pollinators 190, 195
pollution 5
pooter 17

ports of entry 5
potassium cyanide 15
praying mantids 164
prehensile 161
primary reproductives 168
primitive 121
printed labels 84
Procampodeidae 157
professional entomologists 101
Projapygidae 157
proper pin positions 54
propodeum 194
Prostigmata 115
protecting specimens 91
protocols 5, 6
Protura 119, 121, 144, 157
proturans 121, 157
Pseudoscorpionida 113
psocids 123
Psocoptera 123, 138, 148, 172
psyllids 172, 178
Pterygota 118-119, 121, 124
punkies 189
pupa 118, 124, 175, 182-183, 185, 190, 192
puparia 36
Pycnogonida 106

Q

queen 168-169

R

Raphidiodea 137, 152
rapidograph 84
rearing 3–6, 24, 35-36, 38, 41
  cages 24
reference collections 67, 101
refrigeration 45
regulated insects 34
regulating 5
relaxing chamber 48
retinaculum 158

Rhaphidioidea 183
Ricinulei 113
Riker mounts 66, 67, 91
ringing 76
RNA 42
robber flies 189
Robinson trap 26
rock crawlers 123, 166
rock roller 184

S

sample vials 8
sawflies 124, 182, 189, 193-195
scale insects 123, 131, 177-178, 189, 289
Schizomida 112
scientific name 104-105, 158, 170, 172, 196
    italicized 104
Sciomyzidae 190
scorpionflies 124, 182-183, 189
Scorpionida 111
secondary reproductives 168
seed bugs 123
sejugal furrow 115
SEL (SELIS) 299
sensitive situations 5
sheep ked 190
shipping live specimens 95
shore flies 190
Sialidae 183
sifters 20
silverfish 121, 159
Simuliidae 190
Siphonaptera 119, 122, 124, 139, 154, 192-193
size of a label 85
skin beetle 180
skipper 187
skippers 185
Sminthuridae 158
snakeflies 124, 183

snap traps 29
snipe flies 189
snow scorpionflies 183
social system 168
sodium cyanide 15
soft forceps 7
soft scales 178
soil insects 34
soldiers 121, 167-168
solid killing agents 15
Solpugida 113
soothsayers 164
sounds 33
species 100-105, 108, 110, 117, 121, 124, 157-196, 198
specimens in vials 94
spermatophore 157, 158
spiders 106-107, 110
spider wasp 195
spot plate 78
spreading blocks 64
spreading boards 60, 62
springtails 121, 158
stadium 118
staining 72, 78, 80
standard insect pins 53
sticky traps 28
sting 6
stinging insects 12
stink bugs 123
stoneflies 123, 170-171
storage 45, 73, 77, 89-90
storage containers 89-90
Strepsiptera 122, 124, 129, 152, 154, 181-182
student 101-102, 198
styli 145, 157, 159
Stylopidae 182
subfamilies 105
subimago 118, 159-160
submittal 299-301
submitting specimens 299, 303
succession 6

sucking lice 123, 174
sugaring 30-31
swallowtail 186
sweeping net 10, 11
symbiotic bacteria 167
Symphyla 109, 117
Symphyta 193
syrphid fly 189
Systematic Entomology Laboratory 289, 299, 301

T

Tabanidae 28
tachnid fly 189
taxonomic hierarchy 104
taxonomic information 88
temperature 39
temporary mounting 73
teneral 164
teneral adults 161
tephiid wasp 195
termitaria 168
termites 121, 123, 157, 167-169, 173, 180
Tetranychidae 290
Thelyphonida 112
thornbug 178
thread-waisted wasp 194
thrips 123, 175-176
Thysanoptera 123, 132, 153, 175
Thysanura 119, 121, 145, 158-159, 181
ticks 19, 32
tiger beetle 180
tooth 160
tow trap 33
tracing paper 62, 65
trap 21-29, 31, 33, 39, 43, 86
treehoppers 123, 178
Triassomachilidae 158
Trichodectidae 173

INDEX

Trichoptera 122, 124-125, 134, 141-143, 184-185
triungulinid 181
trochanter 158, 181
true bugs 176
true flies, 123-124, 189
twisted-winged parasites 124, 181

U

U.S. Department of Agriculture 95-96
ultraviolet light 26
Uropygi 112

V

Vapona 16, 91-92
vapor degreasing 50
vehicle-mounted net 12
vertical illuminator 73
viviparity 175
viviparous 163
voucher specimens 9, 42

W

walkingsticks 123, 162, 165
washing 71
wasps 124, 154, 179, 182, 193, 195-197
watchmaker forceps 7
waterboatman 176
waterproofing spray 84
water bugs 123
water strider 176
webspinners 123, 169-170
weevils 124, 180
whiteflies 123, 177-178, 289, 300
Wilkinson trap 25
windowpane trap 22
wing-flexion 122
wingless 121–198
wings 108, 118-119, 121-124, 127, 129, 131-138, 157, 159-163, 165-169, 171-177, 179, 181-186, 188-189, 191-196
wings spread 55, 60, 87
wood boring beetles 124
wool-eaters 173
wrigglers 154
written accounts 6

X

Xiphosura 106
xylene 20, 28, 51-52, 54-55, 67

Z

Zoraptera 122-123, 138, 148, 172-173
zorapterans 123, 173
Zygoptera 160